Introduction to Plasma Physics

Introduction to Plasma Physics

Edited by
Eric Buchanan

Larsen & Keller
www.larsen-keller.com

Introduction to Plasma Physics
Edited by Eric Buchanan
ISBN: 978-1-63549-225-5 (Hardback)

☰ Larsen & Keller

Published by Larsen and Keller Education,
5 Penn Plaza,
19th Floor,
New York, NY 10001, USA

Cataloging-in-Publication Data

Introduction to plasma physics / edited by Eric Buchanan.
    p. cm.
Includes bibliographical references and index.
ISBN 978-1-63549-225-5
1. Plasma (Ionized gases). 2. Physics.
I. Buchanan, Eric.
QC718 .I58 2017
530.44--dc23

The publisher's policy is to use permanent paper from mills that operate a sustainable forestry policy. Furthermore, the publisher ensures that the text paper and cover boards used have met acceptable environmental accreditation standards.

Printed and bound in the United States of America.

For more information regarding Larsen and Keller Education and its products, please visit the publisher's website www.larsen-keller.com

Table of Contents

Preface

This book elucidates the concepts and innovative models around prospective developments with respect to plasma physics. It talks about the different concepts and theories related to this subject in detail. Plasma is a fundamental state of matter, along with solid, liquid and gas. It is the most abundantly found matter on our planet. Plasma physics is the building base for nuclear fusion and fusion power. This text is a compilation of chapters that discuss the most vital concepts in the field of plasma physics. Coherent flow of topics, student-friendly language and extensive use of examples make this book an invaluable source of knowledge. It will provide comprehensive insights into the field of plasma physics. The textbook is appropriate for those seeking detailed information in this area.

A detailed account of the significant topics covered in this book is provided below:

Chapter 1- Matter has four fundamental states; one of the four states is plasma. The other three are solid, liquid and gas. A plasma constitutes of properties that are not similar to the other states. This chapter will provide an integrated understanding of plasma physics.

Chapter 2- Guiding center, Debye sheath, double layer, microplasma, plasmon and pinch are the key concepts related to plasma physics. Debye sheath is the layer of plasma that contains within itself a density of positive ions whereas a double layer is a structure that has layers of opposite charged layers. The following section unfolds its crucial aspects in a critical yet systematic manner.

Chapter 3- Plasma diagnostics are techniques and methods used to measure all the properties of plasma. Some of the aspects discussed within this text are plasma parameters, Faraday cup, Langmuir probe, plasma scaling etc. The following text will not only provide an overview, it will also delve deep into the topics related to it.

Chapter 4- The techniques used in the processing of plasma are plasma etching, corona treatment, plasma polymerization, plasma cleaning and plasma cutting. This chapter discusses plasma processing techniques in a critical manner providing key analysis to the subject matter.

Chapter 5- Electric discharge in gases usually arises when current flows through a gaseous medium because of ionization of the gas. Electric arc, glow discharge, Paschen's law and Townsend discharge are the topics elucidates in the chapter. This section on electric discharge in gases offers an insightful focus, keeping in mind the complex subject matter.

Chapter 6- This section focuses on all the themes concerned with fusion power. Fusion power is the energy that is produced by nuclear fusion. Pinch and types of pinch have also been explicated in the text. This chapter has been carefully written to provide an easy understanding of the varied facets of fusion power.

Chapter 7- The applications of plasma physics are inductively coupled plasma mass spectrometry, reactive-ion etching, plasma propulsion engine, plasma display, plasma actuator etc. Plasma displays are flat panel display television sets whereas plasma actuators are actuators that are developed for aerodynamic flow control. The diverse applications of plasma physics in the current scenario have been thoroughly discussed in this section.

Chapter 8- The allied fields of plasma physics are magnetohydrodynamics, fluid dynamics, physical cosmology and electrodynamics. Plasma physics is a vast subject that branches out into significant fields that have been thoroughly discussed in this chapter.

I would like to make a special mention of my publisher who considered me worthy of this opportunity and also supported me throughout the process. I would also like to thank the editing team at the back-end who extended their help whenever required.

Editor

Introduction to Plasma Physics

Matter has four fundamental states; one of the four states is plasma. The other three are solid, liquid and gas. A plasma constitutes of properties that are not similar to the other states. This chapter will provide an integrated understanding of plasma physics.

Plasma is one of the four fundamental states of matter, the others being solid, liquid, and gas. A plasma has properties unlike those of the other states.

A plasma can be created by heating a gas or subjecting it to a strong electromagnetic field, applied with a laser or microwave generator at temperatures above 5000 °C. This decreases or increases the number of electrons in the atoms or molecules, creating positive or negative charged particles called ions, and is accompanied by the dissociation of molecular bonds, if present.

The presence of a significant number of charge carriers makes plasma electrically conductive so that it responds strongly to electromagnetic fields. Like gas, plasma does not have a definite shape or a definite volume unless enclosed in a container. Unlike gas, under the influence of a magnetic field, it may form structures such as filaments, beams and double layers.

Plasma is the most abundant form of ordinary matter in the universe (the properties of dark matter are still mostly unknown; whether it can be equated to ordinary matter has yet to be determined), most of which is in the rarefied intergalactic regions, particularly the intracluster medium, and in stars, including the Sun. A common form of plasma on Earth is produced in neon signs.

Much of the understanding of plasma has come from the pursuit of controlled nuclear fusion and fusion power, for which plasma physics provides the scientific foundation.

Properties and Parameters

Artist's rendition of the Earth's plasma fountain, showing oxygen, helium, and hydrogen ions that gush into space from regions near the Earth's poles. The faint yellow area shown above the north pole represents gas lost from Earth into space; the green area is the aurora borealis, where plasma energy pours back into the atmosphere.

Definition

Plasma is an electrically neutral medium of unbound positive and negative particles (i.e. the overall charge of a plasma is roughly zero). It is important to note that although the particles are unbound, they are not 'free' in the sense of not experiencing forces. When a charged particle moves, it generates an electric current with magnetic fields; in plasma, the movement of a charged particle affects and is affected by the general field created by the movement of other charges. This governs collective behavior with many degrees of variation. Three factors are listed in the definition of a plasma stream:

1. The plasma approximation: Charged particles must be close enough together that each particle influences many nearby charged particles, rather than just interacting with the closest particle (these collective effects are a distinguishing feature of a plasma). The plasma approximation is valid when the number of charge carriers within the sphere of influence (called the *Debye sphere* whose radius is the Debye screening length) of a particular particle is higher than unity to provide collective behavior of the charged particles. The average number of particles in the Debye sphere is given by the plasma parameter, «Λ» (the Greek uppercase letter Lambda).

2. Bulk interactions: The Debye screening length (defined above) is short compared to the physical size of the plasma. This criterion means that interactions in the bulk of the plasma are more important than those at its edges, where boundary effects may take place. When this criterion is satisfied, the plasma is quasineutral.

3. Plasma frequency: The electron plasma frequency (measuring plasma oscillations of the electrons) is large compared to the electron-neutral collision frequency (measuring frequency of collisions between electrons and neutral particles). When this condition is valid, electrostatic interactions dominate over the processes of ordinary gas kinetics.

Ranges of Parameters

The factors of a plasma stream can vary by many orders of magnitude, but the properties of plasmas with apparently disparate parameters may be very similar. The following chart considers only conventional atomic plasmas and not exotic phenomena like quark gluon plasmas:

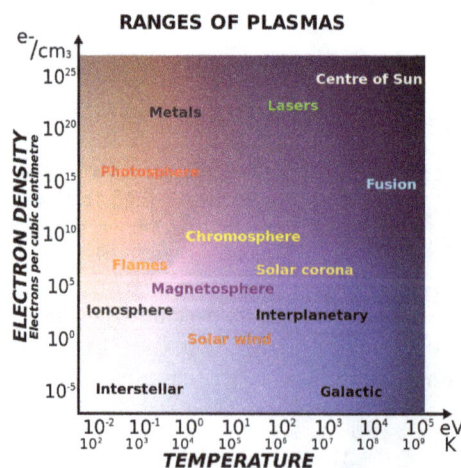

Range of plasmas. Density increases upwards, temperature increases towards the right. The free electrons in a metal may be considered an electron plasma.

Typical ranges of plasma parameters: orders of magnitude (OOM)		
Characteristic	**Terrestrial plasmas**	**Cosmic plasmas**
Size in meters	10^{-6} m (lab plasmas) to 10^{2} m (lightning) (~8 OOM)	10^{-6} m (spacecraft sheath) to 10^{25} m (intergalactic nebula) (~31 OOM)
Lifetime in seconds	10^{-12} s (laser-produced plasma) to 10^{7} s (fluorescent lights) (~19 OOM)	10^{1} s (solar flares) to 10^{17} s (intergalactic plasma) (~16 OOM)
Density in particles per cubic meter	10^{7} m^{-3} to 10^{32} m^{-3} (inertial confinement plasma)	1 m^{-3} (intergalactic medium) to 10^{30} m^{-3} (stellar core)
Temperature in Kelvin	~0 K (crystalline non-neutral plasma) to 10^{8} K (magnetic fusion plasma)	10^{2} K (aurora) to 10^{7} K (solar core)
Magnetic fields in teslas	10^{-4} T (lab plasma) to 10^{3} T (pulsed-power plasma)	10^{-12} T (intergalactic medium) to 10^{11} T (near neutron stars)

Degree of Ionization

For plasma to exist, ionization is necessary. The term "plasma density" by itself usually refers to the "electron density", that is, the number of free electrons per unit volume. The degree of ionization of a plasma is the proportion of atoms that have lost or gained electrons, and is controlled mostly by the temperature. Even a partially ionized gas in which as little as 1% of the particles are ionized can have the characteristics of a plasma (i.e., response to magnetic fields and high electrical conductivity). The degree of ionization, α, is defined as $\alpha = \dfrac{n_i}{n_i + n_n}$, where n_e is the number density of ions and α is the number density of neutral atoms. The *electron density* is related to this by the average charge state $\langle Z \rangle$ of the ions through $n_e = \langle Z \rangle n_i$, where n_e is the number density of electrons.

Temperatures

Plasma temperature is commonly measured in kelvins or electronvolts and is, informally, a measure of the thermal kinetic energy per particle. High temperatures are usually needed to sustain ionization, which is a defining feature of a plasma. The degree of plasma ionization is determined by the electron temperature relative to the ionization energy (and more weakly by the density), in a relationship called the Saha equation. At low temperatures, ions and electrons tend to recombine into bound states—atoms—and the plasma will eventually become a gas.

In most cases the electrons are close enough to thermal equilibrium that their temperature is relatively well-defined, even when there is a significant deviation from a Maxwellian energy distribution function, for example, due to UV radiation, energetic particles, or strong electric fields. Because of the large difference in mass, the electrons come to thermodynamic equilibrium amongst themselves much faster than they come into equilibrium with the ions or neutral atoms. For this reason, the ion temperature may be very different from (usually lower than) the electron temperature. This is especially common in weakly ionized technological plasmas, where the ions are often near the ambient temperature.

Thermal vs. Nonthermal Plasmas

Based on the relative temperatures of the electrons, ions and neutrals, plasmas are classified as "thermal" or "non-thermal". Thermal plasmas have electrons and the heavy particles at the same temperature, i.e. they are in thermal equilibrium with each other. Nonthermal plasmas on the other hand have the ions and neutrals at a much lower temperature (sometimes room temperature), whereas electrons are much "hotter" ($n_e = \langle Z \rangle n_i$).

Complete vs. Incomplete Ionization

A plasma is sometimes referred to as being "hot" if it is nearly fully ionized, or "cold" if only a small fraction (for example 1%) of the gas molecules are ionized, but other definitions of the terms "hot plasma" and "cold plasma" are common. Even in a "cold" plasma, the electron temperature is still typically several thousand degrees Celsius. Plasmas utilized in "plasma technology" ("technological plasmas") are usually cold plasmas in the sense that only a small fraction of the gas molecules are ionized.

Plasma Potential

Since plasmas are very good electrical conductors, electric potentials play an important role. The potential as it exists on average in the space between charged particles, independent of the question of how it can be measured, is called the "plasma potential", or the "space potential". If an electrode is inserted into a plasma, its potential will generally lie considerably below the plasma potential due to what is termed a Debye sheath. The good electrical conductivity of plasmas makes their electric fields very small. This results in the important concept of "quasineutrality", which says the density of negative charges is approximately equal to the density of positive charges over large volumes of the plasma (), but on the scale of the Debye length there can be charge imbalance. In the special case that *double layers* are formed, the charge separation can extend some tens of Debye lengths.

Lightning is an example of plasma present at Earth's surface. Typically, lightning discharges 30,000 amperes at up to 100 million volts, and emits light, radio waves, X-rays and even gamma rays. Plasma temperatures in lightning can approach 28,000 K (28,000 °C; 50,000 °F) and electron densities may exceed 10^{24} m^{-3}.

The magnitude of the potentials and electric fields must be determined by means other than simply finding the net charge density. A common example is to assume that the electrons satisfy the Boltzmann relation:

$$n_e \propto e^{e\Phi/k_B T_e}.$$

Differentiating this relation provides a means to calculate the electric field from the density:

$$\vec{E} = (k_B T_e / e)(\nabla n_e / n_e).$$

It is possible to produce a plasma that is not quasineutral. An electron beam, for example, has only negative charges. The density of a non-neutral plasma must generally be very low, or it must be very small, otherwise it will be dissipated by the repulsive electrostatic force.

In astrophysical plasmas, Debye screening prevents electric fields from directly affecting the plasma over large distances, i.e., greater than the Debye length. However, the existence of charged particles causes the plasma to generate, and be affected by, magnetic fields. This can and does cause extremely complex behavior, such as the generation of plasma double layers, an object that separates charge over a few tens of Debye lengths. The dynamics of plasmas interacting with external and self-generated magnetic fields are studied in the academic discipline of magnetohydrodynamics.

Magnetization

Plasma with a magnetic field strong enough to influence the motion of the charged particles is said to be magnetized. A common quantitative criterion is that a particle on average completes at least one gyration around the magnetic field before making a collision, i.e., $\omega_{ce} / v_{coll} > 1$, where ω_{ce} is the "electron gyrofrequency" and v_{coll} is the "electron collision rate". It is often the case that the electrons are magnetized while the ions are not. Magnetized plasmas are *anisotropic*, meaning that their properties in the direction parallel to the magnetic field are different from those perpendicular to it. While electric fields in plasmas are usually small due to the high conductivity, the electric field associated with a plasma moving in a magnetic field is given by $\mathbf{E} = -v \times \mathbf{B}$ (where $\mathbf{E}$ is the electric field, v is the velocity, and $\mathbf{B}$ is the magnetic field), and is not affected by Debye shielding.

Comparison of Plasma and Gas Phases

Plasma is often called the *fourth state of matter* after solid, liquids and gases, despite plasma typically being an ionized gas. It is distinct from these and other lower-energy states of matter. Although it is closely related to the gas phase in that it also has no definite form or volume, it differs in a number of ways, including the following:

Property	Gas	Plasma
Electrical conductivity	Very low: Air is an excellent insulator until it breaks down into plasma at electric field strengths above 30 kilovolts per centimeter.	Usually very high: For many purposes, the conductivity of a plasma may be treated as infinite.
Independently acting species	One: All gas particles behave in a similar way, influenced by gravity and by collisions with one another.	Two or three: Electrons, ions, protons and neutrons can be distinguished by the sign and value of their charge so that they behave independently in many circumstances, with different bulk velocities and temperatures, allowing phenomena such as new types of waves and instabilities.

Velocity distribution	Maxwellian: Collisions usually lead to a Maxwellian velocity distribution of all gas particles, with very few relatively fast particles.	Often non-Maxwellian: Collisional interactions are often weak in hot plasmas and external forcing can drive the plasma far from local equilibrium and lead to a significant population of unusually fast particles.
Interactions	Binary: Two-particle collisions are the rule, three-body collisions extremely rare.	Collective: Waves, or organized motion of plasma, are very important because the particles can interact at long ranges through the electric and magnetic forces.

Common Plasmas

Plasmas are by far the most common phase of ordinary matter in the universe, both by mass and by volume. Essentially, all of the visible light from space comes from stars, which are plasmas with a temperature such that they radiate strongly at visible wavelengths. Most of the ordinary (or baryonic) matter in the universe, however, is found in the intergalactic medium, which is also a plasma, but much hotter, so that it radiates primarily as X-rays.

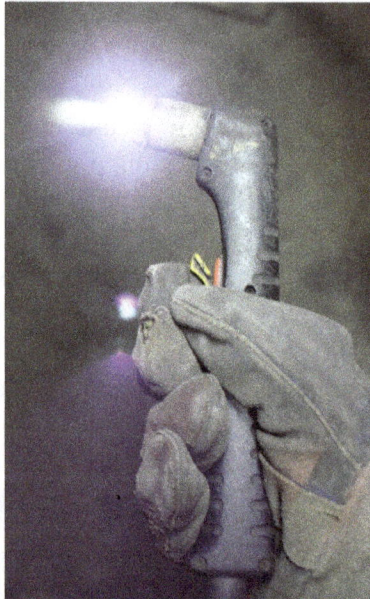

A handheld plasma cutter used to cut steel plates. A jet of gas is heated with an electric arc into a plasma.

In 1937, Hannes Alfvén argued that if plasma pervaded the universe, it could then carry electric currents capable of generating a galactic magnetic field. After winning the Nobel Prize, he emphasized that:

In order to understand the phenomena in a certain plasma region, it is necessary to map not only the magnetic but also the electric field and the electric currents. Space is filled with a network of currents which transfer energy and momentum over large or very large distances. The currents often pinch to filamentary or surface currents. The latter are likely to give space, as also interstellar and intergalactic space, a cellular structure.

By contrast the current scientific consensus is that about 96% of the total energy density in the universe is not plasma or any other form of ordinary matter, but a combination of cold dark matter and dark energy. Our Sun, and all of the other stars, are made of plasma, much of interstellar space is filled with a plasma, albeit a very sparse one, and intergalactic space too. Even black holes, which are not directly visible, are thought to be fuelled by accreting ionising matter (i.e. plasma), and they are associated with astrophysical jets of luminous ejected plasma, such as M87's jet that extends 5,000 light-years.

In our solar system, interplanetary space is filled with the plasma of the Solar Wind that extends from the Sun out to the heliopause. However, the density of ordinary matter is much higher than average and much higher than that of either dark matter or dark energy. The planet Jupiter accounts for most of the *non*-plasma within the orbit of Pluto (about 0.1% by mass, or 10^{-15}% by volume).

Dust and small grains within a plasma will also pick up a net negative charge, so that they in turn may act like a very heavy negative ion component of the plasma.

Common forms of plasma		
Artificially produced	**Terrestrial plasmas**	**Space and astrophysical plasmas**
• Those found in plasma displays, including TV screens. • Inside fluorescent lamps (low energy lighting), neon signs • Rocket exhaust and ion thrusters • The area in front of a spacecraft's heat shield during re-entry into the atmosphere • Inside a corona discharge ozone generator • Fusion energy research • The electric arc in an arc lamp, an arc welder or plasma torch • Plasma ball (sometimes called a plasma sphere or plasma globe) • Arcs produced by Tesla coils (resonant air core transformer or disruptor coil that produces arcs similar to lightning, but with alternating current rather than static electricity) • Plasmas used in semiconductor device fabrication including reactive-ion etching, sputtering, surface cleaning and plasma-enhanced chemical vapor deposition • Laser-produced plasmas (LPP), found when high power lasers interact with materials. • Inductively coupled plasmas (ICP), formed typically in argon gas for optical emission spectroscopy or mass spectrometry • Magnetically induced plasmas (MIP), typically produced using microwaves as a resonant coupling method • Static electric sparks	• Lightning • St. Elmo's fire • Upper-atmospheric lightning (e.g. Blue jets, Blue starters, Gigantic jets, ELVES) • Sprites • The ionosphere • The plasmasphere • The polar aurorae • Some flames • The polar wind, a plasma fountain	• The Sun and other stars (plasmas heated by nuclear fusion) • The solar wind • The interplanetary medium (space between planets) • The interstellar medium (space between star systems) • The Intergalactic medium (space between galaxies) • The Io-Jupiter flux tube • Accretion discs • Interstellar nebulae • Cometary ion tail

Complex Plasma Phenomena

Although the underlying equations governing plasmas are relatively simple, plasma behavior is extraordinarily varied and subtle: the emergence of unexpected behavior from a simple model is a typical feature of a complex system. Such systems lie in some sense on the boundary between ordered and disordered behavior and cannot typically be described either by simple, smooth, mathematical functions, or by pure randomness. The spontaneous formation of interesting spatial features on a wide range of length scales is one manifestation of plasma complexity. The features are interesting, for example, because they are very sharp, spatially intermittent (the distance between features is much larger than the features themselves), or have a fractal form. Many of these features were first studied in the laboratory, and have subsequently been recognized throughout the universe. Examples of complexity and complex structures in plasmas include:

Filamentation

Striations or string-like structures, also known as Birkeland currents, are seen in many plasmas, like the plasma ball, the aurora, lightning, electric arcs, solar flares, and supernova remnants. They are sometimes associated with larger current densities, and the interaction with the magnetic field can form a magnetic rope structure. High power microwave breakdown at atmospheric pressure also leads to the formation of filamentary structures.

Filamentation also refers to the self-focusing of a high power laser pulse. At high powers, the non-linear part of the index of refraction becomes important and causes a higher index of refraction in the center of the laser beam, where the laser is brighter than at the edges, causing a feedback that focuses the laser even more. The tighter focused laser has a higher peak brightness (irradiance) that forms a plasma. The plasma has an index of refraction lower than one, and causes a defocusing of the laser beam. The interplay of the focusing index of refraction, and the defocusing plasma makes the formation of a long filament of plasma that can be micrometers to kilometers in length. One interesting aspect of the filamentation generated plasma is the relatively low ion density due to defocusing effects of the ionized electrons.

Shocks or Double Layers

Plasma properties change rapidly (within a few Debye lengths) across a two-dimensional sheet in the presence of a (moving) shock or (stationary) double layer. Double layers involve localized charge separation, which causes a large potential difference across the layer, but does not generate an electric field outside the layer. Double layers separate adjacent plasma regions with different physical characteristics, and are often found in current carrying plasmas. They accelerate both ions and electrons.

Electric Fields and Circuits

Quasineutrality of a plasma requires that plasma currents close on themselves in electric circuits. Such circuits follow Kirchhoff's circuit laws and possess a resistance and inductance. These circuits must generally be treated as a strongly coupled system, with the behavior in each plasma region dependent on the entire circuit. It is this strong coupling between system elements, together with nonlinearity, which may lead to complex behavior. Electrical circuits in plasmas store inductive (magnetic) energy, and should the circuit be disrupted, for example, by a plasma instability, the inductive energy will be released as plasma heating and acceleration. This is a common explanation for the heating that takes place in the solar corona. Electric currents, and in particular, magnetic-field-aligned electric currents (which are sometimes generically referred to as "Birkeland currents"), are also observed in the Earth's aurora, and in plasma filaments.

Cellular Structure

Narrow sheets with sharp gradients may separate regions with different properties such as magnetization, density and temperature, resulting in cell-like regions. Examples include the magnetosphere, heliosphere, and heliospheric current sheet. Hannes Alfvén wrote: "From the cosmological point of view, the most important new space research discovery is probably the cellular structure of space. As has been seen in every region of space accessible to in situ measurements, there are a number of 'cell walls', sheets of electric currents, which divide space into compartments with different magnetization, temperature, density, etc."

Critical Ionization Velocity

The critical ionization velocity is the relative velocity between an ionized plasma and a neutral gas, above which a runaway ionization process takes place. The critical ionization process is a quite general mechanism for the conversion of the kinetic energy of a rapidly streaming gas into ionization and plasma thermal energy. Critical phenomena in general are typical of complex systems, and may lead to sharp spatial or temporal features.

Ultracold Plasma

Ultracold plasmas are created in a magneto-optical trap (MOT) by trapping and cooling neutral atoms, to temperatures of 1 mK or lower, and then using another laser to ionize the atoms by giving each of the outermost electrons just enough energy to escape the electrical attraction of its parent ion.

One advantage of ultracold plasmas are their well characterized and tunable initial conditions, including their size and electron temperature. By adjusting the wavelength of the ionizing laser, the kinetic energy of the liberated electrons can be tuned as low as 0.1 K, a limit set by the frequency bandwidth of the laser pulse. The ions inherit the millikelvin temperatures of the neutral atoms, but are quickly heated through a process known as disorder induced heating (DIH). This type of non-equilibrium ultracold plasma evolves rapidly, and displays many other interesting phenomena.

One of the metastable states of a strongly nonideal plasma is Rydberg matter, which forms upon condensation of excited atoms.

Non-neutral Plasma

The strength and range of the electric force and the good conductivity of plasmas usually ensure that the densities of positive and negative charges in any sizeable region are equal ("quasineutrality"). A plasma with a significant excess of charge density, or, in the extreme case, is composed of a single species, is called a non-neutral plasma. In such a plasma, electric fields play a dominant role. Examples are charged particle beams, an electron cloud in a Penning trap and positron plasmas.

Dusty Plasma/Grain Plasma

A dusty plasma contains tiny charged particles of dust (typically found in space). The dust particles acquire high charges and interact with each other. A plasma that contains larger particles is called grain plasma. Under laboratory conditions, dusty plasmas are also called *complex plasmas*.

Impermeable Plasma

Impermeable plasma is a type of thermal plasma which acts like an impermeable solid with respect to gas or cold plasma and can be physically pushed. Interaction of cold gas and thermal plasma was briefly studied by a group led by Hannes Alfvén in 1960s and 1970s for its possible applications in insulation of fusion plasma from the reactor walls. However, later it was found that the external magnetic fields in this configuration could induce kink instabilities in the plasma and subsequently lead to an unexpectedly high heat loss to the walls. In 2013, a group of materials scientists reported that they have successfully generated stable impermeable plasma with no magnetic confinement using only an ultrahigh-pressure blanket of cold gas. While spectroscopic data on the characteristics of plasma were claimed to be difficult to obtain due to the high pressure, the passive effect of plasma on synthesis of different nanostructures clearly suggested the effective confinement. They also showed that upon maintaining the impermeability for a few tens of seconds, screening of ions at the plasma-gas interface could give rise to a strong secondary mode of heating (known as viscous heating) leading to different kinetics of reactions and formation of complex nanomaterials.

Mathematical Descriptions

To completely describe the state of a plasma, we would need to write down all the particle locations and velocities and describe the electromagnetic field in the plasma region. However, it is generally not practical or necessary to keep track of all the particles in a plasma. Therefore, plasma physicists commonly use less detailed descriptions, of which there are two main types:

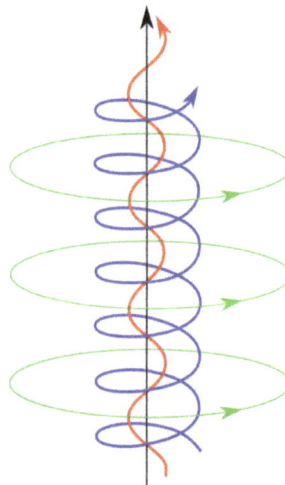

The complex self-constricting magnetic field lines and current paths in a field-aligned Birkeland current that can develop in a plasma.

Fluid Model

Fluid models describe plasmas in terms of smoothed quantities, like density and averaged velocity around each position. One simple fluid model, magnetohydrodynamics, treats the plasma as a single fluid governed by a combination of Maxwell's equations and the Navier–Stokes equations. A more general description is the two-fluid plasma picture, where the ions and electrons are described separately. Fluid models are often accurate when collisionality is sufficiently high to keep the plasma velocity distribution close to a Maxwell–Boltzmann distribution. Because fluid models usually describe the plasma in terms of a single flow at a certain temperature at each spatial location, they can neither capture velocity space structures like beams or double layers, nor resolve wave-particle effects.

Kinetic Model

Kinetic models describe the particle velocity distribution function at each point in the plasma and therefore do not need to assume a Maxwell–Boltzmann distribution. A kinetic description is often necessary for collisionless plasmas. There are two common approaches to kinetic description of a plasma. One is based on representing the smoothed distribution function on a grid in velocity and position. The other, known as the particle-in-cell (PIC) technique, includes kinetic information by following the trajectories of a large number of individual particles. Kinetic models are generally more computationally intensive than fluid models. The Vlasov equation may be used to describe the dynamics of a system of charged particles interacting with an electromagnetic field. In magnetized plasmas, a gyrokinetic approach can substantially reduce the computational expense of a fully kinetic simulation.

Artificial Plasmas

Most artificial plasmas are generated by the application of electric and/or magnetic fields through a gas. Plasma generated in a laboratory setting and for industrial use can be generally categorized by:

- The type of power source used to generate the plasma—DC, RF and microwave

- The pressure they operate at—vacuum pressure (< 10 mTorr or 1 Pa), moderate pressure (~ 1 Torr or 100 Pa), atmospheric pressure (760 Torr or 100 kPa)

- The degree of ionization within the plasma—fully, partially, or weakly ionized

- The temperature relationships within the plasma—thermal plasma $(T_e = T_i = T_{gas})$, non-thermal or "cold" plasma $(T_e \gg T_i = T_{gas})$

- The electrode configuration used to generate the plasma

- The magnetization of the particles within the plasma—magnetized (both ion and electrons are trapped in Larmor orbits by the magnetic field), partially magnetized (the electrons but not the ions are trapped by the magnetic field), non-magnetized (the magnetic field is too weak to trap the particles in orbits but may generate Lorentz forces)

Generation of Artificial Plasma

Just like the many uses of plasma, there are several means for its generation, however, one principle is common to all of them: there must be energy input to produce and sustain it. For this case, plasma is generated when an electric current is applied across a dielectric gas or fluid (an electrically non-conducting material) as can be seen in the image to the right, which shows a discharge tube as a simple example (DC used for simplicity).

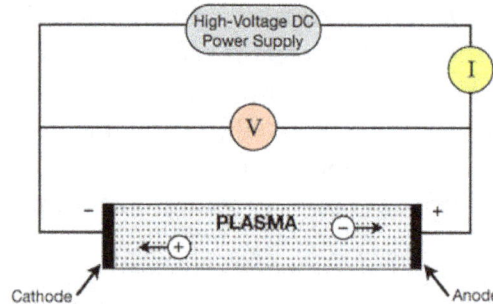

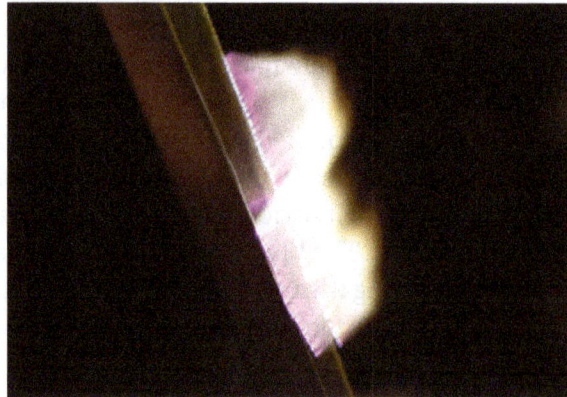

Artificial plasma produced in air by a Jacob's Ladder

The potential difference and subsequent electric field pull the bound electrons (negative) toward the anode (positive electrode) while the cathode (negative electrode) pulls the nucleus. As the voltage increases, the current stresses the material (by electric polarization) beyond its dielectric limit (termed strength) into a stage of electrical breakdown, marked by an electric spark, where the material transforms from being an insulator into a conductor (as it becomes increasingly ionized). The underlying process is the Townsend avalanche, where collisions between electrons and neutral gas atoms create more ions and electrons (as can be seen in the figure on the right). The first impact of an electron on an atom results in one ion and two electrons. Therefore, the number of charged particles increases rapidly (in the millions) only "after about 20 successive sets of collisions", mainly due to a small mean free path (average distance travelled between collisions).

Electric Arc

With ample current density and ionization, this forms a luminous electric arc (a continuous electric discharge similar to lightning) between the electrodes. Electrical resistance along the continuous electric arc creates heat, which dissociates more gas molecules and ionizes the resulting atoms (where degree of ionization is determined by temperature), and as per the sequence: solid-liquid-gas-plasma, the gas is gradually turned into a thermal plasma. A thermal plasma is in thermal equilibrium, which is to say that the temperature is relatively homogeneous throughout the heavy particles (i.e. atoms, molecules and ions) and electrons. This is so because when thermal plasmas are generated, electrical energy is given to electrons, which, due to their great mobility and large numbers, are able to disperse it rapidly and by elastic collision (without energy loss) to the heavy particles.

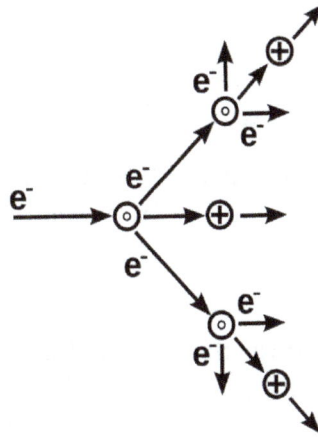

Cascade process of ionization. Electrons are 'e−', neutral atoms 'o', and cations '+'.

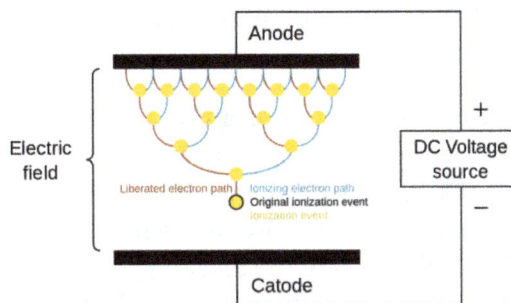

Avalanche effect between two electrodes. The original ionisation event liberates one electron, and each subsequent collision liberates a further electron, so two electrons emerge from each collision: the ionising electron and the liberated electron.

Examples of Industrial/Commercial Plasma

Because of their sizable temperature and density ranges, plasmas find applications in many fields of research, technology and industry. For example, in: industrial and extractive metallurgy, surface treatments such as plasma spraying (coating), etching in microelectronics, metal cutting and welding; as well as in everyday vehicle exhaust cleanup and fluorescent/luminescent lamps, while even playing a part in supersonic combustion engines for aerospace engineering.

Low-pressure Discharges

- *Glow discharge plasmas*: non-thermal plasmas generated by the application of DC or low frequency RF (<100 kHz) electric field to the gap between two metal electrodes. Probably the most common plasma; this is the type of plasma generated within fluorescent light tubes.

- *Capacitively coupled plasma (CCP)*: similar to glow discharge plasmas, but generated with high frequency RF electric fields, typically 13.56 MHz. These differ from glow discharges in that the sheaths are much less intense. These are widely used in the microfabrication and integrated circuit manufacturing industries for plasma etching and plasma enhanced chemical vapor deposition.

- *Cascaded Arc Plasma Source*: a device to produce low temperature (~1eV) high density plasmas (HDP).

- *Inductively coupled plasma (ICP)*: similar to a CCP and with similar applications but the electrode consists of a coil wrapped around the chamber where plasma is formed.

- *Wave heated plasma*: similar to CCP and ICP in that it is typically RF (or microwave). Examples include helicon discharge and electron cyclotron resonance (ECR).

Atmospheric Pressure

- *Arc discharge:* this is a high power thermal discharge of very high temperature (~10,000 K). It can be generated using various power supplies. It is commonly used in metallurgical processes. For example, it is used to smelt minerals containing Al_2O_3 to produce aluminium.

- *Corona discharge:* this is a non-thermal discharge generated by the application of high voltage to sharp electrode tips. It is commonly used in ozone generators and particle precipitators.

- *Dielectric barrier discharge (DBD):* this is a non-thermal discharge generated by the application of high voltages across small gaps wherein a non-conducting coating prevents the transition of the plasma discharge into an arc. It is often mislabeled 'Corona' discharge in industry and has similar application to corona discharges. It is also widely used in the web treatment of fabrics. The application of the discharge to synthetic fabrics and plastics functionalizes the surface and allows for paints, glues and similar materials to adhere.

- *Capacitive discharge:* this is a nonthermal plasma generated by the application of RF power (e.g., 13.56 MHz) to one powered electrode, with a grounded electrode held at a small separation distance on the order of 1 cm. Such discharges are commonly stabilized using a noble gas such as helium or argon.

- "Piezoelectric direct discharge plasma:" is a nonthermal plasma generated at the high-side of a piezoelectric transformer (PT). This generation variant is particularly suited for high efficient and compact devices where a separate high voltage power supply is not desired.

History

Plasma was first identified in a Crookes tube, and so described by Sir William Crookes in 1879 (he called it "radiant matter"). The nature of the Crookes tube "cathode ray" matter was subsequently identified by British physicist Sir J.J. Thomson in 1897. The term "plasma" was coined by Irving Langmuir in 1928, perhaps because the glowing discharge molds itself to the shape of the Crookes tube. Langmuir described his observations as:

Except near the electrodes, where there are *sheaths* containing very few electrons, the ionized gas contains ions and electrons in about equal numbers so that the resultant space charge is very small. We shall use the name *plasma* to describe this region containing balanced charges of ions and electrons.

Research

Plasmas are the object of study of the academic field of *plasma science* or *plasma physics*, including sub-disciplines such as space plasma physics. There are multiple journals devoted to the subject. It involves the following fields of active research:

- Plasma theory
  - Plasma equilibria and stability
  - Plasma interactions with waves and beams
  - Guiding center
  - Adiabatic invariant
  - Debye sheath
  - Coulomb collision
- Plasmas in nature
  - The Earth's ionosphere
  - Northern and southern (polar) lights
  - Space plasmas, e.g. Earth's plasmasphere (an inner portion of the magnetosphere dense with plasma)
  - Astrophysical plasma
  - Interplanetary medium
- Industrial plasmas
  - Plasma chemistry
  - Plasma processing
  - Plasma spray
  - Plasma display
  - Plasma sources
  - Dusty plasmas
- Plasma diagnostics
  - Thomson scattering
  - Langmuir probe
  - Ball-pen probe
  - Faraday cup
  - Spectroscopy
  - Interferometry
  - Ionospheric heating
  - Incoherent scatter radar

- Plasma applications
  - Dielectric barrier discharge
  - Enhanced oil recovery
  - Fusion power
    - Magnetic fusion energy (MFE) — tokamak, stellarator, reversed field pinch, magnetic mirror, dense plasma focus
    - Inertial fusion energy (IFE) (also Inertial confinement fusion — ICF)
    - Plasma weapon
  - Ion implantation
  - Ion thruster
  - MAGPIE (short for Mega Ampere Generator for Plasma Implosion Experiments)
  - Plasma ashing
  - Food processing (nonthermal plasma, aka "cold plasma")
  - Plasma arc waste disposal, convert waste into reusable material with plasma.
  - Plasma acceleration
  - Plasma medicine (e. g. Dentistry)
  - Plasma window

Solar plasma

Plasma spraying

Hall effect thruster. The electric field in a plasma double layer is so effective at accelerating ions that electric fields are used in ion drives.

References

- Sturrock, Peter A. (1994). Plasma Physics: An Introduction to the Theory of Astrophysical, Geophysical & Laboratory Plasmas. Cambridge University Press. ISBN 978-0-521-44810-9.

- Hazeltine, R.D.; Waelbroeck, F.L. (2004). The Framework of Plasma Physics. Westview Press. ISBN 978-0-7382-0047-7.

- Hastings, Daniel & Garrett, Henry (2000). Spacecraft-Environment Interactions. Cambridge University Press. ISBN 978-0-521-47128-2.

- von Engel, A. and Cozens, J.R. (1976) "Flame Plasma" in Advances in electronics and electron physics, L. L. Marton (ed.), Academic Press, ISBN 978-0-12-014520-1, p. 99

- Yaffa Eliezer, Shalom Eliezer, The Fourth State of Matter: An Introduction to the Physics of Plasma, Publisher: Adam Hilger, 1989, ISBN 978-0-85274-164-1, 226 pages, page 5

- Mészáros, Péter (2010) The High Energy Universe: Ultra-High Energy Events in Astrophysics and Cosmology, Publisher Cambridge University Press, ISBN 978-0-521-51700-3, p. 99.

- Raine, Derek J. and Thomas, Edwin George (2010) Black Holes: An Introduction, Publisher: Imperial College Press, ISBN 978-1-84816-382-9, p. 160

- National Research Council (U.S.). Plasma 2010 Committee (2007). Plasma science: advancing knowledge in the national interest. National Academies Press. pp. 190–193. ISBN 978-0-309-10943-7.

- Hippler, R.; Kersten, H.; Schmidt, M.; Schoenbach, K.M., eds. (2008). "Plasma Sources". Low Temperature Plasmas: Fundamentals, Technologies, and Techniques (2nd ed.). Wiley-VCH. ISBN 978-3-527-40673-9.

- National Research Council (1991). Plasma Processing of Materials : Scientific Opportunities and Technological Challenges. National Academies Press. ISBN 978-0-309-04597-1.

- Brown, Sanborn C. (1978). "Chapter 1: A Short History of Gaseous Electronics". In HIRSH, Merle N. e OSKAM, H. J. Gaseous Electronics. 1. Academic Press. ISBN 978-0-12-349701-7.

Key Concepts of Plasma Physics

Guiding center, Debye sheath, double layer, microplasma, plasmon and pinch are the key concepts related to plasma physics. Debye sheath is the layer of plasma that contains within itself a density of positive ions whereas a double layer is a structure that has layers of opposite charged layers. The following section unfolds its crucial aspects in a critical yet systematic manner.

Guiding Center

In physics, the motion of an electrically charged particle (such as an electron or ion in a plasma) in a magnetic field can be treated as the superposition of a relatively fast circular motion around a point called the guiding center and a relatively slow drift of this point. The drift speeds may differ for various species depending on their charge states, masses, or temperatures, possibly resulting in electric currents or chemical separation.

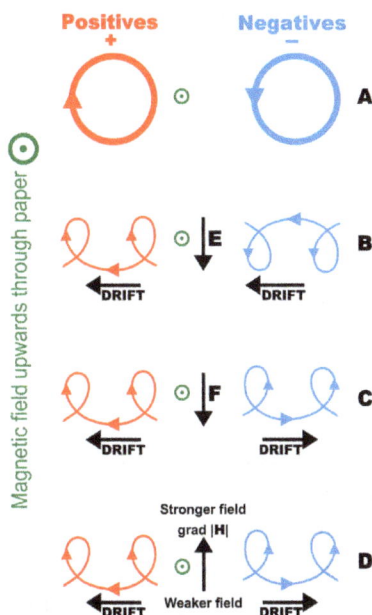

Charged particle drifts in a homogeneous magnetic field. (A) No disturbing force (B) With an electric field, E (C) With an independent force, F (e.g. gravity) (D) In an inhomogeneous magnetic field, grad H

Gyration

If the magnetic field is uniform and all other forces are absent, then the Lorentz force will cause a particle to undergo a constant acceleration perpendicular to both the particle velocity and the magnetic field. This does not affect particle motion parallel to the magnetic field, but results in circular motion at constant speed in the plane perpendicular to the magnetic field. This circular motion is known as the gyromotion. For a particle with mass m and charge B moving in a magnetic field with strength m, it has a frequency, called the gyrofrequency or cyclotron frequency, of

$$\omega_c = |q| B / m.$$

For a speed perpendicular to the magnetic field of $v_\perp$, the radius of the orbit, called the gyroradius or Larmor radius, is

$$\rho_L = v_\perp / \omega_c.$$

Parallel Motion

Since the magnetic Lorentz force is always perpendicular to the magnetic field, it has no influence (to lowest order) on the parallel motion. In a uniform field with no additional forces, a charged particle will gyrate around the magnetic field according to the perpendicular component of its velocity and drift parallel to the field according to its initial parallel velocity, resulting in a helical orbit. If there is a force with a parallel component, the particle and its guiding center will be correspondingly accelerated.

If the field has a parallel gradient, a particle with a finite Larmor radius will also experience a force in the direction away from the larger magnetic field. This effect is known as the magnetic mirror. While it is closely related to guiding center drifts in its physics and mathematics, it is nevertheless considered to be distinct from them.

General Force Drifts

Generally speaking, when there is a force on the particles perpendicular to the magnetic field, then they drift in a direction perpendicular to both the force and the field. If $\vec{F}$ is the force on one particle, then the drift velocity is

$$\vec{v}_f = \frac{1}{q} \frac{\vec{F} \times \vec{B}}{B^2}.$$

These drifts, in contrast to the mirror effect and the non-uniform B drifts, do not depend on finite Larmor radius, but are also present in cold plasmas. This may seem counterintuitive. If a particle is stationary when a force is turned on, where does the motion perpendicular to the force come from and why doesn't the force produce a motion parallel to itself? The answer is the interaction with the magnetic field. The force initially results in an acceleration parallel to itself, but the magnetic field deflects the resulting motion in the drift direction. Once the particle is moving in the drift direction, the magnetic field deflects it back against the external force, so that the average acceleration in the direction of the force is zero. There is, however, a one-time displacement in the direction of the force equal to $(f/m)\omega_c^{-2}$, which should be considered a consequence of the polarization drift while the force is being turned on. The resulting motion is a cycloid. More generally, the superposition of a gyration and a uniform perpendicular drift is a trochoid.

All drifts may be considered special cases of the force drift, although this is not always the most useful way to think about them. The obvious cases are electric and gravitational forces. The grad-B drift can be considered to result from the force on a magnetic dipole in a field gradient. The curvature, inertia, and polarisation drifts result from treating the acceleration of the particle as fictitious forces. The diamagnetic drift can be derived from the force due to a pressure gradient. Finally, other forces such as radiation pressure and collisions also result in drifts.

Gravitational Field

A simple example of a force drift is a plasma in a gravitational field, e.g. the ionosphere. The drift velocity is

$$\vec{v}_g = \frac{m}{q} \frac{\vec{g} \times \vec{B}}{B^2}$$

Because of the mass dependence, the gravitational drift for the electrons can normally be ignored.

The dependence on the charge of the particle implies that the drift direction is opposite for ions as for electrons, resulting in a current. In a fluid picture, it is this current crossed with the magnetic field that provides that force counteracting the applied force.

Electric Field

This drift, often called the $\vec{E} \times \vec{B}$ (E-cross-B) drift, is a special case because the electric force on a particle depends on its charge (as opposed, for example, to the gravitational force considered above). As a result, ions (of whatever mass and charge) and electrons both move in the same direction at the same speed, so there is no net current (assuming quasineutrality). In the context of special relativity, in the frame moving with this velocity, the electric field vanishes. The value of the drift velocity is given by

$$\vec{v}_E = \frac{\vec{E} \times \vec{B}}{B^2}$$

Nonuniform E

If the electric field is not uniform, the above formula is modified to read

$$\vec{v}_E = \left(1 + \frac{1}{4} \rho_L^2 \nabla^2\right) \frac{\vec{E} \times \vec{B}}{B^2}$$

Nonuniform B

Guiding center drifts may also result not only from external forces but also from non-uniformities in the magnetic field. It is convenient to express these drifts in terms of the parallel and perpendicular kinetic energies

$$K_{\parallel} = \frac{1}{2} m v_{\parallel}^2$$

$$K_{\perp} = \frac{1}{2} m v_{\perp}^2$$

In that case, the explicit mass dependence is eliminated. If the ions and electrons have similar temperatures, then they also have similar, though oppositely directed, drift velocities.

Grad-B Drift

When a particle moves into a larger magnetic field, the curvature of its orbit becomes tighter, transforming the otherwise circular orbit into a cycloid. The drift velocity is

$$\vec{v}_{\nabla B} = \frac{K_{\perp}}{qB} \frac{\vec{B} \times \nabla B}{B^2}$$

Curvature Drift

In order for a charged particle to follow a curved field line, it needs a drift velocity out of the plane of curvature to provide the necessary centripetal force. This velocity is

$$\vec{v}_R = \frac{2K_{\parallel}}{qB} \frac{\vec{R}_c \times \vec{B}}{R_c^2 B}$$

where $\vec{R}_c$ is the radius of curvature pointing outwards, away from the center of the circular arc which best approximates the curve at that point.

$$\vec{v}_{\text{inertial}} = \frac{v_{\parallel}}{\omega_c} \vec{b} \times \frac{d\vec{b}}{dt},$$

where $\vec{b} = \vec{B}/B$ is the unit vector in the direction of the magnetic field. This drift can be decomposed into the sum of the curvature drift and the term

$$\frac{v_{\parallel}}{\omega_c} \vec{b} \times \left[\frac{\partial \vec{b}}{\partial t} + (\vec{v}_E \cdot \nabla \vec{b}) \right].$$

In the important limit of stationary magnetic field and weak electric field, the inertial drift is dominated by the curvature drift term.

Curved Vacuum Drift

In the limit of small plasma pressure, Maxwell's equations provide a relationship between gradient and curvature that allows the corresponding drifts to be combined as follows

$$\vec{v}_R + \vec{v}_{\nabla B} = \frac{2K_\parallel + K_\perp}{qB} \frac{\vec{R}_c \times \vec{B}}{R_c^2 B}$$

For a species in thermal equilibrium, $2K_\parallel + K_\perp$ can be replaced by $2k_B T$ ($k_B T >$ for $K_\parallel$ and $k_B T$ for $K_\perp$).

The expression for the grad-B drift above can be rewritten for the case when ∇B is due to the curvature. This is most easily done by realizing that in a vacuum, Ampere's Law is $\nabla \times \vec{B} = 0$. In cylindrical coordinates chosen such that the azimuthal direction is parallel to the magnetic field and the radial direction is parallel to the gradient of the field, this becomes

$$\nabla \times \vec{B} = \frac{1}{r} \frac{\partial}{\partial r} (rB_\theta) \hat{z} = 0$$

Since rB_θ is a constant, this implies that

$$\nabla B = -B \frac{\vec{R}_c}{R_c^2}$$

and the grad-B drift velocity can be written

$$\vec{v}_{\nabla B} = -\frac{K_\perp}{q} \frac{\vec{B} \times \vec{R}_c}{R_c^2 B^2}$$

Polarization Drift

A time-varying electric field also results in a drift given by

$$\vec{v}_p = \frac{m}{qB^2} \frac{d\vec{E}}{dt}$$

Obviously this drift is different from the others in that it cannot continue indefinitely. Normally an oscillatory electric field results in a polarization drift oscillating 90 degrees out of phase. Because of the mass dependence, this effect is also called the inertia drift. Normally the polarization drift can be neglected for electrons because of their relatively small mass.

Diamagnetic Drift

The diamagnetic drift is not actually a guiding center drift. A pressure gradient does not cause any single particle to drift. Nevertheless, the fluid velocity is defined by counting the particles moving through a reference area, and a pressure gradient results in more particles in one direction than in the other. The net velocity of the fluid is given by

$$\vec{v}_D = -\frac{\nabla p \times \vec{B}}{qnB^2}$$

Drift Currents

With the important exception of the E-cross-B drift, the drift velocities of different species will be different. The differential velocity of charged particles results in a current, while the mass dependence of the drift velocity can result in chemical separation.

Debye Sheath

The Debye sheath (also electrostatic sheath) is a layer in a plasma which has a greater density of positive ions, and hence an overall excess positive charge, that balances an opposite negative charge on the surface of a material with which it is in contact. The thickness of such a layer is several Debye lengths thick, a value whose size depends on various characteristics of plasma (e.g. temperature, density, etc.).

A Debye sheath arises in a plasma because the electrons usually have a temperature on the order of magnitude or greater than that of the ions and are much lighter. Consequently, they are faster than the ions by at least a factor of $\sqrt{m_i / m_e}$. At the interface to a material surface, therefore, the electrons will fly out of the plasma, charging the surface negative relative to the bulk plasma. Due to Debye shielding, the scale length of the transition region will be the Debye length λ_D. As the potential increases, more and more electrons are reflected by the sheath potential. An equilibrium is finally reached when the potential difference is a few times the electron temperature.

The Debye sheath is the transition from a plasma to a solid surface. Similar physics is involved between two plasma regions that have different characteristics; the transition between these regions is known as a double layer, and features one positive, and one negative layer.

Description

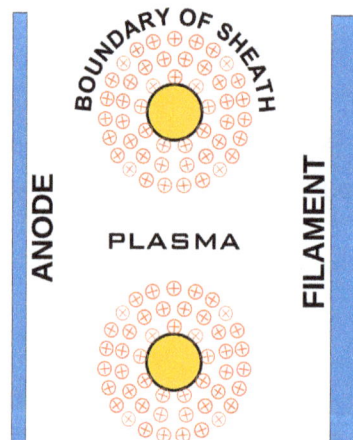

Positive ion **sheaths** around grid wires in a thermionic gas tube, where ⊕ represents a positive charge (not to scale) (After Langmuir, 1929)

Sheaths were first described by American physicist Irving Langmuir. In 1923 he wrote:

"Electrons are repelled from the negative electrode while positive ions are drawn towards it.

Around each negative electrode there is thus a *sheath* of definite thickness containing only pos-itive ions and neutral atoms. [..] Electrons are reflected from the outside surface of the sheath while all *positive* ions which reach the sheath are attracted to the electrode. [..] it follows direct-ly that no change occurs in the positive ion current reaching the electrode. The electrode is in fact perfectly screened from the discharge by the positive ion sheath, and its potential cannot influence the phenomena occurring in the arc, nor the current flowing to the electrode."

Langmuir and co-author Albert W. Hull further described a sheath formed in a thermionic valve:

"Figure 1 shows graphically the condition that exists in such a tube containing mercury vapor. The space between filament and plate is filled with a mixture of electrons and pos-itive ions, in nearly equal numbers, to which has been given the name "plasma". A wire immersed in the plasma, at zero potential with respect to it, will absorb every ion and electron that strikes it. Since the electrons move about 600 times as fast as the ions, 600 times as many electrons will strike the wire as ions. If the wire is insulated it must assume such a negative potential that it receives equal numbers of electrons and ions, that is, such a potential that it repels all but 1 in 600 of the electrons headed for it."

"Suppose that this wire, which we may take to be part of a grid, is made still more negative with a view to controlling the current through the tube. It will now repel all the electrons headed for it, but will receive all the positive ions that fly toward it. There will thus be a region around the wire which contains positive ions and no electrons, as shown diagram-matically in Fig. 1. The ions are accelerated as they approach the negative wire, and there will exist a potential gradient in this sheath, as we may call it, of positive ions, such that the potential is less and less negative as we recede from the wire, and at a certain distance is equal to the potential of the plasma. This distance we define as the boundary of the sheath. Beyond this distance there is no effect due to the potential of the wire."

Mathematical Treatment

The Planar Sheath Equation

The quantitative physics of the Debye sheath is determined by four phenomena:

Energy conservation of the ions: If we assume for simplicity cold ions of mass m_i entering the sheath with a velocity u_0, having charge opposite to the electron, conservation of energy in the sheath potential requires

$$\frac{1}{2}m_i u(x)^2 = \frac{1}{2}m_i u_0^2 - e\varphi(x),$$

where e is the charge of the electron taken positively, i.e. $e = 1.602 \times 10^{-19}$ C.

Ion continuity: In the steady state, the ions do not build up anywhere, so the flux is everywhere the same:

$$n_0 u_0 = n_i(x)u(x).$$

Boltzmann relation for the electrons: Since most of the electrons are reflected, their density is given by

$$n_e(x) = n_0 \exp\left(\frac{e\varphi(x)}{k_B T_e}\right).$$

Poisson's equation: The curvature of the electrostatic potential is related to the net charge density as follows:

$$\frac{d^2\varphi(x)}{dx^2} = \frac{e(n_e(x) - n_i(x))}{\epsilon_0}.$$

Combining these equations and writing them in terms of the dimensionless potential, position, and ion speed,

$$\chi(\xi) = -\frac{e\varphi(\xi)}{k_B T_e}$$

$$\xi = \frac{x}{\lambda_D}$$

$$\mathfrak{M} = \frac{u_0}{(k_B T_e / m_i)^{1/2}}$$

we arrive at the sheath equation:

$$\chi'' = \left(1 + \frac{2\chi}{\mathfrak{M}^2}\right)^{-1/2} - e^{-\chi}.$$

The Bohm Sheath Criterion

The sheath equation can be integrated once by multiplying by χ':

$$\int_0^\xi \chi'\chi'' d\xi_1 = \int_0^\xi \left(1 + \frac{2\chi}{\mathfrak{M}^2}\right)^{-1/2} \chi' d\xi_1 - \int_0^\xi e^{-\chi}\chi' d\xi_1$$

At the sheath edge ($\xi = 0$), we can define the potential to be zero ($\chi = 0$) and assume that the electric field is also zero ($\chi' = 0$). With these boundary conditions, the integrations yield

$$\frac{1}{2}\chi'^2 = \mathfrak{M}^2\left[\left(1 + \frac{2\chi}{\mathfrak{M}^2}\right)^{1/2} - 1\right] + e^{-\chi} - 1$$

This is easily rewritten as an integral in closed form, although one that can only be solved numerically. Nevertheless, an important piece of information can be derived analytically. Since the left-hand-side is a square, the right-hand-side must also be non-negative for every value of χ, , in particular for small values. Looking at the Taylor expansion around $\chi = 0$, we see that the first term that does not vanish is the quadratic one, so that we can require

$$\frac{1}{2}\chi^2 \left(-\frac{1}{\mathfrak{M}^2} + 1 \right) \geq 0,$$

or

$$\mathfrak{M}^2 \geq 1,$$

or

$$u_0 \geq (k_B T_e / m_i)^{1/2}.$$

This inequality is known as the Bohm sheath criterion after its discoverer, David Bohm. If the ions are entering the sheath too slowly, the sheath potential will "eat" its way into the plasma to accelerate them. Ultimately a so-called pre-sheath will develop with a potential drop on the order of $(k_B T_e / 2e)$ and a scale determined by the physics of the ion source (often the same as the dimensions of the plasma). Normally the Bohm criterion will hold with equality, but there are some situations where the ions enter the sheath with supersonic speed.

The Child–Langmuir Law

Although the sheath equation must generally be integrated numerically, we can find an approximate solution analytically by neglecting the $e^{-\chi}$ term. This amounts to neglecting the electron density in the sheath, or only analyzing that part of the sheath where there are no electrons. For a "floating" surface, i.e. one that draws no net current from the plasma, this is a useful if rough approximation. For a surface biased strongly negative so that it draws the ion saturation current, the approximation is very good. It is customary, although not strictly necessary, to further simplify the equation by assuming that $2\chi / \mathfrak{M}^2$ is much larger than unity. Then the sheath equation takes on the simple form

$$\chi'' = \frac{\mathfrak{M}}{(2\chi)^{1/2}}.$$

As before, we multiply by χ' and integrate to obtain

$$\frac{1}{2}\chi'^2 = \mathfrak{M}(2\chi)^{1/2},$$

or

$$\chi^{-1/4}\chi' = 2^{3/4}\mathfrak{M}^{1/2}.$$

This is easily integrated over ξ to yield

$$\frac{4}{3}\chi_{w}^{3/4}=2^{3/4}\mathfrak{M}^{1/2}d,$$

where χ_{w} is the (normalized) potential at the wall (relative to the sheath edge), and d is the thickness of the sheath. Changing back to the variables u_0 and φ and noting that the ion current into the wall is $J=en_0u_0$, we have

$$J=\frac{4}{9}\left(\frac{2e}{m_i}\right)^{1/2}\frac{|\varphi_w|^{3/2}}{4\pi d^2}.$$

This equation is known as Child's Law, after Clement D. Child (1868–1933), who first published it in 1911, or as the Child–Langmuir Law, honoring as well Irving Langmuir, who discovered it independently and published in 1913. It was first used to give the space-charge-limited current in a vacuum diode with electrode spacing d. It can also be inverted to give the thickness of the Debye sheath as a function of the voltage drop by setting $J=j_{ion}^{sat}$:

$$d=\frac{2}{3}\left(\frac{2e}{m_i}\right)^{1/4}\frac{|\varphi_w|^{3/4}}{2\sqrt{\pi}j_{ion}^{sat}}.$$

Double Layer (Plasma Physics)

A double layer is a structure in a plasma which consists of two parallel layers of opposite electrical charge. The sheets of charge, which are not necessarily planar, produce localised excursions of electric potential, resulting in a relatively strong electric field between the layers and weaker but more extensive compensating fields outside, which restore the global potential. Ions and electrons within the double layer are accelerated, decelerated, or deflected by the electric field, depending on their direction of motion.

However, the line integral of electrostatic potential across a double layer is zero. because the contours of equipotential are necessarily closed. As the electrostatic potential introduced by a double layer at distances comparable to the sheet separation rapidly tends to zero, it also follows that any charged particle entering the region of a double layer will experience no net change in energy on passing right through it.

Double layers can be created in discharge tubes, where sustained energy is provided within the layer for electron acceleration by an external power source. Double layers are claimed to have been observed in the aurora and are invoked in astrophysical applications. For many decades, a view has been held that double layers are instrumental in accelerating auroral electrons. This interpretation treats only the double layer's internal electric field, and overlooks the fact that this is neutralised by the double layer's external fields.

Electrostatic double layers are especially common in current-carrying plasmas, and are very thin (typically ten Debye lengths), compared to the sizes of the plasmas that contain them. Other names for a double layer are electrostatic double layer, electric double layer, plasma double layers. The term 'electrostatic shock' in the magnetosphere has been applied to electric fields oriented at an oblique angle to the magnetic field in such a way that the perpendicular electric field is much stronger than the parallel electric field, In laser physics, a double layer is sometimes called an ambipolar electric field.

Double layers are conceptually related to the concept of a 'sheath'. An early review of double layers from laboratory experiment and simulations is provided by Torvén.

Double Layer Classification

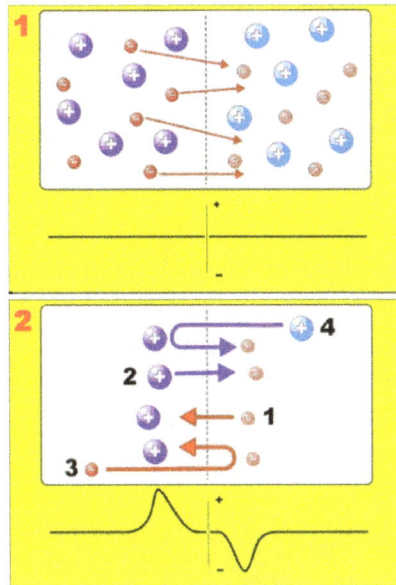

Double layer formation. Formation of a double layer requires electrons to move between two adjacent regions (Diagram 1, top) causing a charge separation. An electrostatic potential imbalance may result (Diagram 2, bottom)

Double layers may be classified in the following ways:

- *Weak* and *strong* double layers. The strength of a double layer is expressed as the ratio of the potential drop in comparison with the plasma's equivalent thermal energy, or in comparison with the rest mass energy of the electrons. A double layer is said to be strong if the potential drop within the layer is greater than the equivalent thermal energy of the plasma's components.

- *Relativistic* or *non-relativistic* double layers. A double layer is said to be relativistic if the potential drop within the layer is comparable to the rest mass energy (~512KeV) of the electron. Double layers of such energy are to be found in laboratory experiments. The charge density is low between the two opposing potential regions and the double layer is similar to the charge distribution in a capacitor in that respect.

- *Current carrying double layers* These double layers may be generated by current-driven plasma instabilities which amplify variations of the plasma density. One example of these instabilities is the Farley–Buneman instability which occurs when the streaming velocity of electrons (basically the current density divided by the electron density) exceeds the electron thermal velocity of the plasma. It occurs in collisional plasmas having a neutral component, and is driven by drift currents.

- *Current-free double layers* These occur at the boundary between plasma regions with different plasma properties. A plasma may have a higher electron temperature, and thermal velocity, on one side of a boundary layer than on the other. The same may apply for plasma densities. Charged particles exchanged between the regions may enable potential differences to be maintained between them locally. The overall charge density, as in all double layers, will be neutral.

Potential imbalance will be neutralised by electron (1&3) and ion (2&4) migration, unless the potential gradients are sustained by an external energy source. Under most laboratory situations, unlike outer space conditions, charged particles may effectively originate within the double layer, by ionization at the anode or cathode, and be sustained.

The figure shows the localised perturbation of potential produced by an idealised double layer consisting of two oppositely charged discs. The perturbation is zero at a distance from the double layer in every direction.

If an incident charged particle, such as a precipitating auroral electron, encounters such a static or quasistatic structure in the magnetosphere, provided that the particle energy exceeds half the electric potential difference within the double layer, it will pass through without any net change in energy. Incident particles with less energy than this will also experience no net change in energy but will undergo more overall deflection.

Hasegawa–Mima Equation

In plasma physics, the Hasegawa–Mima equation, named after Akira Hasegawa and Kunioki Mima, is an equation that describes a certain regime of plasma, where the time scales are very fast, and the distance scale in the direction of the magnetic field is long. In particular the equation is useful for describing turbulence in some tokamaks. The equation was introduced in Hasegawa and Mima's paper submitted in 1977 to *Physics of Fluids*, where they compared it to the results of the ATC tokamak.

Assumptions

- The magnetic field is large enough that:

$$\frac{1}{\omega_{ci}} \frac{\partial}{\partial t} \ll 1$$

for all quantities of interest. When the particles in the plasma are moving through a magnetic field, they spin in a circle around the magnetic field. The frequency of oscillation, ω_{ci} known as the cyclotron frequency or gyrofrequency, is directly proportional to the magnetic field.

- The particle density follows the quasineutrality condition:

$$n_e \approx Z n_i$$

where Z is the number of protons in the ions. If we are talking about hydrogen Z = 1, and n is the same for both species. This condition is true as long as the electrons can shield out electric fields. A cloud of electrons will surround any charge with an approximate radius known as the Debye length. For that reason this approximation means the size scale is much larger than the Debye length. The ion particle density can be expressed by a first order term that is the density defined by the quasineutrality condition equation, and a second order term which is how much it differs from the equation.

- The first order ion particle density is a function of position, but not time. This means that perturbations of the particle density change at a timescale much slower than the scale of interest. The second order particle density which causes a charge density and thus an electric potential can change with time.

- The magnetic field, B must be uniform in space, and not be a function of time. The magnetic field also moves at a timescale much slower than the scale of interest. This allows the time derivative in the momentum balance equation to be neglected.

- The ion temperature must be much smaller than the electron temperature. This means that the ion pressure can be neglected in the ion momentum balance equation.

- The electrons follow a Boltzmann distribution where:

$n = n_0 e^{e\phi/T_e}$

Since the electrons are free to move along the direction of the magnetic field, they screen away electric potentials. This screening causes a Boltzmann distribution of electrons to form around the electric potentials.

The Equation

The Hasegawa–Mima equation is a second order nonlinear partial differential equation that describes the electric potential. The form of the equation is:

$$\frac{\partial}{\partial t}\left(\nabla^2\phi-\phi\right)-\left[(\nabla\phi\times\hat{\mathbf{z}})\cdot\nabla\right]\left[\nabla^2\phi-\ln\left(\frac{n_0}{\omega_{ci}}\right)\right]=0.$$

Although the quasi neutrality condition holds, the small differences in density between the electrons and the ions cause an electric potential. The Hasegawa–Mima equation is derived from the continuity equation:

$$\frac{\partial n}{\partial t}+\nabla\cdot n\mathbf{v}=0.$$

The fluid velocity can be approximated by the E cross B drift:

$$\mathbf{V_E}=\frac{\mathbf{E}\times\mathbf{B}}{cB^2}=\frac{-\nabla\phi\times\hat{\mathbf{z}}}{cB}.$$

Previous models derived their equations from this approximation. The divergence of the E cross B drift is zero, which keeps the fluid incompressible. However, the compressibility of the fluid is very important in describing the evolution of the system. Hasegawa and Mima argued that the assumption was invalid. The Hasegawa–Mima equation introduces a second order term for the fluid velocity known as the polarization drift in order to find the divergence of the fluid velocity. Due to the assumption of large magnetic field, the polarization drift is much smaller than the E cross B drift. Nevertheless, it introduces important physics.

For a two-dimensional incompressible fluid which is not a plasma, the Navier–Stokes equations say:

$$\frac{\partial}{\partial t}\left(\nabla^2\psi\right) - \left[\left(\nabla\psi\times\hat{\mathbf{z}}\right)\cdot\nabla\right]\nabla^2\psi = 0$$

after taking the curl of the momentum balance equation. This equation is almost identical to the Hasegawa–Mima equation except the second and fourth terms are gone, and the electric potential is replaced with the fluid velocity vector potential where:

$$\mathbf{v} = -\nabla\psi\times\hat{\mathbf{z}}.$$

The first and third terms to the Hasegawa–Mima equation, which are the same as the Navier Stokes equation, are the terms introduced by adding the polarization drift. In the limit where the wavelength of a perturbation of the electric potential is much smaller than the gyroradius based on the sound speed, the Hasegawa–Mima equations become the same as the two-dimensional incompressible fluid.

Normalization

One way to understand an equation more fully is to understand what it is normalized to, which gives you an idea of the scales of interest. The time, position, and electric potential are normalized to t', x', and ϕ'

The time scale for the Hasegawa–Mima equation is the inverse ion gyrofrequency:

$$t' = \omega_{ci}t, \qquad \omega_{ci} = \frac{eZB}{m_ic}.$$

From the large magnetic field assumption the normalized time is very small. However, it is still large enough to get information out of it.

The distance scale is the gyroradius based on the sound speed:

$$x' = \frac{x}{\rho_s}, \qquad \rho_s^2 \equiv \frac{T_e}{m_i\omega_{ci}^2}.$$

If you transform to k-space, it is clear that when k, the wavenumber, is much larger than one, the terms that make the Hasegawa–Mima equation differ from the equation derived from Navier-Stokes equation in a two dimensional incompressible flow become much smaller than the rest.

From the distance and time scales we can determine the scale for velocities. This turns out to be the sound speed. The Hasegawa–Mima equation, shows us the dynamics of fast moving sounds as opposed to the slower dynamics such as flows that are captured in the MHD equations. The motion is even faster than the sound speed given that the time scales are much smaller than the time normalization.

The potential is normalized to:

$$\phi' = \frac{e\phi}{T_e}.$$

Since the electrons fit a Maxwellian and the quasineutrality condition holds, this normalized potential is small, but similar order to the normalized time derivative.

The entire equation without normalization is:

$$\frac{1}{\omega_{ci}}\frac{\partial}{\partial t}\left(\rho_s^2\nabla^2\frac{e\phi}{T_e}-\frac{e\phi}{T_e}\right)-\left[\left(\rho_s\nabla\frac{e\phi}{T_e}\times\hat{\mathbf{z}}\right)\cdot\rho_s\nabla\right]\left[\rho_s^2\nabla^2\frac{e\phi}{T_e}-\ln\left(\frac{n_0}{\omega_{ci}}\right)\right]=0.$$

Although the time derivative divided by the cyclotron frequency is much smaller than unity, and the normalized electric potential is much smaller than unity, as long as the gradient is on the order of one, both terms are comparable to the nonlinear term. The unperturbed density gradient can also be just as small as the normalized electric potential and be comparable to the other terms.

Other Forms of the Equation

Often the Hasegawa–Mima equation is expressed in a different form using Poisson brackets. These Poisson brackets are defined as:

$$[A,B]\equiv\frac{\partial A}{\partial x}\frac{\partial B}{\partial y}-\frac{\partial A}{\partial y}\frac{\partial B}{\partial x}.$$

Using these Poisson brackets, the equation can be reexpressed as:

$$\frac{\partial}{\partial t}\left(\nabla^2\phi-\phi\right)+\left[\phi,\nabla^2\phi\right]-\left[\phi,\ln\left(\frac{n_0}{\omega_{ci}}\right)\right]=0.$$

Often the particle density is assumed to vary uniformly just in one direction, and the equation is written in a sightly different form. The Poisson bracket including the density is replaced with the definition of the Poisson bracket, and a constant replaces the derivative of the density dependent term.

Conserved Quantities

There are two quantities that are conserved in a two-dimensional incompressible fluid. The kinetic energy:

$$\int (\nabla \psi)^2 dV = \int v_x^2 + v_y^2 \, dV.$$

And the enstrophy:

$$\int (\nabla^2 \psi)^2 dV = \int (\nabla \times \mathbf{v})^2 dV.$$

For the Hasegawa–Mima equation, there are also two conserved quantities, that are related to the above quantities. The generalized energy:

$$\int \left[\phi^2 + (\nabla \phi)^2 \right] dV.$$

And the generalized enstrophy:

$$\int \left[(\nabla \phi) + (\nabla \phi) \right] dV$$

In the limit where the Hasegawa–Mima equation is the same as an incompressible fluid, the generalized energy, and enstrophy become the same as the kinetic energy and enstrophy.

Microplasma

Microplasmas are plasmas of small dimensions, ranging from tens to thousands of micrometers. They can be generated at a variety of temperatures and pressures, existing as either thermal or non-thermal plasmas. Non-thermal microplasmas that can maintain their state at standard temperatures and pressures are readily available and accessible to scientists as they can be easily sustained and manipulated under standard conditions. Therefore, they can be employed for commercial, industrial, and medical applications, giving rise to the evolving field of microplasmas.

What is a Microplasma?

There are 4 states of matter: solid, liquid, gas, and plasma. Plasmas make up more than 99% of the visible universe. In general, when energy is applied to a gas, internal electrons of gas molecules (atoms) are excited and move up to higher energy levels. If the energy applied is high enough, outermost electron(s) can even be stripped off the molecules (atoms), forming ions. Electrons, molecules (atoms), excited species and ions form a soup of species which involves many interactions between species and demonstrate collective behavior under the influence of external electric and magnetic fields. Light always accompanies plasmas: as the excited species relax and move to lower energy levels, energy is released in the form of light. Microplasma is a subdivision of plasma in which the dimensions of the plasma can range between tens, hundreds, or even thousands of micrometers in size. The majority of microplasmas that are employed in commercial applications are cold plasmas. In a cold plasma, electrons have much higher energy than the accompanying ions and neutrals. Microplasmas are typically generated at elevated pressure to atmospheric pressure or higher.

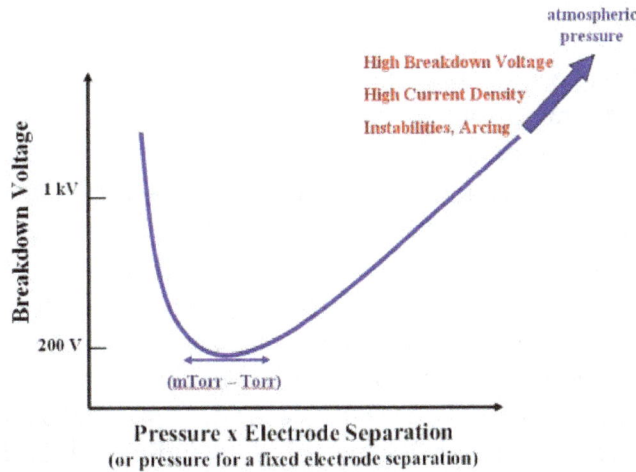

A simplified Paschen Breakdown Curve for most gases

Successful ignition of microplasmas is governed by Paschen's Law, which describes the breakdown voltage (the voltage at which the plasma begins to arc) as a function of the product of electrode distance and pressure,

$$V_b = \frac{B(pd)}{\ln(pd) + \ln(A / \ln(1 + \frac{1}{\gamma}))}$$

where pd is the product of pressure and distance, and A and B are the gas constants for calculating Townsend's first ionization coefficient and γ is the secondary emission coefficient of the material. As the pressure increases, the distance between the electrodes must decrease to achieve the same breakdown voltage. This law is proven to be valid at inter-electrode distances as small as tens of micrometers and pressures higher than atmospheric. However, its validity at even smaller scales (approaching debye length) is still currently under investigation.

Generating Microplasmas

While microplasma devices have been studied experimentally for more than a decade, understanding has been spurred in the past few years as the result of modelling and computational investigations of microplasmas.

Confinement To Small Spaces

When the pressure of the gas medium in which the microplasma is generated increases, the distance between the electrodes must decrease to maintain the same breakdown voltage. In such microhollow cathode discharges, the product of pressure and distance ranges from fractions of Torr cm to about 10 Torr cm. At values below 5 Torr cm, the discharges are called "pre-discharges" and are low intensity glow discharges. Above 10 Torr cm the discharge can become uncontrollable and extend from the anode to random locations within the cavity. Further research by David Staack provided a graph of ideal electrode distances, voltages, and carrier gases tested for microplasma generation.

Dielectric Materials

Dielectrics are poor electrical conductors, but support electrostatic fields and electric polarization. Dielectric barrier discharge microplasmas are typically created between metal plates, which are covered by a thin layer of dielectric or highly resistive material. The dielectric layer plays an important role in suppressing the current: the cathode/anode layer is charged by incoming positive ions/electrons during a positive cycle of AC is applied which reduces the electric field and hinders charge transport towards the electrode. DBD also has a large surface-to-volume ratio, which promotes diffusion losses and maintains a low gas temperature. When a negative cycle of AC is applied, the electrons are repelled off of the anode, and are ready to collide with other particles. Frequencies of 1000 Hz or more are required to move the electrons fast enough to create a microplasma, but excessive frequencies can damage the electrode (~50 kHz). Although dielectric barrier discharge comes in various shapes and dimensions, each individual discharge is in micrometer scale.

Pulsed Power

AC and high frequency power are often used to excite dielectrics, in place of DC. Take AC as an example, there are positive and negative cycles in each period. When the positive cycle occurs, electrons accumulate on the dielectric surface. On the other hand, the negative cycle would repel the accumulated electrons, causing collisions in the gas and creating plasma. During the switch from the negative to positive cycles, the above-mentioned frequency range of 1000 Hz-50,000 Hz is needed in order for a microplasma to be generated. Because of the small mass of the electrons, they are able to absorb the sudden switch in energy and become excited; the larger particles (atoms, molecules, and ions), however aren't able to follow the fast switching, therefore keeping the gas temperature low.

RF- and Microwave Signals

Based on transistor amplifiers low power RF (radio frequency) and microwave sources are used to generate a microplasma. Most of the solutions work at 2.45 GHz. Meanwhile, is a technology developed which provide the ignition on the one hand and the high efficient operation on the other hand with the same electronic and couple network.

Laser Induced

With the use of lasers, solid substrates can be converted directly into microplasmas. Solid targets are struck by high energy lasers, usually gas lasers, which are pulsed at time periods from picoseconds to femtoseconds (mode-locking). Successful experiments have used Ti:Sm, KrF, and YAG lasers, which can be applied to a variety of substrates such as lithium, germanium, plastics, and glass.

History

In 1857, Werner von Siemens, a German scientist, originated ozone generation using a dielectric barrier discharge apparatus for biological decontamination. His observations were explained without the knowledge of "microplasmas", but were later recognized as the first use of microplasmas to date. The early electrical engineers, such as Edison and Tesla, were actually trying to prevent the generation of such "micro-discharges", and used dielectrics to insulate the first electrical infrastructures. Subsequent studies have observed the Paschen breakdown curve as being the prime cause of microplasma generation in an article published in 1916.

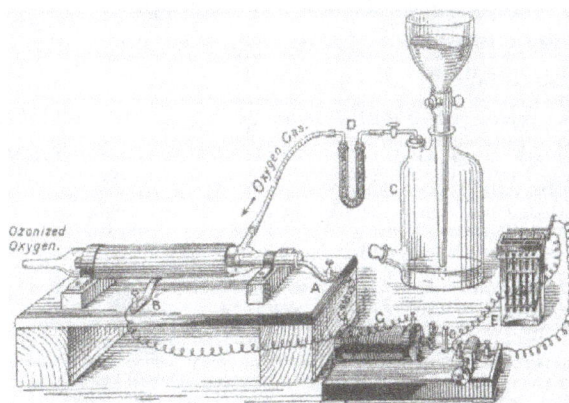

A - terminal connected to the inner surface, B - terminal connected to outer surface, C - gasholder, D - calcium chloride drying tube, E - battery, G - induction coil

Subsequent articles during the course of the 20th century have described the various conditions and specifications that lead to the generation of microplasmas. After Siemens' interactions with microplasma, Ulrich Kogelschatz was the first to identify these "micro-discharges" and define their fundamental properties. Kogelschatz also realized that microplasmas could be used for excimer formation. His experiments spurred the rapid development of the microplasma field.

In February 2003, Kunihide Tachibana, a professor of Kyoto University held the first international workshop on microplasmas (IWM) in Hyogo, Japan., The workshop, titled "The New World of Microplasmas", opened a new era of microplasma research. Tachibana is recognized as one of the founding fathers as he coined the term "microplasma". The Second IWM was organized in October 2004 by Professors K.H. Becker, J.G. Eden, and K.H. Schoenbach at Steven's Institute of Technology in Hoboken, New Jersey. The third international workshop was coordinated by the Institute of Low Temperature Plasma Physics alongside the Institute of Physics of Ernst-Moitz-Arndt-University in Griefswald, Germany, May 2006. Topics discussed were inspiring scientific and arising technological opportunities of microplasmas. The fourth IWM was held in Taiwan in October 2007, the fifth in San Diego, California in March 2009, and the sixth in Paris, France in April 2011. The next (seventh) workshop will be held in China, in approximately May 2013.

Applications

The rapid growth of applications of microplasmas renders it impossible to name all of them within a short space, but some selected applications are listed here.

Plasma Displays

Artificially generated microplasmas are found on the flat panel screen of a plasma display. The technology utilizes small cells and contains electrically charged ionized gases. Across this plasma display panel, there are a millions of tiny cells called pixels that are confined to form a visual image. In the plasma display panels, X and Y grid of electrodes, separated by a MgO dielectric layer and surrounded by a mixture of inert gases - such as argon, neon or xenon, the individual picture elements are addressed. They work on the principle that passing a high voltage through a low-pressure gas generates light. Essentially, a PDP can be viewed as a matrix of tiny fluorescent tubes which are controlled in a sophisticated fashion. Each pixel comprises a small capacitor with three electrodes, one for each primary color (some newer displays include an electrode for yellow). An electrical discharge across the electrodes causes the rare gases sealed in the cell to be converted to plasma form as it ionizes. Being electrically neutral, it contains equal quantities of electrons and ions and is, by definition, a good conductor. Once energized, the plasma cells release ultraviolet (UV) light which then strikes and excites red, green and blue phosphors along the face of each pixel, causing them to glow.

Illumination

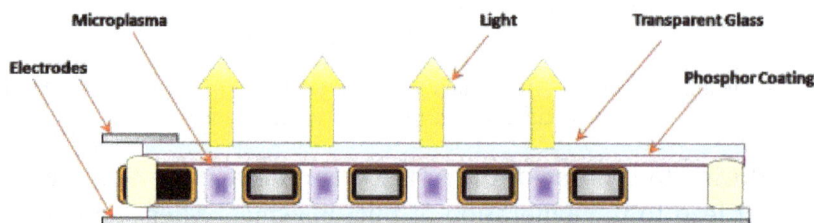

Schematic for device being developed by Eden and Park

The team of Gary Eden and Sung-Jin Park are pioneering the use of microplasmas for general illumination. Their apparatus uses many microplasma generators in a large array, which emit light through a clear, transparent window. Unlike fluorescent lamps, which require the electrodes to be far apart in a cylindrical cavity and vacuum conditions, microplasma light sources can be put into many different shapes and configurations, and generate heat. This is opposed to the more commonly used fluorescent lamps which require a noble gas atmosphere (usually argon), where eximer formation and resulting radiative decomposition strikes a phosphor coating to create light. Excimer light sources are also being produced and researched. The stable, non-equilibrium condition of microplasmas favors three-body collisions which can lead to excimer formation. The excimer, an unstable molecule produced by collisions of excited atoms, is very short lived due to its rapid dissociation. Upon their decomposition, excimers release different kinds of radiation when electrons fall to lower energy levels. One application, which has been pursued by the Hyundai Display Advanced Technology R&D Research Center and the University of Illinois, is to use excimer light sources in flat panel displays.

Destruction of Volatile Organic Compounds (VOC's)

Microplasma are used to destroy volatile organic compounds. For example, capillary plasma electrode (CPE) discharge was used to effectively destroy volatile organic compounds such as benzene, toluene, ethylbenzene, xylene, ethylene, heptane, octane, and ammonia in the surrounding air for use in advanced life support systems designed for enclosed environments. Destruction efficiencies were determined as a function of plasma energy density, initial contaminant concentration, residence time in plasma volume, reactor volume, and the number of contaminants in the gas flow stream. Complete destruction of VOC's can be achieved in the annular reactor for specific energies of 3 J cm−3 and above. Furthermore, specific energies approaching 10 J cm−3 are required to achieve a comparable destruction efficiency in the cross-flow reactor. This indicates that optimization of the reactor geometry is a critical aspect of achieving maximum destruction efficiencies. Koutsospyros *et al.* (2004, 2005) and Yin *et al.* (2003) reported results regarding studies of VOC destruction using CPE plasma reactors. All compounds studied reached maximum VOC destruction efficiencies between 95% and 100%. The VOC destruction efficiency increased initially with the specific energy, but remained at values of the specific energy that are compound-dependent. A similar observation was made for the dependence of the VOC destruction efficiency on the residence time. The destruction efficiency increased with rising initial contaminant concentration. For chemically similar compounds, the maximum destruction efficiency was found to be inversely related to the ionization energy of the compound and directly related to the degree of chemical substitution. This may suggest that chemical substitution sites offer the highest plasma-induced chemical activity.

Environmental Sensors

The small size and modest power required for microplasma devices employ a variety of environmental sensing applications and detect trace concentrations of hazardous species. Microplasmas are sensitive enough to act as detectors, which can distinguish between excessive quantities of complex molecules. C.M. Herring and his colleagues at Caviton Inc. have simulated this system by coupling a microplasma device with a commercial gas chromatography column (GC). The microplasma device is situated at the exit of the GC column, which records the relative fluorescence intensity of specific atomic and molecular dissociation fragments. This apparatus possesses the ability to detect minute concentrations of toxic and environmentally hazardous molecules. It can also detect a wide range of wavelengths and the temporal signature of chromatograms, which identifies the species of interest. For the detection of less complex species, the temporal sorting done by the GC column is not necessary since the direct observation of fluorescence produced in the microplasma is sufficient.

Ozone Generation for Water Purification

Microplasmas are being used for the formation of ozone from atmospheric oxygen. Ozone (O_3) has been shown to be a good disinfectant and water treatment that can cause breakdown of organic and inorganic materials. Ozone is not potable and reverts to diatomic oxygen, with a half-life of about 3 days in air room temperature (about 20 °C). In water, however, ozone has a half-life of only 20 minutes at the same temperature of 20 (°C) . Degremont Technologies (Switzerland) produces microplasma arrays for commercial and industrial production of ozone for water treatment. By passing molecular oxygen through a series of dielectric barriers, using what Degremont calls the Intelligent Gap System (IGS), an increasing concentration of ozone is produced by altering the gap size and coatings used on the electrodes farther down the system. The ozone is then directly bubbled into the water to be made potable (suitable for drinking). Unlike chlorine, which is still used in many water purification systems to treat water, ozone does not remain in the water for extended periods. Because ozone decomposes with a half-life of 20 minutes in water at room temperature, there are no lasting effects that may cause harm.

Current Research

Fuel Cells

Microplasmas serve as energetic sources of ions and radicals, which are desirable for activating chemical reactions. Microplasmas are used as flow reactors that allow molecular gases to flow through the microplasma inducing chemical modifications by molecular decomposition. The high energy electrons of microplasmas accommodate chemical modification and reformation of liquid hydrocarbon fuels to produce fuel for fuel cells. Becker and his co-workers used a single flow-through dc-excited microplasma reactor to generate hydrogen from an atmospheric pressure mixture of ammonia and argon for use in small, portable fuel cells. Lindner and Besser experimented with reforming model hydrocarbons such as methane, methanol, and butane into hydrogen for fuel cell feed. Their novel microplasma reactor was a microhollow cathode discharge with a microfluidic channel. Mass and energy balances on these experiments revealed conversions up to nearly 50%, but the conversion of electrical power input to chemical reaction enthalpy was only on the order of 1%. Although through modeling the reforming reaction it was found that the amount of input electrical power to chemical conversion could increase by improving the device as well as the system parameters.

Nanomaterial Synthesis and Deposition

The use of microplasmas is being lo=oked into for the synthesis of complex macromolecules, as well as the addition of functional groups to the surfaces of other substrates. An article by Klages *et al.* describes the addition of amino groups to the surfaces of polymers after treatment with a pulsed DC discharge apparatus using nitrogen containing gases. It was found that ammonia gas microplasmas add on an average of 2.4 amino groups per square nanometer of a nitrocellulose membrane, and increase the strength at which the layers of the substrate can bind. The treatment can also provide a reactive surface for biomedicine, as amino groups are extremely electron rich and energetic. Mohan Sankaran has done work on the synthesis of nanoparticles using a pulsed DC discharge. His research team has found that by applying a microplasma jet to an electrolytic solution which has either a gold or silver anode is submerged produces the relevant cations. These cations can then capture electrons supplied by the microplasma jet and results in the formation of nanoparticles. The research shows that more nanoparticles of gold and silver are shown in the solution than there are of the resulting salts that form from the acid conducting solution.

Cosmetics

Microplasma uses in research are being considered. The plasma skin regeneration (PSR) device consists of an ultra–high-radiofrequency generator that excites a tuned resonator and imparts energy to a flow of inert nitrogen gas within the handpiece. The plasma generated has an optical emission spectrum with peaks in the visible range (mainly indigo and violet) and near-infrared range. Nitrogen is used as the gaseous source because it is able to purge oxygen from the surface of the skin, minimizing the risk of unpredictable hot spots, charring, and scar formation. As the plasma hits the skin, energy is rapidly transferred to the skin surface, causing instantaneous heating in a controlled uniform manner, without an explosive effect on tissue or epidermal removal. In pretreatment samples, the zone of collagen shows a dense accumulation of elastin, but in posttreatment samples, this zone contains less dense elastin with significant, interlocking new collagen. Repeated low-energy PSR treatment is an effective modality for improving dyspigmentation, smoothness, and skin laxity associated with photoaging. Histologic analysis of posttreatment samples confirms the production of new collagen and remodeling of dermal architecture. Changes consist of erythema and superficial epidermal peeling without complete removal, generally complete by 4 to 5 days.*Bogle, Melissa; et al. (2007). "Evaluation of plasma skin regeneration technology in low energy full-facial rejuvenation". Arch Dermatol. 143 (2): 168–174. doi:10.1001/archderm.143.2.168.*

Plasma Medicine

Dental Treatments

Scientists found that microplasmas are capable of inactivating bacteria that causes tooth decay and periodontal diseases. By directing low temperature microplasma beams at the calcified tissue structure beneath the tooth enamel coating called dentin, it severely reduces the amount of dental bacteria and in turn reduces infection. This aspect of microplasma could allow dentists to use microplasma technology to destroy bacteria in tooth cavities instead of using mechanical means. Developers claim that microplasma devices will enable dentists to effectively treat oral-borne diseases with little pain to their patients. Recent studies show that microplasmas can be a very effective method of controlling oral biofilms. Biofilms (also known as slime) are highly organized, three-dimensional bacterial communities. Dental plaque is a common example of oral biofilms. It is the main cause of both tooth decay and periodontal diseases such as Gingivitis and Periodontitis. At the University of Southern California, Parish Sedghizadeh, Director of the USC Center for Biofilms and Chunqi Jiang, assistant research professor in the Ming Hsieh Department of Electrical Engineering-Electrophysics, work with researchers from Viterbi School of Engineering searching for new ways to fight off these bacterial infections. Sedghizadeh explained that the biofilms' slimy matrix acts as extra protection against traditional antibiotics. However, the centers' study confirms that biofilms cultivated in the root canal of extracted human teeth can be easily destroyed by the application of microplasma. The plasma emission microscopy obtained during each experiment suggests that the atomic oxygen produced by the microplasma is responsible for the inactivation of bacteria. Sedghizadeh then suggested that the oxygen free radicals could disrupt the biofilms cellular membrane and cause them to break down. According to their ongoing research at USC, Sedghizadeh and Jiang have found that microplasma is not harmful to surrounding healthy tissues and they are confident that microplasma technology will soon become a groundbreaking tool in the medical industry.J.K. Lee along with other scientists in this field have found that microplasma can also be used for teeth bleaching. This reactive species can effectively bleach teeth along with saline or whitening gels that consist of hydrogen peroxide. Lee and his colleagues experimented with this method, examining how microplasma along with hydrogen peroxide effects blood stained human teeth. These scientists took forty extracted single-root, blood stained human teeth and randomly divided them into two groups of twenty. Group one received 30% hydrogen peroxide activated by microplasma for thirty minutes in a pulp chamber, while group two received 30% hydrogen peroxide alone for thirty minutes in the pulp chamber and the temperature was maintained at thirty seven degrees Celsius for both groups. After the tests had been performed, they found that microplasma treatment with 30% hydrogen peroxide had a significant effect on the whiteness of the teeth in group one. Lee and his associates concluded that the application of microplasma along with hydrogen peroxide is an efficient method in the bleaching of stained teeth due to its ability to remove proteins on the surface of teeth and the increased production of hydroxide.

Wound Care

Microplasma that is sustained near room temperature can destroy bacteria, viruses, and fungi deposited on the surfaces of surgical instruments and medical devices. Researchers discovered that bacteria cannot survive in the harsh environment created by microplasmas. They consist of chemically reactive species such as hydroxyl (OH) and atomic oxygen (O) that can kill harmful bacteria through oxidation. Oxidation of the lipids and proteins that compose a cell's membrane can lead to the breakdown of the membrane and deactivate the bacteria. Microplasma can contact skin without harming it, making it ideal for disinfecting wounds. "Medical plasmas are said to be in the 'Goldilocks' range—hot enough to produce and be an effective treatment, but cold enough to leave tissues unharmed" (Larousi, Kong 1). Researchers have found that microplasmas can be applied directly to living tissues to deactivate pathogens. Scientists have also discovered that microplasmas stop bleeding without damaging healthy tissue, disinfect wounds, accelerate wound healing, and selectively kill some types of cancer cells. At moderate doses, microplasmas can destroy pathogens. At low doses, they can accelerate the replication of cells—an important step in the wound healing process. The ability of microplasma to kill bacteria cells and accelerate the replication of healthy tissue cells is known as the "plasma kill/plasma heal" process, this led scientists to further experiment with the use of microplasmas for wound care. Preliminary tests have also demonstrated successful treatments of some types of chronic wounds.

Cancer Treatments

Since microplasmas deactivate bacteria they may have the ability to destroy cancer cells. Jean Michel Pouvesle has been working at the University of Orléans in France, in the Group for Research and Studies on Mediators of Inflammation (GREMI), experimenting with the effects of microplasma on cancer cells. Pouvesle along with other scientists has created a dielectric barrier discharge and plasma gun for cancer treatment, in which microplasma will be applied to both in vitro and in vivo experiments. This application will reveal the role of ROS (Reactive Oxygen Species), DNA damage, cell cycle modification, and apoptosis induction. Studies show that microplasma treatments are able to induce programmed death (apoptosis) among cancer cells—stopping the rapid reproduction of cancerous cells, with little damage to living human tissues. GREMI performs many experiments with microplasmas in cancerology, their first experiment applies microplasma to mice tumors growing beneath the skin's surface. During this experiment, scientists found no changes or burns on the surface of the skin. After a five-day microplasma treatment, the results displayed a significant decrease in the growth of U87 glioma cancer (brain tumor), compared to the control group where microplasma was not applied. GREMI performed further in vitro studies regarding U87 gliomal cancer (brain tumors) and HCT116 (colon tumor) cell lines where microplasma was applied. This microplasma treatment was proven to be an efficient method in destroying cancer cells after being applied over periods of a few tens of seconds. Further studies are being conducted on the effects of microplasma treatment in oncology; this application of microplasma will impact the medical field significantly.

Plasmon

In physics, a plasmon is a quantum of plasma oscillation. Just as light [optical oscillation] consists of photons, the plasma oscillation consists of plasmons. The plasmon can be considered as a quasiparticle since it arises from the quantization of plasma oscillations, just like phonons are quantizations of mechanical vibrations. Thus, plasmons are collective (a discrete number) oscillations of the free electron gas density. For example, at optical frequencies, plasmons can couple with a photon to create another quasiparticle called a plasmon polariton.

Derivation

The plasmon was initially proposed in 1952 by David Pines and David Bohm and was shown to arise from a Hamiltonian for the long-range electron-electron correlations.

Since plasmons are the quantization of classical plasma oscillations, most of their properties can be derived directly from Maxwell's equations.

Explanation

Plasmons can be described in the classical picture as an oscillation of free electron density with respect to the fixed positive ions in a metal. To visualize a plasma oscillation, imagine a cube of metal placed in an external electric field pointing to the right. Electrons will move to the left side (uncovering positive ions on the right side) until they cancel the field inside the metal. If the electric field is removed, the electrons move to the right, repelled by each other and attracted to the positive ions left bare on the right side. They oscillate back and forth at the plasma frequency until the energy is lost in some kind of resistance or damping. Plasmons are a quantization of this kind of oscillation.

Role of Plasmons

Plasmons play a large role in the optical properties of metals and semiconductors. Light of frequencies below the plasma frequency is reflected by a material because the electrons in the material screen the electric field of the light. Light of frequencies above the plasma frequency is transmitted by a material because the electrons in the material cannot respond fast enough to screen it. In most metals, the plasma frequency is in the ultraviolet, making them shiny (reflective) in the visible range. Some metals, such as copper and gold, have electronic interband transitions in the visible range, whereby specific light energies (colors) are absorbed, yielding their distinct color. In semiconductors, the valence electron plasmon frequency is usually in the deep ultraviolet, while their electronic interband transitions are in the visible range, whereby specific light energies (colors) are absorbed, yielding their distinct color which is why they are reflective. It has been shown that the plasmon frequency may occur in the mid-infrared and near-infrared region when semiconductors are in the form of nanoparticles with heavy doping.

The plasmon energy can often be estimated in the free electron model as

$$E_p = \hbar \sqrt{\frac{ne^2}{m\epsilon_0}} = \hbar\,\omega_p,$$

where n is the conduction electron density, n is the elementary charge, m is the electron mass, ϵ_0 the permittivity of free space, $\hbar$ the reduced Planck constant and ω_p the plasmon frequency.

Surface Plasmons

Surface plasmons are those plasmons that are confined to surfaces and that interact strongly with light resulting in a polariton. They occur at the interface of a material exhibiting positive real part of their relative permittivity, i.e. dielectric constant, (e.g. vacuum, air, glass and other dielectrics) and a material whose real part of permittivity is negative at the given frequency of light, typically a metal or heavily doped semiconductors. In addition to opposite sign of the real part of the permittivity, the magnitude of the real part of the permittivity in the negative permittivity region should typically be larger than the magnitude of the permittivity in the positive permittivity region, otherwise the light is not bound to the surface (i.e. the surface plasmons do not exist) as shown in the famous book by Raether. At visible wavelengths of light, e.g. 632.8 nm wavelength provided by a He-Ne laser, interfaces supporting surface plasmons are often formed by metals like silver or gold (negative real part permittivity) in contact with dielectrics such as air or silicon dioxide. The particular choice of materials can have a drastic effect on the degree of light confinement and propagation distance due to losses. Surface plasmons can also exist on interfaces other than flat surfaces, such as particles, or rectangular strips, v-grooves, cylinders, and other structures. Many structures have been investigated due to the capability of surface plasmons to confine light below the diffraction limit of light.

Surface plasmons can play a role in surface-enhanced Raman spectroscopy and in explaining anomalies in diffraction from metal gratings (Wood's anomaly), among other things. Surface plasmon resonance is used by biochemists to study the mechanisms and kinetics of ligands binding to receptors (i.e. a substrate binding to an enzyme). Multi-Parametric Surface Plasmon Resonance can be used not only to measure molecular interactions, but also nanolayer properties or structural changes in the adsorbed molecules, polymer layers or graphene, for instance.

Surface plasmons may also be observed in the X-ray emission spectra of metals. A dispersion relation for surface plasmons in the X-ray emission spectra of metals has been derived (Harsh and Agarwal).

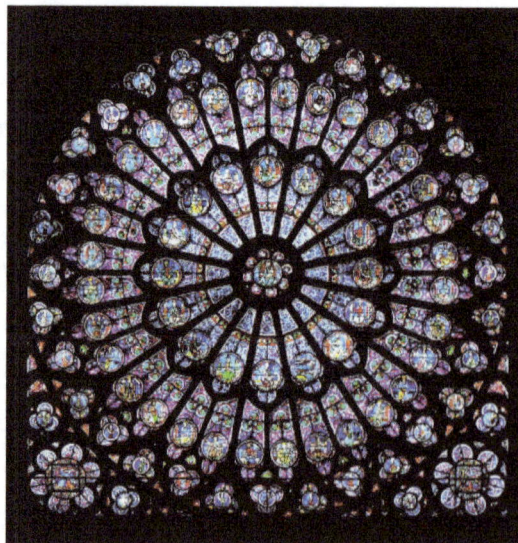

Gothic stained glass rose window of Notre-Dame de Paris. The colors were achieved by colloids of gold nano-particles.

More recently surface plasmons have been used to control colors of materials. This is possible since controlling the particle's shape and size determines the types of surface plasmons that can couple to it and propagate across it. This in turn controls the interaction of light with the surface. These effects are illustrated by the historic stained glass which adorn medieval cathedrals. In this case, the color is given by metal nanoparticles of a fixed size which interact with the optical field to give the glass its vibrant color. In modern science, these effects have been engineered for both visible light and microwave radiation. Much research goes on first in the microwave range because at this wavelength material surfaces can be produced mechanically as the patterns tend to be of the order a few centimeters. To produce optical range surface plasmon effects involves producing surfaces which have features <400 nm. This is much more difficult and has only recently become possible to do in any reliable or available way.

Recently, graphene has also been shown to accommodate surface plasmons, observed via near field infrared optical microscopy techniques and infrared spectroscopy. Potential applications of graphene plasmonics mainly addressed the terahertz to midinfrared frequencies, such as optical modulators, photodetectors, biosensors.

Possible Applications

The position and intensity of plasmon absorption and emission peaks are affected by molecular adsorption, which can be used in molecular sensors. For example, a fully operational device detecting casein in milk has been prototyped, based on detecting a change in absorption of a gold layer. Localized surface plasmons of metal nanoparticles can be used for sensing different types molecules, proteins, etc.

Plasmons are being considered as a means of transmitting information on computer chips, since plasmons can support much higher frequencies (into the 100 THz range, whereas conventional wires become very lossy in the tens of GHz). However, for plasmon-based electronics to be practical, a plasmon-based amplifier analogous to the transistor, called a plasmonstor, needs to be created.

Plasmons have also been proposed as a means of high-resolution lithography and microscopy due to their extremely small wavelengths; both of these applications have seen successful demonstrations in the lab environment.

Finally, surface plasmons have the unique capacity to confine light to very small dimensions, which could enable many new applications.

Surface plasmons are very sensitive to the properties of the materials on which they propagate. This has led to their use to measure the thickness of monolayers on colloid films, such as screening and quantifying protein binding events. Companies such as Biacore have commercialized instruments that operate on these principles. Optical surface plasmons are being investigated with a view to improve makeup by L'Oréal and others.

In 2009, a Korean research team found a way to greatly improve organic light-emitting diode efficiency with the use of plasmons.

A group of European researchers led by IMEC has begun work to improve solar cell efficiencies and costs through incorporation of metallic nanostructures (using plasmonic effects) that can enhance absorption of light into different types of solar cells: crystalline silicon (c-Si), high-performance III-V, organic, and dye-sensitized. However, for plasmonic photovoltaic devices to function optimally, ultra-thin transparent conducting oxides are necessary. Full color holograms using *plasmonics* have been demonstrated.

"The silicon absorbs only a certain fraction of the spectrum, and it's transparent to the rest. If I put a photo voltaic module on my roof, the silicon absorbs that portion of the spectrum, and some of that light gets converted into power, but the rest of it ends up just heating the roof and waste. The new technique they've developed is based on a phenomenon observed in metallic structures known as plasmon resonance. Plasmons are coordinated waves, or ripples, of electrons that exist on the surfaces of metals at the point where the metal meets the air.

Plasma Oscillation

Plasma oscillations, also known as "Langmuir waves" (after Irving Langmuir), are rapid oscillations of the electron density in conducting media such as plasmas or metals. The oscillations can be described as an instability in the dielectric function of a free electron gas. The frequency only depends weakly on the wavelength of the oscillation. The quasiparticle resulting from the quantization of these oscillations is the plasmon.

Langmuir waves were discovered by American physicists Irving Langmuir and Lewi Tonks in the 1920s. They are parallel in form to Jeans instability waves, which are caused by gravitational instabilities in a static medium.

Mechanism

Consider an electrically neutral plasma in equilibrium, consisting of a gas of positively charged ions and negatively charged electrons. If one displaces by a tiny amount all of the electrons with respect to the ions, the Coulomb force pulls the electrons back, acting as a restoring force.

'Cold' Electrons

If the thermal motion of the electrons is ignored, it is possible to show that the charge density oscillates at the *plasma frequency*

$$\omega_{pe} = \sqrt{\frac{n_e e^2}{m^* \varepsilon_0}}, [\text{rad} / \text{s}] (\text{SI units}),$$

$$\omega_{pe} = \sqrt{\frac{4\pi n_e e^2}{m^*}}, (\text{cgs units}),$$

where n_e is the number density of electrons, e is the electric charge, m^* is the effective mass of the electron, and ε_0 is the permittivity of free space. Note that the above formula is derived under the approximation that the ion mass is infinite. This is generally a good approximation, as the electrons are so much lighter than ions. (One must modify this expression in the case of electron-positron plasmas, often encountered in astrophysics). Since the frequency is independent of the wavelength, these oscillations have an infinite phase velocity and zero group velocity.

Note that, if m^* is electron mass ($m^* = m_e$), , plasma frequency ω_{pe} depends only on physical constants and concentration of electrons n_e. The numeric expression for plasma ordinary frequency

$$f_{pe} = \omega_{pe} / 2\pi$$

is

$$f_{pe} \approx 8980\sqrt{n_e} \text{ Hz}$$

with number density n_e in cm^{-3}.

Metals are only transparent to light with higher frequency than the metal's plasma frequency. For typical metals such as copper or silver, n_e is approximately 10^{23} cm^{-3}, which brings the plasma frequency into the ultraviolet region. This is why most metals reflect visible light and appear shiny.

'Warm' Electrons

When the effects of the electron thermal speed $v_{e,th} = \sqrt{\dfrac{k_B T_e}{m_e}}$ are taken into account, the electron

pressure acts as a restoring force as well as the electric field and the oscillations propagate with frequency and wavenumber related by the longitudinal Langmuir wave:

$$\omega^2 = \omega_{pe}^2 + \frac{3k_B T_e}{m_e}k^2 = \omega_{pe}^2 + 3k^2 v_{e,th}^2$$

called the Bohm-Gross dispersion relation. If the spatial scale is large compared to the Debye length, the oscillations are only weakly modified by the pressure term, but at small scales the pressure term dominates and the waves become dispersionless with a speed of $\sqrt{3} \cdot v_{e,th}$. For such waves, however, the electron thermal speed is comparable to the phase velocity, i.e.,

$$v \sim v_{ph} \stackrel{\text{def}}{=} \frac{\omega}{k},$$

so the plasma waves can accelerate electrons that are moving with speed nearly equal to the phase velocity of the wave. This process often leads to a form of collisionless damping, called Landau damping. Consequently, the large-k portion in the dispersion relation is difficult to observe and seldom of consequence.

In a bounded plasma, fringing electric fields can result in propagation of plasma oscillations, even when the electrons are cold.

In a metal or semiconductor, the effect of the ions' periodic potential must be taken into account. This is usually done by using the electrons' effective mass in place of m.

Magnetic Reconnection

Magnetic Reconnection: This view is a cross-section through four magnetic domains undergoing separator reconnection. Two separatrices divide space into four magnetic domains with a separator at the center of the figure. Field lines (and associated plasma) flow inward from above and below the separator, reconnect, and spring outward horizontally. A current sheet may be present but is not required for reconnection to occur. This process is not well understood: once started, it proceeds many orders of magnitude faster than predicted by standard models.

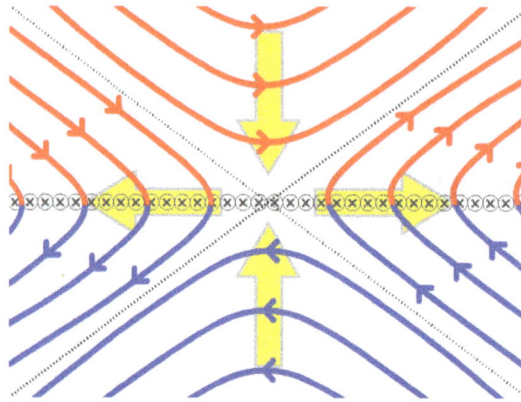

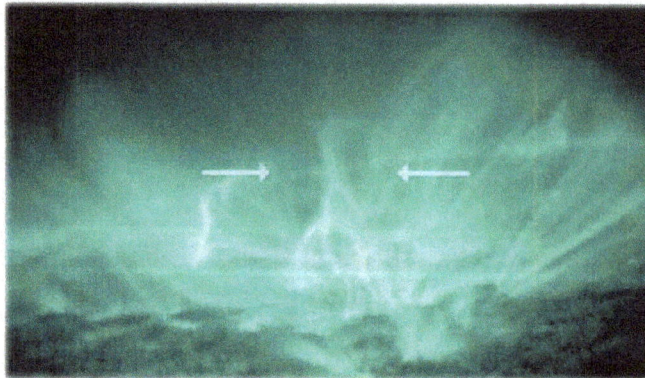

A magnetic reconnection event on the sun.

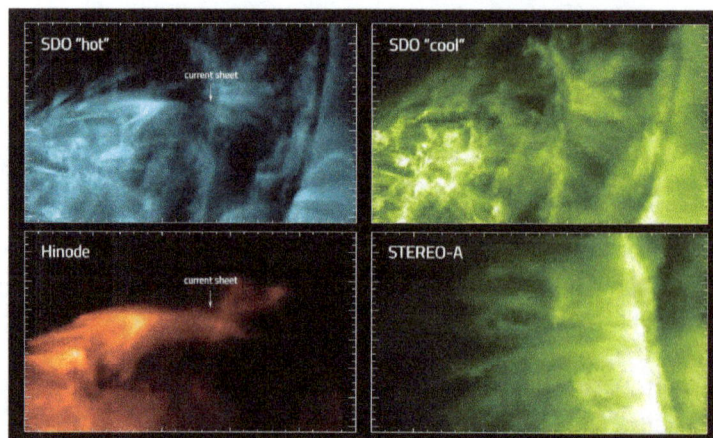

The evolution of magnetic reconnection during a solar flare.

Magnetic reconnection is a physical process in highly conducting plasmas in which the magnetic topology is rearranged and magnetic energy is converted to kinetic energy, thermal energy, and particle acceleration. Magnetic reconnection occurs on timescales intermediate between slow resistive diffusion of the magnetic field and fast Alfvénic timescales.

The qualitative description of the reconnection process is such that magnetic field lines from different magnetic domains (defined by the field line connectivity) are spliced to one another, changing their patterns of connectivity with respect to the sources. It is a violation of an approximate conservation law in plasma physics, called the Alfvén's Theorem, and can concentrate mechanical or magnetic energy in both space and time. Solar flares, the largest explosions in the Solar System, may involve the reconnection of large systems of magnetic flux on the Sun, releasing, in minutes, energy that has been stored in the magnetic field over a period of hours to days. Magnetic reconnection in Earth's magnetosphere is one of the mechanisms responsible for the aurora, and it is important to the science of controlled nuclear fusion because it is one mechanism preventing magnetic confinement of the fusion fuel.

In an electrically conductive plasma, magnetic field lines are grouped into 'domains'—bundles of field lines that connect from a particular place to another particular place, and that are topologically distinct from other field lines nearby. This topology is approximately preserved even when the magnetic field itself is strongly distorted by the presence of variable currents or motion of magnetic sources, because effects that might otherwise change the magnetic topology instead induce eddy currents in the plasma; the eddy currents have the effect of canceling out the topological change.

In two dimensions, the most common type of magnetic reconnection is separator reconnection, in which four separate magnetic domains exchange magnetic field lines. Domains in a magnetic plasma are separated by *separatrix surfaces*: curved surfaces in space that divide different bundles of flux. Field lines on one side of the separatrix all terminate at a particular magnetic pole, while field lines on the other side all terminate at a different pole of similar sign. Since each field line generally begins at a north magnetic pole and ends at a south magnetic pole, the most general way of dividing simple flux systems involves four domains separated by two separatrices: one separatrix surface divides the flux into two bundles, each of which shares a south pole, and the other separatrix surface divides the flux into two bundles, each of which shares a north pole. The intersection of the separatrices forms a *separator*, a single line that is at the boundary of the four separate domains. In separator reconnection, field lines enter the separator from two of the domains, and are spliced one to the other, exiting the separator in the other two domains.

According to simple resistive magnetohydrodynamics (MHD) theory, reconnection happens because the plasma's electrical resistivity near the boundary layer opposes the currents necessary to sustain the change in the magnetic field. The need for such a current can be seen from one of Maxwell's equations,

$$\nabla \times \mathbf{B} = \mu \mathbf{J} + \mu \epsilon \frac{\partial \mathbf{E}}{\partial t}.$$

The resistivity of the current layer allows magnetic flux from either side to diffuse through the current layer, cancelling out flux from the other side of the boundary. When this happens, the plasma is pulled out by magnetic tension along the direction of the magnetic field lines. The resulting drop in pressure pulls more plasma and magnetic flux into the central region, yielding a self-sustaining process.

A current problem in plasma physics is that observed reconnection happens much faster than predicted by MHD in high Lundquist number plasmas: solar flares, for example, proceed 13-14 orders of magnitude faster than a naive calculation would suggest, and several orders of magnitude faster than current theoretical models that include turbulence and kinetic effects. One possible mechanism to explain the discrepancy is that the electromagnetic turbulence in the boundary layer is sufficiently strong to scatter electrons, raising the plasma's local resistivity. This would allow the magnetic flux to diffuse faster.

Theoretical Descriptions of Magnetic Reconnection

The Sweet-Parker Model

At a conference in 1956, Peter Sweet pointed out that by pushing two plasmas with oppositely directed magnetic fields together, resistive diffusion is able to occur on a length scale much shorter than a typical equilibrium length scale. Eugene Parker was in attendance at this conference and developed scaling relations for this model during his return travel.

The Sweet-Parker model describes time-independent magnetic reconnection in the resistive MHD framework when the reconnecting magnetic fields are antiparallel (oppositely directed) and effects related to viscosity and compressibility are unimportant. The ideal Ohm's law then yields the relation

$$E_y = v_{in} B_{in}$$

where E_y is the out-of-plane electric field, v_{in} is the characteristic inflow velocity, and B_{in} is the characteristic upstream magnetic field strength. By neglecting displacement current, the low-frequency Ampere's law, $\mathbf{J} = \dfrac{\nabla \times \mathbf{B}}{\mu_0}$, , gives the relation

$$J_y \sim \frac{B_{in}}{\mu_0 \delta},$$

where δ is the current sheet half-thickness. This relation uses that the magnetic field reverses over a distance of $\sim 2\delta$. By matching the ideal electric field outside of the layer with the resistive electric field $\mathbf{E} = \dfrac{1}{\sigma} \mathbf{J}$ inside the layer (using Ohm's law), we find that

$$v_{in} = \frac{E_y}{B_{in}} \sim \frac{1}{\mu_0 \sigma \delta} = \frac{\eta}{\delta},$$

where η is the magnetic diffusivity. When the inflow density is comparable to the outflow density, conservation of mass yields the relationship

$$v_{in} L \sim v_{out} \delta,$$

where L is the half-length of the current sheet and v_{out} is the outflow velocity. The left and right hand sides of the above relation represent the mass flux into the layer and out of the layer, respectively. Equating the upstream magnetic pressure with the downstream dynamic pressure gives

$$\frac{B_{in}^2}{2\mu_0} \sim \frac{\rho v_{out}^2}{2}$$

where ρ is the mass density of the plasma. Solving for the outflow velocity then gives

$$v_{out} \sim \frac{B_{in}}{\sqrt{\mu_0 \rho}} \equiv v_A$$

where v_A is the Alfvén velocity. With the above relations, the dimensionless reconnection rate R can then be written in two forms, the first in terms of (δ, L) using the result earlier derived from Ohm's law, the second in terms of (η, δ, v_A) from the conservation of mass as

$$R = \frac{v_{in}}{v_{out}} \sim \frac{\eta}{v_A \delta} \sim \frac{\delta}{L}.$$

Since the dimensionless Lundquist number S is given by

$$S \equiv \frac{L v_A}{\eta},$$

the two different expressions of R are multiplied by each other and then square-rooted, giving a simple relation between the reconnection rate R and the Lundquist number S

$$R \sim \sqrt{\frac{\eta}{v_A L}} = \frac{1}{S^{1/2}}.$$

Sweet-Parker reconnection allows for reconnection rates much faster than global diffusion, but is not able to explain the fast reconnection rates observed in solar flares, the Earth's magnetosphere, and laboratory plasmas. Additionally, Sweet-Parker reconnection neglects three-dimensional effects, collisionless physics, time-dependent effects, viscosity, compressibility, and downstream pressure. Numerical simulations of two-dimensional magnetic reconnection typically show agreement with this model. Results from the Magnetic Reconnection Experiment (MRX) of collisional reconnection show agreement with a generalized Sweet-Parker model which incorporates compressibility, downstream pressure, and anomalous resistivity.

The Petschek Model

One of the reasons why Sweet-Parker reconnection is slow is that the aspect ratio of the reconnection layer is very large in high Lundquist number plasmas. The inflow velocity, and thus the reconnection rate, must then be very small. In 1964, Harry Petschek proposed a mechanism where the inflow and outflow regions are separated by stationary slow mode shocks. The aspect ratio of the diffusion region is then of order unity and the maximum reconnection rate becomes

$$\frac{v_{in}}{v_A} \approx \frac{\pi}{8 \ln S}.$$

This expression allows fast reconnection almost independent of the Lundquist number.

Simulations of resistive MHD reconnection with uniform resistivity showed the development of elongated current sheets in agreement with the Sweet-Parker model rather than the Petschek model. When a localized anomalously large resistivity is used, however, Petschek reconnection can be realized in resistive MHD simulations. Because the use of an anomalous resistivity is only appropriate when the particle mean free path is large compared to the reconnection layer, it is likely that other collisionless effects become important before Petschek reconnection can be realized.

Collisionless Reconnection

On length scales shorter than the ion inertial length c / ω_{pi} (where $\omega_{pi} \equiv \sqrt{\dfrac{n_i Z^2 e^2}{\epsilon_0 m_i}}$ is the ion plasma frequency), ions decouple from electrons and the magnetic field becomes frozen into the electron fluid rather than the bulk plasma. On these scales the Hall effect becomes important. Two-fluid simulations show the formation of an X-point geometry rather than the double Y-point geometry characteristic of resistive reconnection. The electrons are then accelerated to very high speeds by Whistler waves. Because the ions can move through a wider "bottleneck" near the current layer and because the electrons are moving much faster in Hall MHD than in standard MHD, reconnection may proceed more quickly. Two-fluid/collisionless reconnection is particularly important in the Earth's magnetosphere.

Anomalous Resistivity and Bohm Diffusion

In the Sweet-Parker model, the common assumption is that the magnetic diffusivity is constant. This can be estimated using the equation of motion for an electron with mass m and electric charge e:

$$\frac{d\mathbf{v}}{dt} = \frac{e}{m}\mathbf{E} - \nu\mathbf{v},$$

where ν is the collision frequency. Since in the steady state, $d\mathbf{v}/dt = 0$, then the above equation along with the definition of electric current, ν, where n is the electron number density, yields

$$\eta = \nu \frac{c^2}{\omega_{pi}^2}.$$

Nevertheless, if the drift velocity of electrons exceeds the thermal velocity of plasma, a steady state cannot be achieved and magnetic diffusivity should be much larger than what is given in the above. This is called anomalous resistivity, η_{anom}, which can enhance the reconnection rate in the Sweet-Parker model by a factor of $\eta_{anom}/\eta.$.

Another proposed mechanism is known as the Bohm diffusion across the magnetic field. This replaces the Ohmic resistivity with $v_A^2 (mc / eB)$, however, its effect, similar to the anomalous resistivity, is still too small compared with the observations.

Stochastic Reconnection

Lazarian and Vishniac (1999) considered the magnetic reconnection in the presence of a random component of magnetic field in a totally ionized and inviscid plasma assuming that the resistive effects could be described with an Ohmic resistivity. For the turbulent flow in the reconnection region a model for magnetohydrodynamic turbulence should be used such as the model developed by Goldreich and Sridhar in 1995. One can imagine that within small scales of the turbulent flow, the Sweet-Parker model is applicable. Lazarian and Vishniac showed that, in general, this cannot affect the final result. In fact, their model is independent of small scale physics which determines the local reconnection rate. According to this model, for a current sheet of the length L, the upper limit for reconnection velocity is given by

$$v = v_{turb} \; min\Big[\Big(\frac{L}{l}\Big)^{1/2}, \Big(\frac{l}{L}\Big)^{1/2}\Big],$$

where $v_{turb} = v_l^2 / v_A$. Here l, and v_l are turbulence injection length scale and velocity respectively and v_A is the Alfvén velocity. This model has been successfully tested by numerical simulations.

Observations of Magnetic Reconnection in Nature and the Laboratory

The Solar Atmosphere

Magnetic reconnection occurs during solar flares, coronal mass ejections, and many other events in the solar atmosphere. The observational evidence for solar flares includes observations of inflows/outflows, downflowing loops, and changes in the magnetic topology. In the past, observations of the solar atmosphere were done using remote imaging; consequently, the magnetic fields were inferred or extrapolated rather than observed directly. However, the first direct observations of solar magnetic reconnection were gathered in 2012 (and released in 2013) by the High Resolution Coronal Imager.

The Earth's Magnetosphere

New measurements from the Cluster mission for the first time now can determine unambiguously the scale sizes of magnetic reconnection in the Earth's magnetosphere, both on the dayside magnetopause and in the magnetotail. Cluster is a four-spacecraft mission, with the four space-craft in a tetrahedron arrangement, to separate spatial from temporal changes as the suite flies through space. Cluster has now also unambiguously discovered 'reverse reconnection' near the polar cusps. 'Dayside reconnection' allows interconnection of the Earth's magnetic field with that of the Sun (the Interplanetary Magnetic Field), allowing particle and energy entry into the Earth's vicinity. Tail reconnection allows release of energy stored in the Earth's magnetic tail, injecting particles deep into the magnetosphere, causing auroral substorms. 'Reverse reconnection' is reconnection of Earth's tail magnetic fields with northward Interplanetary Magnetic Fields, causing sunward convection in the Earth's ionosphere. The Magnetospheric Multiscale Mission, launched on 13 March 2015, is improving on Cluster results by having a tighter constellation of spacecraft, allowing finer spatial measurements and finer time detail. In this way the behavior of the electrical currents in the electron diffusion region will be better understood.

On 26 February 2008, THEMIS probes were able to determine, for the first time, the triggering event for the onset of magnetospheric substorms. Two of the five probes, positioned approximately one third the distance to the Moon, measured events suggesting a magnetic reconnection event 96 seconds prior to Auroral intensification. Dr. Vassilis Angelopoulos of the University of California, Los Angeles, who is the principal investigator for the THEMIS mission, claimed, "Our data show clearly and for the first time that magnetic reconnection is the trigger.".

Laboratory Plasma Experiments

The process of magnetic reconnection has been studied in detail by dedicated laboratory experiments, such as the Magnetic Reconnection Experiment (MRX) at Princeton Plasma Physics Laboratory (PPPL). Experiments such as these have confirmed many aspects of magnetic reconnection, including the Sweet-Parker model in regimes where this model is applicable.

The confinement of plasma in devices such as tokamaks, spherical tokamaks, and reversed field pinches requires the presence of closed magnetic flux surfaces. By changing the magnetic topology, magnetic reconnection degrades confinement by disrupting these closed flux surfaces, allowing the hot central plasma to mix with cooler plasma closer to the wall.

Alfvén Wave

In plasma physics, an Alfvén wave, named after Hannes Alfvén, is a type of magnetohydrodynamic wave in which ions oscillate in response to a restoring force provided by an effective tension on the magnetic field lines.

Definition

An Alfvén wave in a plasma is a low-frequency (compared to the ion cyclotron frequency) travelling oscillation of the ions and the magnetic field. The ion mass density provides the inertia and the magnetic field line tension provides the restoring force.

The wave propagates in the direction of the magnetic field, although waves exist at oblique incidence and smoothly change into the magnetosonic wave when the propagation is perpendicular to the magnetic field.

The motion of the ions and the perturbation of the magnetic field are in the same direction and transverse to the direction of propagation. The wave is dispersionless.

Alfvén Velocity

The low-frequency relative permittivity ϵ of a magnetized plasma is given by

$$\epsilon = 1 + \frac{1}{B^2}c^2\mu_0\rho$$

where B is the magnetic field strength, c is the speed of light, μ_0 is the permeability of the vacuum, and $\rho = \Sigma n_s m_s$ is the total mass density of the charged plasma particles. Here, s goes over all plasma species, both electrons and (few types of) ions.

Therefore, the phase velocity of an electromagnetic wave in such a medium is

$$v = \frac{c}{\sqrt{\epsilon}} = \frac{c}{\sqrt{1 + \frac{1}{B^2} c^2 \mu_0 \rho}}$$

or

$$v = \frac{v_A}{\sqrt{1 + \frac{1}{c^2} v_A^2}}$$

where

$$v_A = \frac{B}{\sqrt{\mu_0 \rho}}$$

is the Alfvén velocity. If $v_A \ll c$, then $v \approx v_A$. On the other hand, when $v_A \gg c$, then $v \approx c$. That is, at high field or low density, the velocity of the Alfvén wave approaches the speed of light, and the Alfvén wave becomes an ordinary electromagnetic wave.

Neglecting the contribution of the electrons to the mass density and assuming that there is a single ion species, we get

$$v_A = \frac{B}{\sqrt{\mu_0 n_i m_i}} \quad \text{in SI}$$

$$v_A = \frac{B}{\sqrt{4\pi n_i m_i}} \quad \text{in Gauss}$$

$$v_A \approx (2.18 \times 10^{11} \, \text{cm/s})(m_i / m_p)^{-1/2} (n_i / \text{cm}^{-3})^{-1/2} (B / \text{gauss})$$

where n_i is the ion number density and m_i is the ion mass.

Alfvén Time

In plasma physics, the Alfvén time τ_A is an important timescale for wave phenomena. It is related to the Alfvén velocity by:

$$\tau_A = \frac{a}{v_A}$$

where a denotes the characteristic scale of the system, for example a is the minor radius of the torus in a tokamak.

Relativistic Case

The general Alfvén wave velocity is defined by Gedalin (1993):

$$v = \frac{c}{\sqrt{1 + \dfrac{e+P}{2P_m}}}$$

where

e is the total energy density of plasma particles, P is the total plasma pressure, and $P_m = \dfrac{1}{2\mu_0}B^2$ is the magnetic field pressure. In the non-relativistic limit $P \ll e \approx \rho c^2$, and we immediately get the expression from the previous section.

Heating the Corona

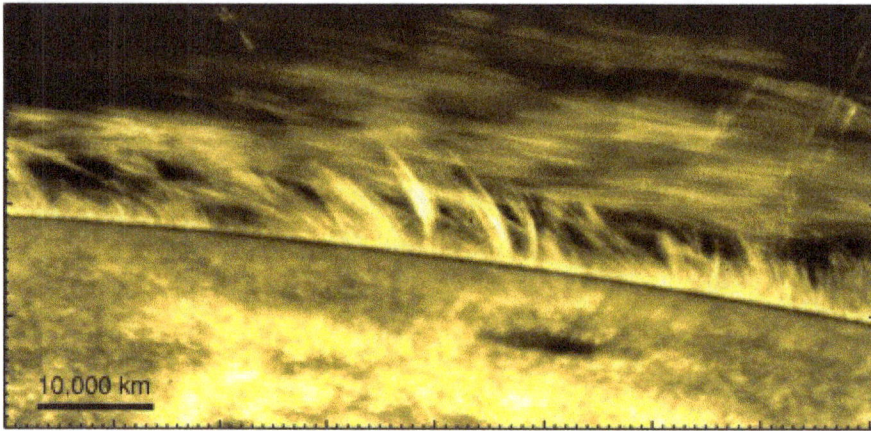

Cold plasma floating in the corona above the solar limb. Alfvén waves were observed for the first time, extrapolated from fluctuations of the plasma.

The coronal heating problem is a longstanding question in heliophysics. It is unknown why the sun's corona lives in a temperature range higher than one million degrees while the sun's surface (photosphere) is only a few thousand degrees in temperature. Natural intuition would predict a decrease in temperature while getting farther away from a heat source, but it is theorized that the photosphere, influenced by the sun's magnetic fields, emits certain waves which carry energy (i.e. heat) to the corona and solar wind. It is important to note that because the density of the corona is quite a bit smaller than the photosphere, the heat and energy level of the photosphere is much higher than the corona. Temperature depends only on the average speed of a species, and less energy is required to heat fewer particles to higher temperatures in the coronal atmosphere. Alfvén first proposed the existence of an electromagnetic-hydrodynamic wave in 1942 in Nature. He claimed the sun had all necessary criteria to support these waves and that they may in turn be responsible for sun spots. From his paper:

If a conducting liquid is placed in a constant magnetic field, every motion of the liquid gives rise to an E.M.F. which produces electric currents. Owing to the magnetic field, these currents give mechanical forces which change the state of motion of the liquid. Thus a kind of combined electromagnetic-hydrodynamic wave is produced.

— Hannes Alfvén, Existence of Electromagnetic-Hydrodynamic Waves,

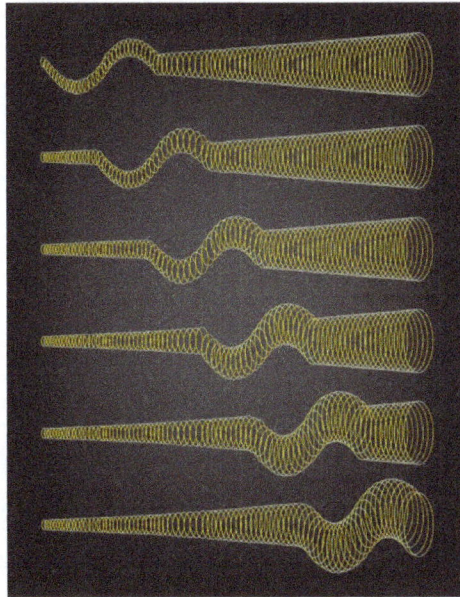

Magnetic waves, called Alfvén S-waves, flow from the base of black hole jets.

Beneath the sun's photosphere lies the convection zone. The rotation of the sun, as well as varying pressure gradients beneath the surface, produces the periodic electromagnetism in the convection zone which can be observed on the sun's surface. This random motion of the surface gives rise to Alfvén waves. The waves travel through the chromosphere and transition zone and interact with much of the ionized plasma. The wave itself carries energy as well as some of the electrically charged plasma. De Pontieu and Haerendel suggested in the early 1990s that Alfven waves may also be associated with the plasma jets known as spicules. It was theorized these brief spurts of superheated gas were carried by the combined energy and momentum of their own upward velocity, as well as the oscillating transverse motion of the Alfven waves. In 2007, Alfven waves were reportedly observed for the first time traveling towards the corona by Tomcyzk et al., but their predictions could not conclude that the energy carried by the Alfven waves were sufficient to heat the corona to its enormous temperatures, for the observed amplitudes of the waves were not high enough. However, in 2011, McIntosh et al. reported the observation of highly energetic Alfven waves combined with energetic spicules which could sustain heating the corona to its million Kelvin temperature. These observed amplitudes (20.0 km/s against 2007's observed 0.5 km/s) contained over one hundred times more energy than the ones observed in 2007. The short period of the waves also allowed more energy transfer into the coronal atmosphere. The 50,000 km long spicules may also play a part in accelerating the solar wind past the corona.

History

How this phenomenon became understood

- 1942: Alfvén suggests the existence of *electromagnetic-hydromagnetic* waves in a paper published in *Nature*.

- 1949: Laboratory experiments by S. Lundquist produce such waves in magnetized mercury, with a velocity that approximated Alfvén's formula.

- 1949: Enrico Fermi uses Alfvén waves in his theory of cosmic rays. According to Alex Dessler in a 1970 *Science* journal article, Fermi had heard a lecture at the University of Chicago, Fermi nodded his head exclaiming "of course" and the next day, the physics world said "of course".

- 1950: Alfvén publishes the first edition of his book, *Cosmical Electrodynamics*, detailing hydromagnetic waves, and discussing their application to both laboratory and space plasmas.

- 1952: Additional confirmation appears in experiments by Winston Bostick and Morton Levine with ionized helium

- 1954: Bo Lehnert produces Alfvén waves in liquid sodium

- 1958: Eugene Parker suggests hydromagnetic waves in the interstellar medium

- 1958: Berthold, Harris, and Hope detect Alfvén waves in the ionosphere after the Argus nuclear test, generated by the explosion, and traveling at speeds predicted by Alfvén formula.

- 1958: Eugene Parker suggests hydromagnetic waves in the Solar corona extending into the Solar wind.

- 1959: D. F. Jephcott produces Alfvén waves in a gas discharge

- 1959: C. H. Kelley and J. Yenser produce Alfvén waves in the ambient atmosphere.

- 1960: Coleman, *et al.*, report the measurement of Alfvén waves by the magnetometer aboard the Pioneer and Explorer satellites

- 1960: Sugiura suggests evidence of hydromagnetic waves in the Earth's magnetic field

- 1961: Normal Alfvén modes and resonances in liquid sodium are studied by Jameson

- 1966: R.O.Motz generates and observes Alfven waves in mercury

- 1970 Hannes Alfvén wins the 1970 Nobel Prize in physics for "fundamental work and discoveries in magneto-hydrodynamics with fruitful applications in different parts of plasma physics"

- 1973: Eugene Parker suggests hydromagnetic waves in the intergalactic medium

- 1974: Hollweg suggests the existence of hydromagnetic waves in interplanetary space

- 1974: Ip and Mendis suggests the existence of hydromagnetic waves in the coma of Comet Kohoutek.

- 1984: Roberts et al. predict the presence of standing MHD waves in the solar corona, thus leading to the field of coronal seismology.

- 1999: Aschwanden, *et al.* and Nakariakov, *et al.* report the detection of damped transverse oscillations of solar coronal loops observed with the EUV imager on board the Transition Region And Coronal Explorer (TRACE), interpreted as standing kink (or "Alfvénic") oscillations of the loops. This fulfilled the prediction of Roberts et al. (1984).

- 2007: Tomczyk, *et al.*, report the detection of Alfvénic waves in images of the solar corona with the Coronal Multi-Channel Polarimeter (CoMP) instrument at the National Solar Observatory, New Mexico. These waves were interpreted as propagating kink waves by Van Doorsselaere et al. (2008)

- 2007: Alfvén wave discoveries appear in articles by Jonathan Cirtain and colleagues, Takenori J. Okamoto and colleagues, and Bart De Pontieu and colleagues. De Pontieu's team proposed that the energy associated with the waves is sufficient to heat the corona and accelerate the solar wind. These results appear in a special collection of 10 articles, by scientists in Japan, Europe and the United States, in the 7 December issue of the journal *Science*. It was demonstrated that those waves should be interpreted in terms of kink waves of coronal plasma structures by Van Doorsselaere, *et al.* (2008); Ofman and Wang (2008); and Vasheghani Farahani, *et al.* (2009).

- 2008: Kaghashvili *et al.* proposed how the detected oscillations can be used to deduct properties of Alfven waves. The mechanism is based on the formalism developed by the Kaghashvili and his collaborators.

- 2011: Experimental evidence of Alfvén wave propagation in a Gallium alloy

References

- Becker, Kurt H. (1998). Novel Aspects of Electron-Molecule Collisions. World Scientific Publishing Company. p. 550. ISBN 978-981-02-3469-0.

- Jackson, J. D. (1975) [1962]. "10.8 Plasma Oscillations". Classical Electrodynamics (2nd ed.). New York: John Wiley & Sons. ISBN 978-0-471-30932-1. OCLC 535998.

- Raether, Heinz (1988). Surface Plasmons on Smooth and Rough Surfaces and on Gratings. Springer. p. 119. ISBN 3540173633.

- Michael G. Cottam & David R. Tilley (1989). Introduction to Surface and Superlattice Excitations. Cambridge University Press. ISBN 0-521-32154-9.

- Heinz Raether (1980). Excitation of plasmons and interband transitions by electrons. Springer-Verlag. ISBN 0-387-09677-9.

- *Andreev, A. A. (2000), An Introduction to Hot Laser Plasma Physics, Huntington, New York: Nova Science Publishers, Inc., ISBN 1-56072-803-5

- Eric Priest, Terry Forbes, Magnetic Reconnection, Cambridge University Press 2000, ISBN 0-521-48179-1, contents and sample chapter online

- Biskamp, Dieter (1997) Nonlinear Magnetohydrodynamics Cambridge University Press, Cambridge, England, page 130, ISBN 0-521-59918-0

- Xin Liu; Mark T. Swihart (2014). "Heavily-doped colloidal semiconductor and metal oxide nanocrystals: an emerging new class of plasmonic nanomaterials". Chem. Soc. Rev. 43: 3908–3920. doi:10.1039/c3cs60417a.

Plasma Diagnostics: An Integrated Study

Plasma diagnostics are techniques and methods used to measure all the properties of plasma. Some of the aspects discussed within this text are plasma parameters, Faraday cup, Langmuir probe, plasma scaling etc. The following text will not only provide an overview, it will also delve deep into the topics related to it.

Plasma Diagnostics

Plasma diagnostics are a pool of methods, instruments, and experimental techniques used to measure properties of a plasma, such as plasma components' density, distribution function over energy (temperature), their spatial profiles and dynamics, which enable to derive plasma parameters.

Invasive Probe Methods

Langmuir Probe

Measurements with electric probes, called Langmuir probes, are the oldest and most often used procedures for low-temperature plasmas. The method was developed by Irving Langmuir and his co-workers in the 1920s, and has since been further developed in order to extend its applicability to more general conditions than those presumed by Langmuir. Langmuir probe measurements are based on the estimation of current versus voltage characteristics of a circuit consisting of two metallic electrodes that are both immersed in the plasma under study. Two cases are of interest: (a) The surface areas of the two electrodes differ by several orders of magnitude. This is known as the *single-probe* method. (b) The surface areas are very small in comparison with the dimensions of the vessel containing the plasma and approximately equal to each other. This is the *double-probe* method.

Conventional Langmuir probe theory assumes collisionless movement of charge carriers in the space charge sheath around the probe. Further it is assumed that the sheath boundary is well-defined and that beyond this boundary the plasma is completely undisturbed by the presence of the probe. This means that the electric field caused by the difference between the potential of the probe and the plasma potential at the place where the probe is located is limited to the volume inside the probe sheath boundary.

The general theoretical description of a Langmuir probe measurement requires the simultaneous solution of the Poisson equation, the collision-free Boltzmann equation or Vlasov equation, and the continuity equation with regard to the boundary condition at the probe surface and requiring that, at large distances from the probe, the solution approaches that expected in an undisturbed plasma.

Ball-pen Probe

A ball-pen probe is novel technique used to measure directly the plasma potential in magnetized plasmas. The probe was invented by Jiří Adámek in the Institute of Plasma Physics AS CR in 2004. The ball-pen probe balances the electron saturation current to the same magnitude as that of the ion saturation current. In this case, its floating potential becomes identical to the plasma potential. This goal is attained by a ceramic shield, which screens off an adjustable part of the electron current from the probe collector due to the much smaller gyro–radius of the electrons. The electron temperature is proportional to the difference of ball-pen probe(plasma potential) and Langmuir probe (floating potential) potential. Thus, the electron temperature can be obtained directly with high temporal resolution without additional power supply.

Self Excited Electron Plasma Resonance Spectroscopy (SEERS)

Nonlinear effects like the I-V characteristic of the boundary sheath are utilized for Langmuir probe measurements but they are usually neglected for modelling of RF discharges due to their very inconvenient mathematical treatment. The Self Excited Electron Plasma Resonance Spectroscopy (SEERS) utilizes exactly these nonlinear effects and known resonance effects in RF discharges. The nonlinear elements, in particular the sheaths, provide harmonics in the discharge current and excite the plasma and the sheath at their series resonance characterized by the so-called geometric resonance frequency.

SEERS provides the spatially and reciprocally averaged electron plasma density and the effective electron collision rate. The electron collision rate reflects stochastic (pressure) heating and ohmic heating of the electrons.

The model for the plasma bulk is based on 2d-fluid model (zero and first order moments of Boltzmann equation) and the full set of the Maxwellian equations leading to the Helmholtz equation for the magnetic field. The sheath model is based additionally on the Poisson equation.

Magnetic (B-dot) Probe

If the magnetic field in the plasma is not stationary, either because the plasma as a whole is transient or because the fields are periodic (radio-frequency heating), the rate of change of the magnetic field with time (, read "B-dot") can be measured locally with a loop or coil of wire. Such coils exploit Faraday's Law, whereby a changing magnetic field induces an electric field. The induced voltage can be measured and recorded with common instruments. Also, by Ampere's Law, the magnetic field is proportional to the currents that produce it, so the measured magnetic field gives information about the currents flowing in the plasma. Both currents and magnetic fields are important in understanding fundamental plasma physics.

Faraday Cup in Plasma Diagnostics

The conventional Faraday cup is applied for measurements of ion (or electron) flows from plasma boundaries and for mass spectrometry.

Passive Spectroscopy

Passive spectroscopic methods simply observe the radiation emitted by the plasma.

Doppler Shift

If the plasma (or one ionic component of the plasma) is flowing in the direction of the line of sight to the observer, emission lines will be seen at a different frequency due to the Doppler effect.

Doppler Broadening

The thermal motion of ions will result in a shift of emission lines up or down, depending on whether the ion is moving toward or away from the observer. The magnitude of the shift is proportional to the velocity along the line of sight. The net effect is a characteristic broadening of spectral lines, known as Doppler broadening, from which the ion temperature can be determined.

Stark Effect

The splitting of some emission lines due to the Stark effect can be used to determine the local electric field.

Stark Broadening

Even if the macroscopic electric field is zero, any single ion will experience an electric field due to the neighboring charged particles in the plasma. This results in a broadening of some lines that can be used to determine the density of the plasma.

Spectral Line Ratios

The brightness of an Atomic spectral line emitted by atoms and ions in a gas (or plasma) can depend on the gas's temperature and pressure.

Due to the completeness and accuracy of modern collisional radiative models the temperature and density of plasmas can be measured by taking ratios of the emission intensities of various Atomic spectral lines

Zeeman Effect

The presence of a magnetic field splits the atomic energy levels due to the Zeeman effect. This leads to broadening or splitting of spectral lines. Analyzing these lines can, therefore, yield the magnetic field strength in the plasma.

Active Spectroscopy

Active spectroscopic methods stimulate the plasma atoms in some way and observe the result (emission of radiation, absorption of the stimulating light or others).

Absorption Spectroscopy

By shining through the plasma a laser with a wavelength, tuned to a certain transition of one of the species present in the plasma, the absorption profile of that transition could be obtained. This profile provides information not only for the plasma parameters, that could be obtained from the emission profile, but also for the line-integrated number density of the absorbing species.

Beam Emission Spectroscopy

A beam of neutral atoms is fired into a plasma. Some atoms are excited by collisions within the plasma and emit radiation. This can be used to probe density fluctuations in a turbulent plasma.

Charge Exchange Recombination Spectroscopy

In very hot plasmas (as in magnetic fusion experiments), light elements are fully ionized and don't emit line radiation. When a beam of neutral atoms is fired into the plasma, electrons from beam atoms are transferred to hot plasma ions, which form hydrogenic ions which promptly emit line radiation. This radiation is analyzed for ion density, temperature, and velocity.

Laser-induced Fluorescence

If the plasma is not fully ionized but contains ions that fluoresce, laser-induced fluorescence can provide very detailed information on temperature, density, and flows.

Motional Stark Effect

If an atom is moving in a magnetic field, the Lorentz force will act in opposite directions on the nucleus and the electrons, just as an electric field does. In the frame of reference of the atom, there *is* an electric field, even if there is none in the laboratory frame. Consequently, certain lines will be split by the Stark effect. With an appropriate choice of beam species and velocity and of geometry, this effect can be used to determine the magnetic field in the plasma.

Two-photon Laser-induced Fluorescence

The two-photon laser-induced fluorescence (TALIF) is a modification of the laser-induced fluorescence technique. In this approach the upper level is excited by absorbing two photons and registering the resulting emission from the excited state. The advantage of this approach is that the registered light from the fluorescence is with a different wavelength from the exciting laser beam, which leads to improved signal to noise ratio.

Optical Effects From Free Electrons

The optical diagnostics above measure line radiation from atoms. Alternatively, the effects of free charges on electromagnetic radiation can be used as a diagnostic.

Electron Cyclotron Emission

In magnetized plasmas, electrons will gyrate around magnetic field lines and emit Cyclotron radiation. The frequency of the emission is given by the Cyclotron resonance condition. In a sufficiently thick and dense plasma, the intensity of the emission will follow Planck's law, and only depend on the electron temperature.

Thomson Scattering

Scattering of laser light from the electrons in a plasma is known as Thomson scattering. The electron temperature can be determined very reliably from the Doppler broadening of the laser line. The electron density can be determined from the intensity of the scattered light, but a careful absolute calibration is required. Although Thomson scattering is dominated by scattering from electrons, since the electrons interact with the ions, in some circumstances information on the ion temperature can also be extracted.

Interferometry

If a plasma is placed in one arm of an interferometer, the phase shift will be proportional to the plasma density integrated along the path.

Faraday Rotation

The Faraday effect will rotate the plane of polarization of a beam passing through a plasma with a magnetic field in the direction of the beam. This effect can be used as a diagnostic of the magnetic field, although the information is mixed with the density profile and is usually an integral value only.

Neutron Diagnostics

Fusion plasmas produces 3.5 MeV and 14 MeV neutrons. By measuring the neutron flux, plasma properties such as ion temperature and fusion power can be determined.

Plasma Parameters

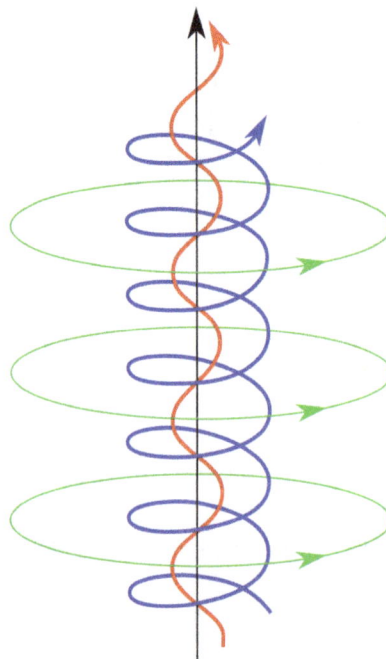

The complex self-constricting magnetic field lines and current paths in a Birkeland current that may develop in a plasma (*Evolution of the Solar System*, 1976)

Plasma parameters define various characteristics of a plasma, an electrically conductive collection of charged particles that responds *collectively* to electromagnetic forces. Plasma typically takes the form of neutral gas-like clouds or charged ion beams, but may also include dust and grains. The behaviour of such particle systems can be studied statistically.

Fundamental Plasma Parameters

All quantities are in Gaussian (cgs) units except energy and temperature expressed in eV and ion mass expressed in units of the proton mass $\mu = m_i / m_p$; Z is charge state; k is Boltzmann's constant; K is wavenumber; $\ln \Lambda$ is the Coulomb logarithm.

Frequencies

- electron gyrofrequency, the angular frequency of the circular motion of an electron in the plane perpendicular to the magnetic field:

$$\omega_{ce} = eB / m_e c = 1.76 \times 10^7 \, B \, \text{rad/s}$$

- ion gyrofrequency, the angular frequency of the circular motion of an ion in the plane perpendicular to the magnetic field:

$$\omega_{ci} = ZeB / m_i c = 9.58 \times 10^3 \, Z \mu^{-1} \, B \, \text{rad/s}$$

- electron plasma frequency, the frequency with which electrons oscillate (plasma oscillation):

$$\omega_{pe} = (4\pi n_e e^2 / m_e)^{1/2} = 5.64 \times 10^4 \, n_e^{1/2} \, \text{rad/s}$$

- ion plasma frequency:

$$\omega_{pi} = (4\pi n_i Z^2 e^2 / m_i)^{1/2} = 1.32 \times 10^3 \, Z \mu^{-1/2} n_i^{1/2} \, \text{rad/s}$$

- electron trapping rate:

$$v_{Te} = (eKE / m_e)^{1/2} = 7.26 \times 10^8 \, K^{1/2} E^{1/2} \, \text{s}^{-1}$$

- ion trapping rate:

$$v_{Ti} = (ZeKE / m_i)^{1/2} = 1.69 \times 10^7 \, Z^{1/2} K^{1/2} E^{1/2} \mu^{-1/2} \, \text{s}^{-1}$$

- electron collision rate in completely ionized plasmas:

$$_e = 2.91 \times 10^{-6} n_e \ln \Lambda T_e^{-3/2} \, \text{s}^{-1}$$

- ion collision rate in completely ionized plasmas:

$$v_i = 4.80 \times 10^{-8} Z^4 \mu^{-1/2} n_i \ln \Lambda T_i^{-3/2} \, \text{s}^{-1}$$

- electron (ion) collision rate in slightly ionized plasmas:

$$v_{e,i} = N\overline{\sigma_{e,i}v} = N\int_0^\infty \sigma(v)_{e,i} f(v)v\,dv$$

where $\sigma(v)_{e,i}$ is a collision crossection of the electron (ion) on the operating gas atoms (molecules), $f(v)$ is the electron (ion) distribution function in plasma, and N is an operating gas concentration.

Lengths

- Electron thermal de Broglie wavelength, approximate average de Broglie wavelength of electrons in a plasma:

$$\Lambda_e = \sqrt{\frac{h^2}{2\pi m_e kT_e}} = 6.919\times10^{-8}\,T_e^{-1/2}\,\mathrm{cm}$$

- classical distance of closest approach, the closest that two particles with the elementary charge come to each other if they approach head-on and each have a velocity typical of the temperature, ignoring quantum-mechanical effects:

$$e^2/kT = 1.44\times10^{-7}\,T^{-1}\,\mathrm{cm}$$

- electron gyroradius, the radius of the circular motion of an electron in the plane perpendicular to the magnetic field:

$$r_e = v_{Te}/\omega_{ce} = 2.38 T_e^{1/2} B^{-1}\,\mathrm{cm}$$

- ion gyroradius, the radius of the circular motion of an ion in the plane perpendicular to the magnetic field:

$$r_i = v_{Ti}/\omega_{ci} = 1.02\times10^2\,\mu^{1/2} Z^{-1} T_i^{1/2} B^{-1}\,\mathrm{cm}$$

- plasma skin depth, the depth in a plasma to which electromagnetic radiation can penetrate:

$$c/\omega_{pe} = 5.31\times10^5\,n_e^{-1/2}\,\mathrm{cm}$$

- Debye length, the scale over which electric fields are screened out by a redistribution of the electrons:

$$\lambda_D = (kT/4\pi ne^2)^{1/2} = 7.43\times10^2\,T^{1/2} n^{-1/2}\,\mathrm{cm}$$

- Ion inertial length, the scale at which ions decouple from electrons and the magnetic field becomes frozen into the electron fluid rather than the bulk plasma:

$$d_i = c/\omega_{pi} = 2.28\times10^7\,Z^{-1}(\mu/n_i)^{1/2}\,\mathrm{cm}$$

- Free path is the average distance between two subsequent collisions of the electron (ion) with plasma components:

$$\lambda_{e,i} = \frac{\overline{v_{e,i}}}{v_{e,i}}$$

where $\overline{v_{e,i}}$ is an average velocity of the electron (ion), and $v_{e,i}$ is the electron or ion collision rate.

Velocities

- electron thermal velocity, typical velocity of an electron in a Maxwell–Boltzmann distribution:

$$v_{Te} = (kT_e / m_e)^{1/2} = 4.19 \times 10^7 T_e^{1/2} \, \text{cm/s}$$

- ion thermal velocity, typical velocity of an ion in a Maxwell–Boltzmann distribution:

$$v_{Ti} = (kT_i / m_i)^{1/2} = 9.79 \times 10^5 \mu^{-1/2} T_i^{1/2} \, \text{cm/s}$$

- ion sound velocity, the speed of the longitudinal waves resulting from the mass of the ions and the pressure of the electrons:

$$c_s = (\gamma Z k T_e / m_i)^{1/2} = 9.79 \times 10^5 (\gamma Z T_e / \mu)^{1/2} \, \text{cm/s},$$

where $\gamma = 1 + 2/n$ is the adiabatic index, and here n is the number of degrees of freedom

- Alfvén velocity, the speed of the waves resulting from the mass of the ions and the restoring force of the magnetic field:

$$v_A = B / (4\pi n_i m_i)^{1/2} = 2.18 \times 10^{11} \mu^{-1/2} n_i^{-1/2} B \, \text{cm/s}$$

Dimensionless

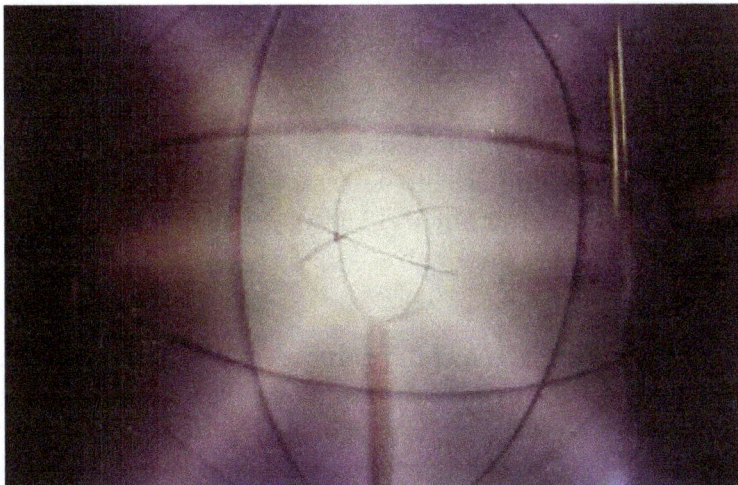

A 'sun in a test tube'. The Farnsworth-Hirsch Fusor during operation in so called "star mode" characterized by "rays" of glowing plasma which appear to emanate from the gaps in the inner grid.

- square root of electron/proton mass ratio

  $$(m_e / m_p)^{1/2} = 2.33 \times 10^{-2} = 1/42.9$$

- number of particles in a Debye sphere

  $$(4\pi / 3)n\lambda_D^3 = 1.72 \times 10^9 T^{3/2} n^{-1/2}$$

- Alfvén velocity/speed of light

  $$v_A / c = 7.28 \mu^{-1/2} n_i^{-1/2} B$$

- electron plasma/gyrofrequency ratio

  $$\omega_{pe} / \omega_{ce} = 3.21 \times 10^{-3} n_e^{1/2} B^{-1}$$

- ion plasma/gyrofrequency ratio

  $$\omega_{pi} / \omega_{ci} = 0.137 \mu^{1/2} n_i^{1/2} B^{-1}$$

- thermal/magnetic pressure ratio ("beta")

  $$\beta = 8\pi nkT / B^2 = 4.03 \times 10^{-11} nTB^{-2}$$

- magnetic/ion rest energy ratio

  $$B^2 / 8\pi n_i m_i c^2 = 26.5 \mu^{-1} n_i^{-1} B^2$$

- Coulomb logarithm is an average coefficient taking into account far Coulomb interactions of charged particles in plasma. Its value is evaluated in the nonrelativistic case approximately

for electrons $\ln \Lambda \approx 13.6$,

for ions $\ln \Lambda \approx 6.8$

Faraday Cup

A Faraday cup is a metal (conductive) cup designed to catch charged particles in vacuum. The resulting current can be measured and used to determine the number of ions or electrons hitting the cup. The Faraday cup is named after Michael Faraday who first theorized ions around 1830.

Principle of Operation

When a beam or packet of ions hits the metal it gains a small net charge while the ions are neutralized. The metal can then be discharged to measure a small current equivalent to the number of impinging ions. Essentially the Faraday cup is part of a circuit where ions are the charge carriers in vacuum and the faraday cup is the interface to the solid metal where electrons act as the charge carriers (as in most circuits). By measuring the electric current (the number of electrons flowing through the circuit per second) in the metal part of the circuit the number of charges being carried by the ions in the vacuum part of the circuit can be determined. For a continuous beam of ions (each with a single charge)

Faraday cup with an electron-suppressor plate in front

$$\frac{N}{t} = \frac{I}{e}$$

where N is the number of ions observed in a time t (in seconds), I is the measured current (in amperes) and e is the elementary charge (about 1.60×10^{-19} C). Thus, a measured current of one nanoamp (10^{-9} A) corresponds to about 6 billion ions striking the faraday cup each second.

Similarly, a Faraday cup can act as a collector for electrons in a vacuum (for instance from an electron beam). In this case electrons simply hit the metal plate/cup and a current is produced. Faraday cups are not as sensitive as electron multiplier detectors, but are highly regarded for accuracy because of the direct relation between the measured current and number of ions.This device is considered a universal charge detector because of its independence from the energy, mass, chemistry, etc. of the analyte.

Faraday Cup in Plasma Diagnostics

Faraday cup utilizes a physical principle according which the electrical charges delivered to the inner surface of a hollow conductor are redistributed around its outer surface due to mutual self-repelling of charges of the same sign – phenomenon discovered by Faraday.

The conventional Faraday cup is applied for measurements of ion (or electron) flows from plasma boundaries and comprises a metallic cylindrical receiver-cap – 1 (Fig. 1) closed with, and insulated from, a washer-type metallic electron-suppressor lid - 2 provided with the round axial through enter-hollow of an aperture with a surface area $S_F = \pi D_F^2 / 4$. Both the receiver cup and the electron-suppressor lid are enveloped in, and insulated from, a grounded cylindrical shield - 3 having an axial round hole coinciding with the hole in the electron-suppressor lid - 2. The electron-suppressor lid is connected by 50 Ω RF cable with the source B_{es} of variable DC voltage U_{es}. The receiver-cup is connected by 50 Ω RF cable through the load resistor R_F with a sweep generator producing saw-type pulses $U_g(t)$. Electric capacity C_F is formed of the capacity of the receiver-cup - 1 to the grounded shield - 3 and the capacity of the RF cable. The signal from R_F enables an observer to acquire an I-V characteristic of the Faraday cup by oscilloscope. Proper operating conditions: $h \geq D_F$ (due to possible potential sag) and $h \ll \lambda_i$, where λ_i is the ion free path. Signal from R_F is the Faraday cup I-V characteristic which can be observed and memorized by oscilloscope

$$i_\Sigma(U_g) = i_i(U_g) - C_F \frac{dU_g}{dt} . \quad (1)$$

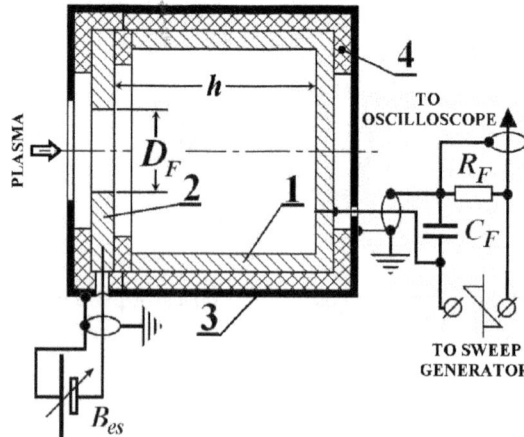

Fig. 1. Faraday Cup for Plasma Diagnostics

In Fig. 1: 1 – cup-receiver, metal (stainless steel). 2 – electron-suppressor lid, metal (stainless steel). 3 – grounded shield, metal (stainless steel). 4 - insulator (teflon, ceramic). C_F - capacity of Faraday cup. R_F - load resistor.

Thus we measure the sum i_Σ of the electric currents through the load resistor R_F: i_i (Faraday cup current) plus the current $i_c(U_g) = -C_F(dU_g >$ induced through the capacitor C_F by the saw-type voltage U_g of the sweep-generator: The current component $i_c(U_g)$ can be measured at the absence of the ion flow and can be subtracted further from the total current $i_\Sigma(U_g)$ measured with plasma to obtain the actual Faraday cup I-V characteristic $i_i(U_g)$ for processing. All of the Faraday cup elements and their assembly that interact with plasma are fabricated usually of temperature-resistant materials (often these are stainless steel and teflon or ceramic for insulators). For processing of the Faraday cup I-V characteristic, we are going to assume that the Faraday cup is installed far enough away from an investigated plasma source where the flow of ions could be considered as the flow of particles with parallel velocities directed exactly along the Faraday cup axis. In this case, the elementary particle current di_i corresponding to the ion density differential i_0 in the range of velocities between v and $v + dv$ of ions flowing in through operating aperture S_F of the electron-suppressor can be written in the form

$$di_i = eZ_i S_F v dn(v) , \quad (2)$$

where

$$dn(v) = nf(v)dv , \quad (3)$$

e is elementary electric charge, Z_i is the ion charge state, and $f(v)$ is the one-dimensional distribution function of ions over velocity v. Therefore the ion current at the ion-decelerating voltage U_g of the Faraday cup can be calculated by integrating Eq. (2) after substituting in it Eq. (3)

$$i_i(U_g) = eZ_i n_i S_F \int\limits_{\sqrt{2eZ_iU_g/M_i}}^{\infty} f(v)v\,dv \quad , (4)$$

where the lower integration limit is defined from the obvious equation $M_i v_{i,s}^2 / 2 = eZ_iU_g$ where $v_{i,s}$ is the velocity of the ion stopped by the decelerating potential U_g, and M_i is the ion mass. Thus the expression (4) represents the I-V characteristic of the Faraday cup. Differentiating Eq. (4) with respect to U_g, one can obtain the relation

$$\frac{di_i(U_g)}{dU_g} = -en_i S_F \frac{eZ_i}{M_i} f\left(\sqrt{2eZ_iU_g/M_i}\right), (5)$$

where the value $-n_i S_F (eZ_i >$ is an invariable constant for each measurement. Therefore the average velocity $\langle v_i \rangle$ of ions arriving into the Faraday cup and their average energy $\langle \mathcal{E}_i \rangle$ can be calculated (under the assumption that we operate with a single type of ion) by the expressions

$$\langle v_i \rangle = 1.389 \times 10^6 \sqrt{\frac{Z_i}{M_A}} \int\limits_0^{\infty} i_i'(U_g)dU_g \left(\int\limits_0^{\infty} \frac{i_i'}{\sqrt{U_g}}dU_g \right)^{-1} \text{[cm/s], (6)}$$

$$\langle \mathcal{E}_i \rangle = \int\limits_0^{\infty} i_i'(U_g)\sqrt{U_g}\,dU_g \left(\int\limits_0^{\infty} \frac{i_i'}{\sqrt{U_g}}dU_g \right)^{-1} \text{[eV], (7)}$$

where M_A is the ion mass in atomic units. The ion concentration n_i in the ion flow at the Faraday cup vicinity can be calculated by the formula

$$n_i = \frac{i_i(0)}{eZ_i \langle v_i \rangle S_F} (8)$$

which follows from Eq. (4) at $U_g = 0$,

$$\int\limits_0^{\infty} f(v)v\,dv = \langle v \rangle , (9)$$

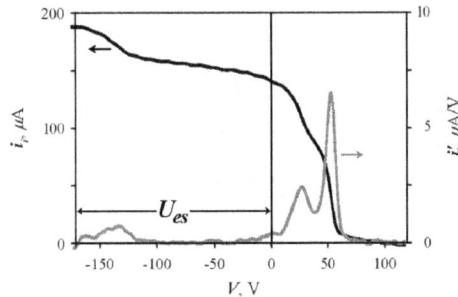

Fig, 2. Faraday cup I-V characteristic

and from the conventional condition for distribution function normalizing

$$\int_{0}^{\infty} f(v)dv = 1. \text{ (10)}$$

Fig. 2 illustrates the I-V characteristic $i_i(V)$ and its first derivative $i_i'(V)$ of the Faraday cup with $S_F = 0.5 cm^2$ installed at output of the Inductively coupled plasma source powered with RF 13.56 MHz and operating at 6 mTorr of H2. The value of the electron-suppressor voltage (accelerating the ions) was set experimentally at $U_{es} = -170V$, near the point of suppression of the secondary electron emission from the inner surface of the Faraday cup.

Error Sources

The counting of charges collected per unit time is impacted by two error sources: 1) the emission of low-energy secondary electrons from the surface struck by the incident charge and 2) backscattering (~180 degree scattering) of the incident particle, which causes it to leave the collecting surface, at least temporarily. Especially with electrons, it is fundamentally impossible to distinguish between a fresh new incident electron and one that has been backscattered or even a fast secondary electron.

Langmuir Probe

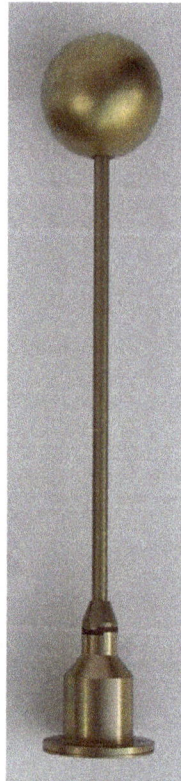

One of two Langmuir probes from the Swedish Institute of Space Physics in Uppsala on board ESA's space vehicle Rosetta, due for a comet. The probe is the spherical part, 50 mm in diameter and made from titanium with a surface coating of titanium nitride.

A Langmuir probe is a device used to determine the electron temperature, electron density, and electric potential of a plasma. It works by inserting one or more electrodes into a plasma, with a constant or time-varying electric potential between the various electrodes or between them and the surrounding vessel. The measured currents and potentials in this system allow the determination of the physical properties of the plasma.

I-V Characteristic of the Debye Sheath

The beginning of Langmuir probe theory is the *I-V* characteristic of the Debye sheath, that is, the current density flowing to a surface in a plasma as a function of the voltage drop across the sheath. The analysis presented here indicates how the electron temperature, electron density, and plasma potential can be derived from the *I-V* characteristic. In some situations a more detailed analysis can yield information on the ion density (n_i), the ion temperature T_i, or the electron energy distribution function (EEDF) or $f_e(v)$.

Ion Saturation Current Density

Consider first a surface biased to a large negative voltage. If the voltage is large enough, essentially all electrons (and any negative ions) will be repelled. The ion velocity will satisfy the Bohm sheath criterion, which is, strictly speaking, an inequality, but which is usually marginally fulfilled. The Bohm criterion in its marginal form says that the ion velocity at the sheath edge is simply the sound speed given by

$$c_s = \sqrt{k_B(ZT_e + \gamma_i T_i)/m_i}.$$

The ion temperature term is often neglected, which is justified if the ions are cold. Even if the ions are known to be warm, the ion temperature is usually not known, so it is usually assumed to be simply equal to the electron temperature. In that case, consideration of finite ion temperature only results in a small numerical factor. Z is the (average) charge state of the ions, and is the adiabatic coefficient for the ions. The proper choice of γ_i is a matter of some contention. Most analyses use $\gamma_i = 1$, corresponding to isothermal ions, but some kinetic theory suggests that $\gamma_i = 3$, corresponding to one degree of freedom is more appropriate. For $Z = 1$ and $T_i = T_e$, using the larger value results in the conclusion that the density is $\sqrt{2}$ times smaller. Uncertainties of this magnitude arise several places in the analysis of Langmuir probe data and are very difficult to resolve.

The charge density of the ions depends on the charge state Z, but quasineutrality allows one to write it simply in terms of the electron density as en_e.

Using these results we have the current density to the surface due to the ions. The current density at large negative voltages is due solely to the ions and, except for possible sheath expansion effects, does not depend on the bias voltage, so it is referred to as the ion saturation current density and is given by

$j_i^{max} = q_e n_e c_s$ where q_e is the charge of an electron, n_e is the number density of electrons, and c_s is as defined above.

The plasma parameters, in particular, the density, are those at the sheath edge.

Exponential Electron Current

As the voltage of the Debye sheath is reduced, the more energetic electrons are able to overcome the potential barrier of the electrostatic sheath. We can model the electrons at the sheath edge with a Maxwell–Boltzmann distribution, i.e.,

$$f(v_x)dv_x \propto e^{-\frac{1}{2}m_e v_x^2/k_B T_e},$$

except that the high energy tail moving away from the surface is missing, because only the lower energy electrons moving toward the surface are reflected. The higher energy electrons overcome the sheath potential and are absorbed. The mean velocity of the electrons which are able to overcome the voltage of the sheath is

$$\langle v_e \rangle = \frac{\int_{v_{e0}}^{\infty} f(v_x) v_x dv_x}{\int_{-\infty}^{\infty} f(v_x) dv_x},$$

where the cut-off velocity for the upper integral is

$$v_{e0} = \sqrt{2q_e \Delta V / m_e}.$$

ΔV is the voltage across the Debye sheath, that is, the potential at the sheath edge minus the potential of the surface. For a large voltage compared to the electron temperature, the result is

$$\langle v_e \rangle = \sqrt{\frac{k_B T_e}{2\pi m_e}} e^{-q_e \Delta V/k_B T_e},$$

With this expression, we can write the electron contribution to the current to the probe in terms of the ion saturation current as

$$j_e = j_i^{max} \sqrt{m_i / 2\pi m_e}\, e^{-q_e \Delta V/k_B T_e},$$

valid as long as the electron current is not more than two or three times the ions current.

Floating Potential

The total current, of course, is the sum of the ion and electron currents:

$$j = j_i^{max} \left(-1 + \sqrt{m_i / 2\pi m_e}\, e^{-q_e \Delta V/k_B T_e} \right).$$

We are using the convention that current *from* the surface into the plasma is positive. An interesting and practical question is the potential of a surface to which no net current flows. It is easily seen from the above equation that

$$\Delta V = (k_B T_e / e)(1/2) \ln(m_i / 2\pi m_e).$$

If we introduce the ion reduced mass $\mu_i = m_i / m_e$, we can write

$$\Delta V = (k_B T_e / e)(2.8 + 0.5 \ln \mu_i)$$

Since the floating potential is the experimentally accessible quantity, the current (below electron saturation) is usually written as

$$j = j_i^{max} \left(-1 + e^{q_e(V_0 - \Delta V)/k_B T_e} \right)..$$

Electron Saturation Current

When the electrode potential is equal to or greater than the plasma potential, then there is no longer a sheath to reflect electrons, and the electron current saturates. Using the Boltzmann expression for the mean electron velocity given above with $v_{e0} = 0$ and setting the ion current to zero, the electron saturation current density would be

$$j_e^{max} = j_i^{max} \sqrt{m_i / \pi m_e} = j_i^{max} \left(24.2 \sqrt{\mu_i} \right)$$

Although this is the expression usually given in theoretical discussions of Langmuir probes, the derivation is not rigorous and the experimental basis is weak. The theory of double layers typically employs an expression analogous to the Bohm criterion, but with the roles of electrons and ions reversed, namely

$$j_e^{max} = q_e n_e \sqrt{k_B (\gamma_e T_e + T_i) / m_e} = j_i^{max} \sqrt{m_i / m_e} = j_i^{max} \left(42.8 \sqrt{\mu_i} \right)$$

where the numerical value was found by taking $T_i = T_e$ and $\gamma_i = \gamma_e$.

In practice, it is often difficult and usually considered uninformative to measure the electron saturation current experimentally. When it is measured, it is found to be highly variable and generally much lower (a factor of three or more) than the value given above. Often a clear saturation is not seen at all. Understanding electron saturation is one of the most important outstanding problems of Langmuir probe theory.

Effects of the Bulk Plasma

The Debye sheath theory explains the basic behavior of Langmuir probes, but is not complete. Merely inserting an object like a probe into a plasma changes the density, temperature, and potential at the sheath edge and perhaps everywhere. Changing the voltage on the probe will also, in general, change various plasma parameters. Such effects are less well understood than sheath physics, but they can at least in some cases be roughly accounted.

Pre-sheath

The Bohm criterion requires the ions to enter the Debye sheath at the sound speed. The potential drop that accelerates them to this speed is called the pre-sheath. It has a spatial scale that depends on the physics of the ion source but which is large compared to the Debye length and often of the order of the plasma dimensions. The magnitude of the potential drop is equal to (at least)

$$\Phi_{pre} = \frac{\frac{1}{2}m_i c_s^2}{Ze} = k_B(T_e + Z\gamma_i T_i)/(2Ze)$$

The acceleration of the ions also entails a decrease in the density, usually by a factor of about 2 depending on the details.

Resistivity

Collisions between ions and electrons will also affect the *I-V* characteristic of a Langmuir probe. When an electrode is biased to any voltage other than the floating potential, the current it draws must pass through the plasma, which has a finite resistivity. The resistivity and current path can be calculated with relative ease in an unmagnetized plasma. In a magnetized plasma, the problem is much more difficult. In either case, the effect is to add a voltage drop proportional to the current drawn, which shears the characteristic. The deviation from an exponential function is usually not possible to observe directly, so that the flattening of the characteristic is usually misinterpreted as a larger plasma temperature. Looking at it from the other side, any measured *I-V* characteristic can be interpreted as a hot plasma, where most of the voltage is dropped in the Debye sheath, or as a cold plasma, where most of the voltage is dropped in the bulk plasma. Without quantitative modeling of the bulk resistivity, Langmuir probes can only give an upper limit on the electron temperature.

Sheath Expansion

It is not enough to know the current *density* as a function of bias voltage since it is the *absolute* current which is measured. In an unmagnetized plasma, the current-collecting area is usually taken to be the exposed surface area of the electrode. In a magnetized plasma, the projected area is taken, that is, the area of the electrode as viewed along the magnetic field. If the electrode is not shadowed by a wall or other nearby object, then the area must be doubled to account for current coming along the field from both sides. If the electrode dimensions are not small in comparison to the Debye length, then the size of the electrode is effectively increased in all directions by the sheath thickness. In a magnetized plasma, the electrode is sometimes assumed to be increased in a similar way by the ion Larmor radius.

The finite Larmor radius allows some ions to reach the electrode that would have otherwise gone past it. The details of the effect have not been calculated in a fully self-consistent way.

If we refer to the probe area including these effects as A_{eff} (which may be a function of the bias voltage) and make the assumptions

- $T_i = T_e$,

- $Z = 1$

- $\gamma_i = 3$, and

- $n_{e,sh} = 0.5 n_e$,

and ignore the effects of

- bulk resistivity, and

- electron saturation,

then the *I-V* characteristic becomes

$$I = I_i^{max}\left(-1 + e^{q_e(V_{pr}-V_{fl})/(k_B T_e)}\right),$$

where

$$I_i^{max} = q_e n_e \sqrt{k_B T_e / m_i}\, A_{eff}.$$

Magnetized Plasmas

The theory of Langmuir probes is much more complex when the plasma is magnetized. The simplest extension of the unmagnetized case is simply to use the projected area rather than the surface area of the electrode. For a long cylinder far from other surfaces, this reduces the effective area by a factor of $\pi/2 = 1.57$. As mentioned before, it might be necessary to increase the radius by about the thermal ion Larmor radius, but not above the effective area for the unmagnetized case.

The use of the projected area seems to be closely tied with the existence of a magnetic sheath. Its scale is the ion Larmor radius at the sound speed, which is normally between the scales of the Debye sheath and the pre-sheath. The Bohm criterion for ions entering the magnetic sheath applies to the motion along the field, while at the entrance to the Debye sheath it applies to the motion normal to the surface. This results in a reduction of the density by the sine of the angle between the field and the surface. The associated increase in the Debye length must be taken into account when considering ion non-saturation due to sheath effects.

Especially interesting and difficult to understand is the role of cross-field currents. Naively, one would expect the current to be parallel to the magnetic field along a flux tube. In many geometries, this flux tube will end at a surface in a distant part of the device, and this spot should itself exhibit an *I-V* characteristic. The net result would be the measurement of a double-probe characteristic; in other words, electron saturation current equal to the ion saturation current.

When this picture is considered in detail, it is seen that the flux tube must charge up and the surrounding plasma must spin around it. The current into or out of the flux tube must be associated with a force that slows down this spinning. Candidate forces are viscosity, friction with neutrals, and inertial forces associated with plasma flows, either steady or fluctuating. It is not known which force is strongest in practice, and in fact it is generally difficult to find any force that is powerful enough to explain the characteristics actually measured.

It is also likely that the magnetic field plays a decisive role in determining the level of electron saturation, but no quantitative theory is as yet available.

Electrode Configurations

Once one has a theory of the *I-V* characteristic of an electrode, one can proceed to measure it and then fit the data with the theoretical curve to extract the plasma parameters. The straightforward way to do this is to sweep the voltage on a single electrode, but, for a number of reasons, configurations using multiple electrodes or exploring only a part of the characteristic are used in practice.

Single Probe

The most straightforward way to measure the I-V characteristic of a plasma is with a single probe, consisting of one electrode biased with a voltage ramp relative to the vessel. The advantages are simplicity of the electrode and redundancy of information, i.e. one can check whether the I-V characteristic has the expected form. Potentially additional information can be extracted from details of the characteristic. The disadvantages are more complex biasing and measurement electronics and a poor time resolution. If fluctuations are present (as they always are) and the sweep is slower than the fluctuation frequency (as it usually is), then the I-V is the *average* current as a function of voltage, which may result in systematic errors if it is analyzed as though it were an instantaneous I-V. The ideal situation is to sweep the voltage at a frequency above the fluctuation frequency but still below the ion cyclotron frequency. This, however, requires sophisticated electronics and a great deal of care.

Double Probe

An electrode can be biased relative to a second electrode, rather than to the ground. The theory is similar to that of a single probe, except that the current is limited to the ion saturation current for both positive and negative voltages. In particular, if V_{bias} is the voltage applied between two identical electrodes, the current is given by;

$$I = I_i^{max}\left(-1+e^{q_e(V_2-V_{fl})/k_BT_e}\right) = -I_i^{max}\left(-1+e^{q_e(V_1-V_{fl})/k_BT_e}\right),$$

which can be rewritten using $V_{bias} = V_2 - V_1$ as a hyperbolic tangent:

$$I = I_i^{max}\tanh\left(\frac{1}{2}\frac{q_eV_{bias}}{k_BT_e}\right).$$

One advantage of the double probe is that neither electrode is ever very far above floating, so the theoretical uncertainties at large electron currents are avoided. If it is desired to sample more of the exponential electron portion of the characteristic, an asymmetric double probe may be used, with one electrode larger than the other. If the ratio of the collection areas is larger than the square root of the ion to electron mass ratio, then this arrangement is equivalent to the single tip probe. If the ratio of collection areas is not that big, then the characteristic will be in-between the symmetric double tip configuration and the single-tip configuration. If A_1 is the area of the larger tip then:

$$I = A_1 J_i^{max}\left[\coth\left(\frac{q_eV_{bias}}{2k_BT_e}\right) + \frac{\left(\dfrac{A_1}{A_2}-1\right)e^{-q_eV_{bias}/2k_BT_e}}{2\sinh\left(\dfrac{q_eV_{bias}}{2k_BT_e}\right)}\right]^{-1}$$

Another advantage is that there is no reference to the vessel, so it is to some extent immune to the disturbances in a radio frequency plasma. On the other hand, it shares the limitations of a single probe concerning complicated electronics and poor time resolution. In addition, the second electrode not only complicates the system, but it makes it susceptible to disturbance by gradients in the plasma.

Triple Probe

An elegant electrode configuration is the triple probe, consisting of two electrodes biased with a fixed voltage and a third which is floating. The bias voltage is chosen to be a few times the electron temperature so that the negative electrode draws the ion saturation current, which, like the floating potential, is directly measured. A common rule of thumb for this voltage bias is 3/e times the expected electron temperature. Because the biased tip configuration is floating, the positive probe can draw at most an electron current only equal in magnitude and opposite in polarity to the ion saturation current drawn by the negative probe, given by :

$$-I_+ = I_- = I_i^{max}$$

and as before the floating tip draws effectively no current:

$$I_{fl} = 0.$$

Assuming that: 1.) The electron energy distribution in the plasma is Maxwellian, 2.) The mean free path of the electrons is greater than the ion sheath about the tips and larger than the probe radius, and 3.) the probe sheath sizes are much smaller than the probe separation, then the current to any probe can be considered composed of two parts – the high energy tail of the Maxwellian electron distribution, and the ion saturation current:

$$I_{probe} = -I_e e^{-q_e V_{probe}/(kT_e)} + I_i^{max}$$

where the current I_e is thermal current. Specifically,

$$I_e = SJ_e = Sn_e q_e \sqrt{kT_e / 2\pi m_e},$$

where S is surface area, J_e is electron current density, and n_e is electron density.

Assuming that the ion and electron saturation current is the same for each probe, then the formulas for current to each of the probe tips take the form

$$I_+ = -I_e e^{-q_e V_+/(kT_e)} + I_i^{max}$$

$$I_- = -I_e e^{-q_e V_-/(kT_e)} + I_i^{max}$$

$$I_{fl} = -I_e e^{-q_e V_{fl}/(kT_e)} + I_i^{max}.$$

It is then simple to show

$$\left(I_{+}-I_{fl}\right)/\left(I_{+}-I_{-}\right)=\left(1-e^{-q_e(V_{fl}-V_{+})/(kT_e)}\right)/\left(1-e^{-q_e(V_{-}-V_{+})/(kT_e)}\right)$$

but the relations from above specifying that $I_{+}=-I_{-}$ and $I_{fl}=0$ give

$$1/2=\left(1-e^{-q_e(V_{fl}-V_{+})/(kT_e)}\right)/\left(1-e^{-q_e(V_{-}-V_{+})/(kT_e)}\right),$$

a transcendental equation in terms of applied and measured voltages and the unknown T_e that in the limit $q_e V_{Bias} = q_e(V_{+}-V) >> k\,T_e$, becomes

$$(V_{+}-V_{fl})=(k_B T_e / q_e)\ln 2.$$

That is, the voltage difference between the positive and floating electrodes is proportional to the electron temperature. (This was especially important in the sixties and seventies before sophisticated data processing became widely available.)

More sophisticated analysis of triple probe data can take into account such factors as incomplete saturation, non-saturation, unequal areas.

Triple probes have the advantage of simple biasing electronics (no sweeping required), simple data analysis, excellent time resolution, and insensitivity to potential fluctuations (whether imposed by an rf source or inherent fluctuations). Like double probes, they are sensitive to gradients in plasma parameters.

Special Arrangements

Arrangements with four (tetra probe) or five (penta probe) have sometimes been used, but the advantage over triple probes has never been entirely convincing. The spacing between probes must be larger than the Debye length of the plasma to prevent an overlapping Debye sheath.

A pin-plate probe consists of a small electrode directly in front of a large electrode, the idea being that the voltage sweep of the large probe can perturb the plasma potential at the sheath edge and thereby aggravate the difficulty of interpreting the I-V characteristic. The floating potential of the small electrode can be used to correct for changes in potential at the sheath edge of the large probe. Experimental results from this arrangement look promising, but experimental complexity and residual difficulties in the interpretation have prevented this configuration from becoming standard.

Various geometries have been proposed for use as ion temperature probes, for example, two cylindrical tips that rotate past each other in a magnetized plasma. Since shadowing effects depend on the ion Larmor radius, the results can be interpreted in terms of ion temperature. The ion temperature is an important quantity that is very difficult to measure. Unfortunately, it is also very difficult to analyze such probes in a fully self-consistent way.

Emissive probes use an electrode heated either electrically or by the exposure to the plasma. When the electrode is biased more positive than the plasma potential, the emitted electrons are pulled back to the surface so the I-V characteristic is hardly changed. As soon as the electrode is biased negative with respect to the plasma potential, the emitted electrons are repelled and contribute a large negative current. The onset of this current or, more sensitively, the onset of a discrepancy between the characteristics of an unheated and a heated electrode, is a sensitive indicator of the plasma potential.

To measure fluctuations in plasma parameters, arrays of electrodes are used, usually one – but occasionally two-dimensional. A typical array has a spacing of 1 mm and a total of 16 or 32 electrodes. A simpler arrangement to measure fluctuations is a negatively biased electrode flanked by two floating electrodes. The ion-saturation current is taken as a surrogate for the density and the floating potential as a surrogate for the plasma potential. This allows a rough measurement of the turbulent particle flux

$$\Phi_{turb} = \langle \tilde{n}_e \tilde{v}_{E\times B} \rangle \propto \langle \tilde{I}_i^{max}(\tilde{V}_{fl,2} - \tilde{V}_{fl,1}) \rangle$$

Cylindrical Langmuir Probe in Electron Flow

Most often, the Langmuir probe is a small size electrode inserted in plasma and connected through an external (with respect to plasma) electric circuit with the electrode of a large surface area contacting with the same plasma (very often it is metallic wall of a chamber containing plasma) to obtain the probe I-V characteristic $i(V)$. The characteristic $i(V)$ is measured by sweeping the voltage V of scanning generator (inserted in the probe circuit) with simultaneous measuring of the probe current.

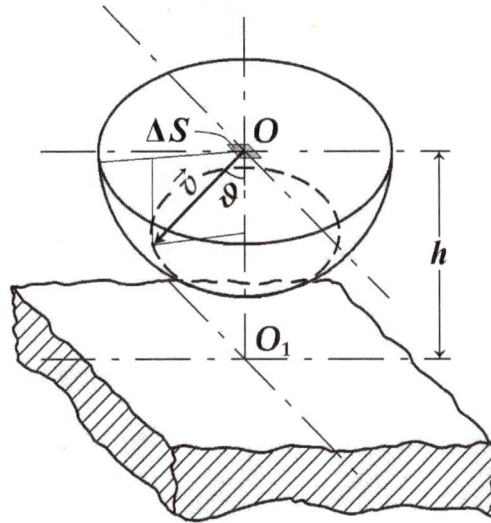

Fig. 1. Illustration to Langmuir Probe I-V Characteristic Derivation

Relations between the probe I-V characteristic and parameters of isotropic plasma were found by the Irving Langmuir and they can be derived most elementary for the planar probe of a large surface area S_z (ignoring the edge effects problem). Let us chose the point O in plasma at the distance from the probe surface where electric field of the probe is negligible while each electron of plasma passing this point could reach the probe surface without collisions with plasma components: $\lambda_D \ll \lambda_{Te}$, λ_D is the Debye length and λ_{Te} is the electron free path calculated for its total cross section with plasma components. In the vicinity of the point O we can imagine a small element of the surface area ΔS parallel to the probe surface. The elementary current di of plasma electrons passing throughout ΔS in a direction of the probe surface can be written in the form

$$di = q_e \Delta S dn(v, \vartheta)v \cos \vartheta, \ (1)$$

where v is a scalar of the electron thermal velocity vector $\vec{v}$,

$$dn(v, \vartheta) = nf(v) \frac{2\pi \sin \vartheta}{4\pi} dv d\vartheta, \ (2)$$

$2\pi \sin \vartheta d\vartheta$ is the element of the solid angle with its relative value $2\pi \sin \vartheta d\vartheta / 4\pi$, $2\pi \sin \vartheta d\vartheta / 4\pi$ is the angle between perpendicular to the probe surface recalled from the point O and the radius-vector of the electron thermal velocity $\vec{v}$ forming a spherical layer of thickness dv in velocity space, and $f(v)$ is the electron distribution function normalized to unity

$$\int_0^\infty f(v)dv = 1. \ (3)$$

Taking into account uniform conditions along the probe surface (boundaries are excluded), $\Delta S \rightarrow S_z$, we can take double integral with respect to the angle ϑ, and with respect to the velocity v, from the expression (1), after substitution Eq. (2) in it, to calculate a total electron current on the probe

$$i(v) = q_e n S_z \frac{1}{4\pi} \int_{\sqrt{2q_eV/m}}^\infty f(v)dv \int_0^\zeta v \cos \vartheta 2\pi \sin \vartheta d\vartheta. \ (4)$$

where V is the probe potential with respect to the potential of plasma $V = 0$, $\sqrt{2q_eV/m}$ is the lowest electron velocity value at which the electron still could reach the probe surface charged to the potential V, ζ is the upper limit of the angle ϑ at which the electron having initial velocity   can still reach the probe surface with a zero-value of its velocity at this surface. That means the value ζ is defined by the condition

$$v \cos \zeta = \sqrt{2q_eV/m}. \ (5)$$

Deriving the value ζ from Eq. (5) and substituting it in Eq. (4), we can obtain the probe I-V characteristic (neglecting the ion current) in the range of the probe potential $-\infty < V \leq 0$ in the form

$$i(V) = \frac{q_e n S_z}{4} \int_{\sqrt{2q_eV/m}}^\infty f(v) \left(1 - \frac{2q_eV}{mv^2} \right) v dv. \ (6)$$

Differentiating Eq. (6) twice with respect to the potential V, one can find the expression describing the second derivative of the probe I-V characteristic (obtained firstly by M. J. Druyvestein)

$$i''(V) = \frac{q_e^2 n S_z}{4m} \frac{1}{V} f\left(\sqrt{2q_eV/m} \right) (7)$$

defining the electron distribution function over velocity $f\left(\sqrt{2q_eV/m} \right)$ in the evident form. M. J. Druyvestein has shown in particular that Eqs. (6) and (7) are valid for description of operation of the probe of any arbitrary convex geometrical shape. Substituting the Maxwellian distribution function:

$$f^{(0)}(v) = \frac{4}{\sqrt{\pi}} \frac{v^2}{v_p^3} \exp\left(-v^2 / v_p^2\right), (8)$$

where $v_p = \langle v \rangle \sqrt{\pi} / 2$ is the most probable velocity, in Eq. (6) we obtain the expression

$$i^{(0)}(V) = \frac{q_e n \langle v \rangle}{4} S_z \exp\left(-q_e V / \mathcal{E}_p\right). (9)$$

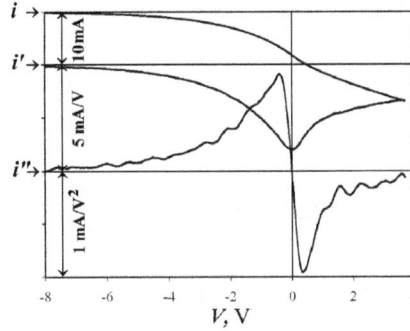

I-V Characteristic of Langmuir Probe in Isotropic Plasma

From which the very useful in practice relation follows

$$\ln\left(i^{(0)}(V) / i^{(0)}(0)\right) = -q_e V / \mathcal{E}_p. (10)$$

allowing one to derive the electron energy $\mathcal{E}_p = k_B T$ (for Maxwellian distribution function only!) by a slop of the probe I-V characteristic in a semilogarithmic scale. Thus in plasmas with isotropic electron distributions, the electron current $i_{th}(0)$ on a surface $S_z = 2\pi r_z l_z$ of the cylindrical Langmuir probe at plasma potential $V = 0$ is defined by the average electron thermal velocity $\langle v \rangle$ and can be written down as equation (see Eqs. (6), (9) at $V = 0$)

$$i_{th}(0) = q_e n \langle v \rangle \frac{1}{4} \times 2\pi r_z l_z, (11)$$

where n is the electron concentration, r_z is the probe radius, and l_z is its length. It is obvious that if plasma electrons form an electron *wind* (*flow*) *across* the *cylindrical* probe axis with a velocity $v_d \gg \langle v \rangle$, the expression

$$i_d = e n v_d \times 2 r_z l_z (12)$$

holds true. In plasmas produced by gas-discharge arc sources as well as inductively coupled sources, the electron wind can develop the Mach number $M^{(0)} = v_d / \langle v \rangle = (\sqrt{\pi} / 2)\alpha \gtrsim 1$. Here the parameter α is introduced along with the Mach number for simplification of mathematical expressions. Note that $(\sqrt{\pi} / 2)\langle v \rangle = v_p$, where v_p is the most probable velocity for the Maxwellian distribution function, so that $\alpha = v_d / v_p$. Thus the general case where $\alpha \gtrsim 1$ is of the theoretical and practical interest. Corresponding physical and mathematical considerations presented in Refs. [9,10] has shown that at the Maxwellian distribution function of the electrons in a reference system moving with the velocity v_d *across axis of the cylindrical* probe set at plasma potential $V = 0$, the electron current on the probe can be written down in the form

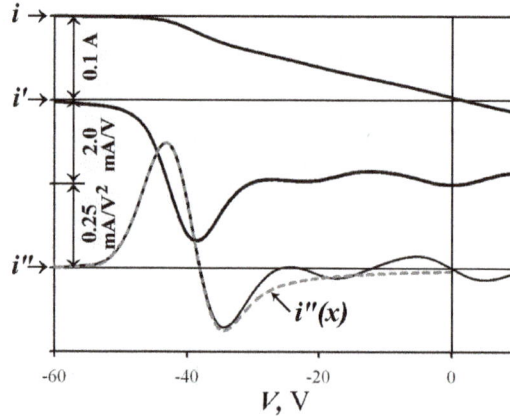

Fig. I-V Characteristic of the cylindrical probe in crossing electron wind

$$\frac{i(0)}{enS_z} = \frac{\langle v \rangle}{4} \exp(-\alpha^2/2) I_0(\alpha^2/2)\left(1+\alpha^2\left(1+I_1(\alpha^2/2)/I_0(\alpha^2/2)\right)\right), \quad (13)$$

where I_0 and I_1 are Bessel functions of imaginary arguments and Eq. (13) is reduced to Eq. (11) at $\alpha \to 0$ being reduced to Eq. (12) at $\alpha \to \infty$. The second derivative of the probe I-V characteristic $i''(V)$ with respect to the probe potential can be presented in this case in the form.

$$i''(x) = enS_z \frac{v_p}{2\pi^{3/2}(\mathcal{E}_p/e)^2} \frac{1}{\sqrt{x}} \int_0^\pi (\sqrt{x}-\cos\varphi)\exp\left(-\alpha^2(\sqrt{x}-\cos\varphi)\right)d\varphi, \quad (14)$$

where

$$x = \frac{1}{\alpha^2}\frac{V}{\mathcal{E}_p/e} \quad (15)$$

and the electron energy $\mathcal{E}_p/e$ is expressed in eV.

All parameters of the electron population: n, α, $\langle v \rangle$ and v_p in plasma can be derived from the experimental probe I-V characteristic second derivative $i''(V)$ by its least square best fitting with the theoretical curve expressed by Eq. (14). For detail and for problem of the general case of none-Maxwellian electron distribution functions see.

Practical Considerations

For laboratory and technical plasmas, the electrodes are most commonly tungsten or tantalum wires several thousandths of an inch thick, because they have a high melting point but can be made small enough not to perturb the plasma. Although the melting point is somewhat lower, molybdenum is sometimes used because it is easier to machine and solder than tungsten. For fusion plasmas, graphite electrodes with dimensions from 1 to 10 mm are usually used because they can withstand the highest power loads (also sublimating at high temperatures rather than melting), and result in reduced bremsstrahlung radiation (with respect to metals) due to the low atomic number of carbon. The electrode surface exposed to the plasma must be defined, e.g. by insulating all but the tip of a wire electrode. If there can be significant deposition of conducting materials (metals or graphite), then the insulator should be separated from the electrode by a meander to prevent short-circuiting.

In a magnetized plasma, it appears to be best to choose a probe size a few times larger than the ion Larmor radius. A point of contention is whether it is better to use proud probes, where the angle between the magnetic field and the surface is at least 15°, or flush-mounted probes, which are embedded in the plasma-facing components and generally have an angle of 1 to 5 °. Many plasma physicists feel more comfortable with proud probes, which have a longer tradition and possibly are less perturbed by electron saturation effects, although this is disputed. Flush-mounted probes, on the other hand, being part of the wall, are less perturbative. Knowledge of the field angle is necessary with proud probes to determine the fluxes to the wall, whereas it is necessary with flush-mounted probes to determine the density.

In very hot and dense plasmas, as found in fusion research, it is often necessary to limit the thermal load to the probe by limiting the exposure time. A reciprocating probe is mounted on an arm that is moved into and back out of the plasma, usually in about one second by means of either a pneumatic drive or an electromagnetic drive using the ambient magnetic field. Pop-up probes are similar, but the electrodes rest behind a shield and are only moved the few millimeters necessary to bring them into the plasma near the wall.

A Langmuir probe can be purchased off the shelf for on the order of 15,000 U.S. dollars, or they can be built by an experienced researcher or technician. When working at frequencies under 100 MHz, it is advisable to use blocking filters, and take necessary grounding precautions.

In low temperature plasmas, in which the probe does not get hot, surface contamination may become an issue. This effect can cause hysteresis in the I-V curve and may limit the current collected by the probe. A heating mechanism or a glow discharge plasma may be used to clean the probe and prevent misleading results.

Ball-pen Probe

A ball-pen probe is novel technique used to measure directly the plasma potential in strongly as well as weakly magnetized plasmas. The probe was invented by Jiří Adámek[16] in the Institute of Plasma Physics [17] AS CR in 2004. The ball-pen probe balances the electron saturation current to the same magnitude as that of the ion saturation current. In this case, its floating potential becomes identical to the plasma potential. This goal is attained by a ceramic shield, which screens off an adjustable part of the electron current from the probe collector due to the much smaller gyro-radius of the electrons. First systematic measurements have been performed on the CASTOR tokamak. The probe has been already used at different fusion devices as ASDEX Upgrade, COMPASS, ISTTOK, MAST, TJ-K, RFX [20], H-1 Heliac, IR-T1 as well as low temperature devices as DC cylindrical magnetron in Prague and linear magnetized plasma devices in Nancy and Ljubljana.

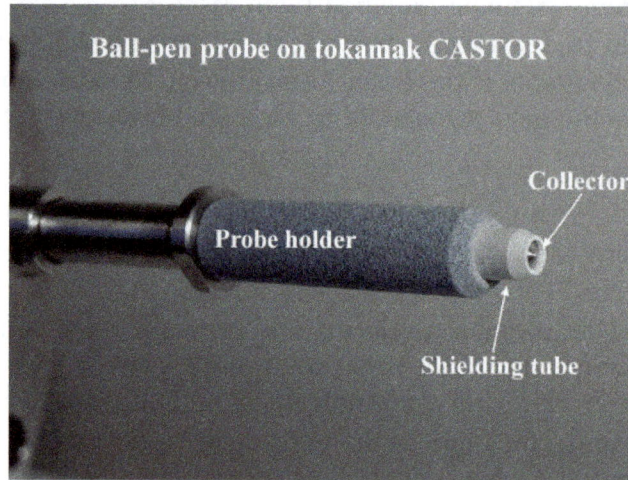

A single Ball-pen probe used on tokamak CASTOR in 2004. It consists of stainless steel collector, which is movable inside the ceramic (boron nitride) shielding tube.

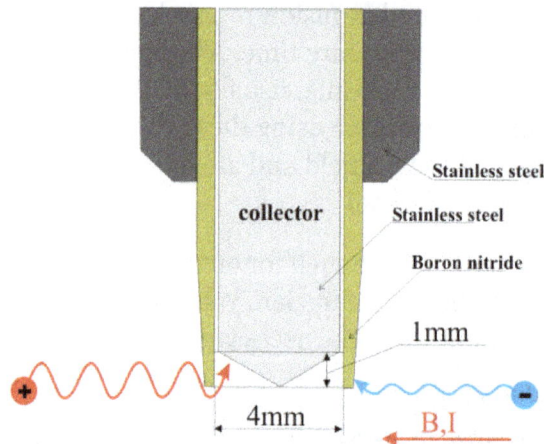

Schematic picture of a single ball-pen probe.

How the Ball-pen Probe Measures the Plasma Potential

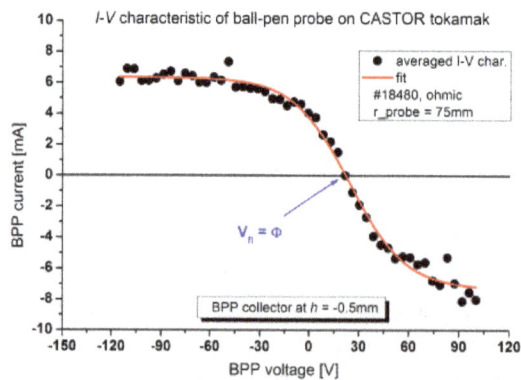

The I-V characteristic of Ball-pen probe on tokamak CASTOR.

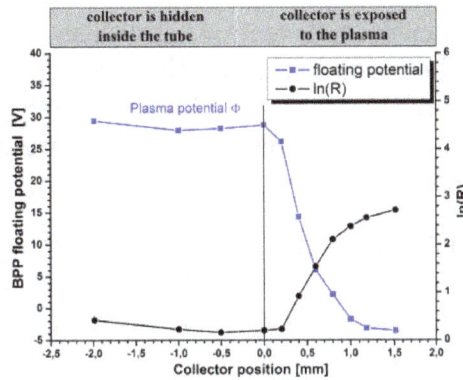

The potential and ln(R) of the Ball-pen probe for different position of collector on tokamak CASTOR.

If a Langmuir probe (electrode) is inserted into a plasma, its potential generally lies considerably below the plasma potential Φ due to what is termed a Debye sheath. Thus, the potential of Langmuir probe is named as floating potential V_{fl}. Therefore, it is impossible to measure directly the plasma potential by simple Langmuir probe. The difference between plasma and floating potential is given by the electron temperature T_e [eV]:

$$V_{fl} = \phi - \alpha * T_e$$

and the coefficient α. The coefficient is given by the ratio of the electron and ion saturation current density (j_e^{sat} and j_i^{sat}) and collecting areas for electrons and ions (A_e and A_i)

$$\alpha = ln(\frac{A_e j_e^{sat}}{A_i j_i^{sat}}) = ln(R)$$

The ball-pen probe, in magnetized plasma, modifies the collecting areas for electrons and ions and makes the ratio R equal to one. Thus, the coefficient α is equal to zero and floating potential of ball-pen probe is equal to the plasma potential independently on electron temperature

$$V_{fl} = \Phi$$

The ball-pen probe inserted into the magnetized plasma is directly on the plasma potential without additional power supplies or electronics.

The Ball-pen Probe Design

The design of the ball-pen probe is shown in the schematic picture. The probe consists of a conically shaped collector (non-magnetic stainless steel, tungsten, copper, molybdenum), which is shielded by an insulating tube (boron nitride, Alumina). The collector is fully shielded and the whole probe head must be oriented perpendicularly to the magnetic field lines. It is necessary to find the sufficient retraction of the ball-pen probe collector in order to reach $R = 1$, which strongly depends on the magnetic field's value. The physics of the ball-pen probe are not yet fully understood, but the collector retraction should be roughly below the ion's Larmor radius. This "calibration" can be done in two different ways:

1) the ball-pen probe collector is biased by swept voltage (low frequency) to provide the I-V characteristics and see the saturation current of electrons as well as ions. The ball-pen probe collector is systematically retracted until the I-V characteristics become symmetric. In this case, the ratio R is close to one. However, the experimental observation at different fusion devices confirmed that the ratio R is close, but not equal, to one. The I-V characteristics remain symmetric for deeper positions of the ball-pen probe collector too.

2) the ball-pen probe collector is fully floating. The ball-pen probe collector is systematically retracted until its potential saturates at some value, which is above Langmuir probe potential. The floating potential of the ball-pen probe remains almost constant for deeper positions too.

The Electron Temperature Measurements Without Power Supply

The probe head with three ball-pen probes and two Langmuir probes used on COMPASS tokamak.

The probe head with four ball-pen probes and four Langmuir probes used on ASDEX Upgrade tokamak.

The array of ball-pen probes installed directly to the divertor target on the COMPASS tokamak

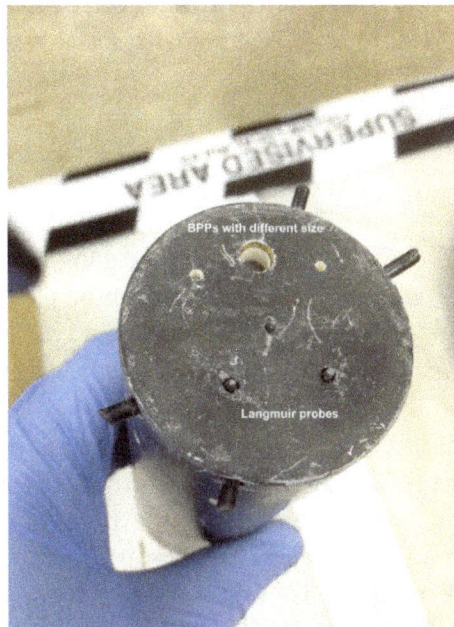

The probe head with three ball-pen probes with different size and set of Langmuir probes used on the MAST tokamak.

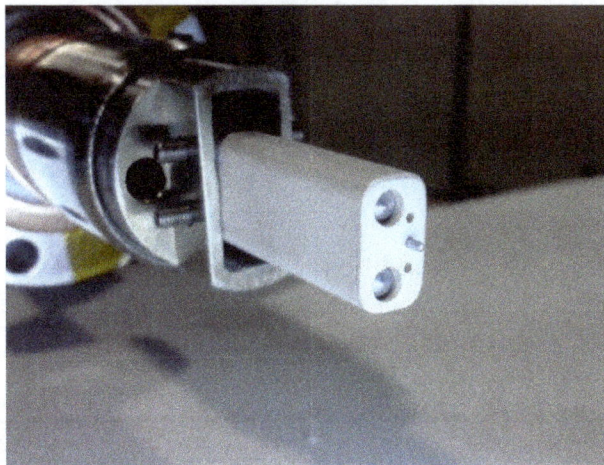

The probe head with two ball-pen probes and one Langmuir probe used on ISTTOK tokamak.

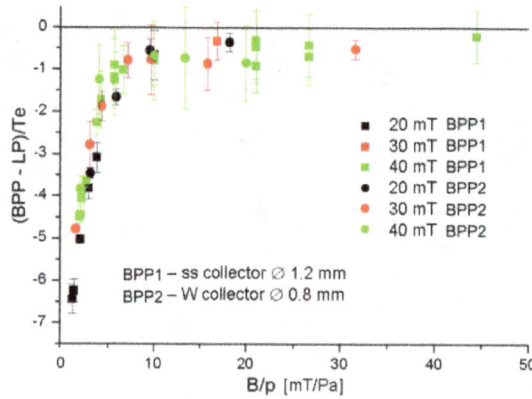

The example of the plasma potential measurements using a ball-pen probe in low-temperature and weekly magnetized plasmas in DC cylindrical magnetron in Prague [15].

The electron temperature can be measured by using ball-pen probe and common Langmuir probe with high temporal resolution in magnetized plasma without any external power supply. The electron temperature can be obtain from previous equation, assuming Maxwellian plasma

$$T_e = \frac{\Phi - V_{fl}}{\alpha}$$

The value of coefficient α is given by the Langmuir probe geometry, plasma gas (Hydrogen, Deuterium, Helium, Argon, Neon,...) and magnetic field. It can be partially effected by other features like secondary electron emission, sheath expansion etc. The coefficient α can be calculated theoretically, and its value is around 3 for non-magnetized hydrogen plasma. This value is obtained under assumption that the ion and electron temperatures are equal and there are no other above mentioned effects (sheath expansion, ...). It should be also taken into account that ratio R of ball-pen probe is close to one, but not equal to one as it is assumed in previous chapter. Therefore, the difference between ball-pen probe and Langmuir probe potential is given by the electron temperature, coefficient α of Langmuir probe and empirical value of the ratio R of ball-pen probe (if there is no empirical value of R it can be used an approximation $R=1$). Therefore, the electron temperature can be simply measured by using formula

$$T_e = \frac{\Phi_{BPP} - V_{fl}}{\bar{\alpha}},$$

where

$$\bar{\alpha} = \alpha - ln(R).$$

The coefficient $\bar{\alpha}$ for different plasma condition:

Device	Magnetic field	gas	
COMPASS	1.15 T	Deuterium	2.2
COMPASS(divertor region)	1.15T	Deuterium	2.2
ASDEX Upgrade	2.5 T	Deuterium	2.2

MAST	0.6 T	Deuterium	2.2
H-1 Heliac	< 1 T	Helium	3.76
CASTOR	1 T	Hydrogen	2.8
ISTTOK	0.6 T	Hydrogen	2.3
TJ-K	0.07 T	Hydrogen	3.0
IR-T1	0.7 T	Hydrogen	2.8
Linear magnetized plasma device, Ljubljana	0.01 T	Argon	5.2
DC cylindrical magnetron, Prague	0.02 T - 0.04 T	Argon	5.2
Linear device Mirabelle, Nancy	0.08 T	Argon	4.1
Linear device Mirabelle, Nancy	0.08 T	Helium	2.9

Plasma Scaling

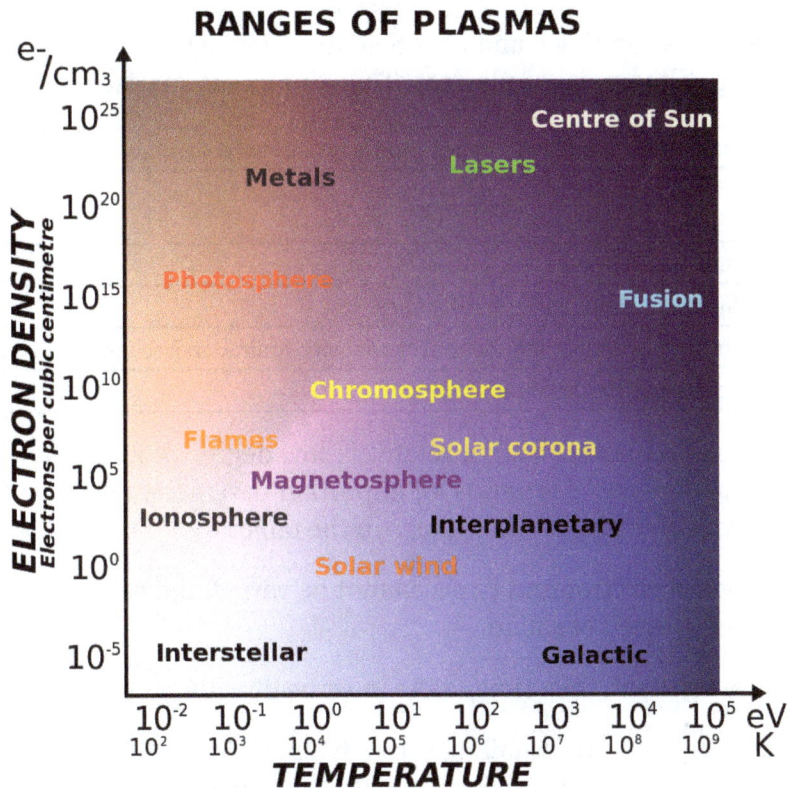

Range of plasmas. Density increases upwards, temperature increases towards the right. The free electrons in a metal may be considered an electron plasma

The parameters of plasmas, including their spatial and temporal extent, vary by many orders of magnitude. Nevertheless, there are significant similarities in the behaviors of apparently disparate plasmas. Understanding the scaling of plasma behavior is of more than theoretical value. It allows the results of laboratory experiments to be applied to larger natural or artificial plasmas of interest. The situation is similar to testing aircraft or studying natural turbulent flow in wind tunnels with smaller-scale models.

Similarity transformations (also called similarity laws) help us work out how plasma properties change in order to retain the same characteristics. A necessary first step is to express the laws governing the system in a nondimensional form. The choice of nondimensional parameters is never unique, and it is usually only possible to achieve by choosing to ignore certain aspects of the system.

One dimensionless parameter characterizing a plasma is the ratio of ion to electron mass. Since this number is large, at least 1836, it is commonly taken to be infinite in theoretical analyses, that is, either the electrons are assumed to be massless or the ions are assumed to be infinitely massive. In numerical studies the opposite problem often appears. The computation time would be intractably large if a realistic mass ratio were used, so an artificially small but still rather large value, for example 100, is substituted. To analyze some phenomena, such as lower hybrid oscillations, it is essential to use the proper value.

A Commonly Used Similarity Transformation

One commonly used similarity transformation was derived for gas discharges by James Dillon Cobine (1941), Alfred Hans von Engel and Max Steenbeck (1934), and further applied by Hannes Alfvén and Carl-Gunne Fälthammar to plasmas. They can be summarised as follows:

Similarity transformations applied to gaseous discharges and some plasmas	
Property	**Scale factor**
length, time, inductance, capacitance	x^1
particle energy, velocity, potential, current, resistance	$x^0 = 1$
electric and magnetic fields, conductivity, neutral gas density, ionization fraction	x^{-1}
current density, electron and ion densities	x^{-2}

This scaling applies best to plasmas with a relatively low degree of ionization. In such plasmas, the ionization energy of the neutral atoms is an important parameter and establishes an absolute *energy* scale, which explains many of the scalings in the table:

- Since the masses of electrons and ions cannot be varied, the *velocities* of the particles are also fixed, as is the speed of sound.

- If velocities are constant, then *time scales* must be directly proportional to distance scales.

- In order that charged particles falling through an *electric potential* gain the same energy, the potentials must be invariant, implying that the *electric field* scales inversely with the distance.

- Assuming that the magnitude of the E-cross-B drift is important and should be invariant, the *magnetic field* must scale like the electric field, namely inversely with the size. This is also the scaling required by Faraday's law of induction and Ampère's law.

- Assuming that the speed of the Alfvén wave is important and must remain invariant, the *ion density* (and with it the electron density) must scale with B^2, that is, inversely with the square of the size. Considering that the temperature is fixed, this also ensures that the ratio of thermal to magnetic energy, known as beta, remains constant. Furthermore, in regions where quasineutrality is violated, this scaling is required by Gauss's law.

- Ampère's law also requires that *current density* scales inversely with the square of the size, and therefore that current itself is invariant.

- The electrical *conductivity* is current density divided by electric field and thus scales inversely with the length.

- In a partially ionized plasma, the electrical conductivity is proportional to the electron density and inversely proportional to the *neutral gas density*, implying that the neutral density must scale inversely with the length, and ionization fraction scales inversely with the length.

Limitations

While these similarity transformations capture some basic properties of plasmas, not all plasma phenomena scale in this way. Consider, for example, the degree of ionization, which is dimensionless and thus would ideally remain unchanged when the system is scaled. The number of charged particles per unit volume is proportional to the current density, which scales as x^{-2}, whereas the number of neutral particles per unit volume scales as x^{-1} in this transformation, so the degree of ionization does not remain unchanged but scales as x^{-1}.

Astrophysical Application

As an example, take an auroral sheet with a thickness of 1 km. A laboratory simulation might have a thickness of 10 cm, a factor of 10^4 smaller. To satisfy the condition of this similarity transformation, the gaseous density would have to be increased by a factor of 10^4 from 10^4 m^{-3} to 10^8 m^{-3} (10^{10} cm^{-3} to 10^{14} cm^{-3}), and the magnetic field would have to be increased by the same factor from 50 microteslas to 500 milliteslas (0.5 gauss to 5 kilogauss). These values are large but within the range of technology. If the experiment captures the essential features of the aurora, the processes will be 10^4 times faster so that a pulse that takes 100 s in nature would take only 10 ms in the laboratory.

Region	Characteristic dimension (cm)			Density (particles/cm³)		Magnetic field (gauss)		Characteristic time		
	Actual	Scaled	Scale Factor	Actual	Scaled	Actual	Scaled	Actual	Scaled	Description
Ionosphere	10^6 - 10^7	10	10^{-5} - 10^{-6}	10^{10}	10^{15} - 10^{16}	0.5	5×10^4 - 5×10^5	100 s	0.1 - 1 ms	Period of giant pulsation
Exosphere	10^9	10	10^{-8}	10^5 - 10	10^{13} - 10^9	0.5 - 5×10^{-4}	5×10^7 - 5×10^4	10^5 s	1 ms	One day
Interplanetary space	10^{13}	10	10^{-12}	1 - 10	10^{12} - 10^{13}	10^{-4}	10^8	2×10^6 s	2 µs	One solar rotation
Interstellar medium	3×10^{22}	10	3×10^{-22}	1	3×10^{21}	10^{-6} - 10^{-5}	3×10^{15} - 3×10^{16}	10^{16} s	3 µs	Period of galactic rotation

Similarity transformations applied to some astrophysical plasmas. Actual plasma properties compared to a laboratory plasma if the scale length is reduced to 10 cm.

Intergalactic space	>3x10^{27}	10	<3x10^{-27}	10^{-4}?	>3x10^{22}	10^{-7}?	>3x10^{19}	4x10^{17} s	10^{-9} s	Age of the universe
Solar chromosphere	10^8	10	10^{-7}	10^{11} - 10^{14}	10^{18} - 10^{21}	10^3 - 1	10^{10} - 10^7	10^3 s	100 μs	Life of solar flare
								10^5 s	10 ms	Life of solar prominence
Solar corona	10^{10} - 10^{11}	10	10^{-9} - 10^{-10}	10^8 - 10^6	10^{17} - 10^{16}	10^2 - 10^{-1}	10^{11} - 10^9	10^3 s	10^{-1} to 1 μs	Life of coronal arc
								22 years	70 to 700 ms	Solar cycle

Particle density of the Earth's atmosphere at sea level is 10^{19} per cm^3. Small bar magnet = 100 milliteslas. Big electromagnet = 2 teslas. 10^9 cm = 10,000 km

The table shows the properties of some actual space plasma. It also shows how other plasma properties would need to be changed, if (a) the characteristic length of a plasma were reduced to just 10 cm, and (b) the characteristics of the plasma were to remain unchanged.

The first thing to notice is that many cosmic phenomena cannot be reproduced in the laboratory because the necessary magnetic field strength is beyond the technological limits. Of the phenomena listed, only the ionosphere and the exosphere can be scaled to laboratory size. Another problem is the ionization fraction. When the size is varied over many orders of magnitude, the assumption of a partially ionized plasma may be violated in the simulation. A final observation is that the plasma densities needed in the laboratory are sizeable, up to 10^{16} cm^{-3} for the ionosphere, compared to the atmospheric density of about 10^{19} particles per cm^3. In other words, the laboratory analogy of a low density space plasma is not a "vacuum chamber", but laboratory plasma with a pressure, when the higher temperature is taken into consideration, which can approach atmospheric pressure.

Dimensionless Parameters in Tokamaks

One of the central questions in fusion power research is to predict the energy confinement time in machines that are larger than any that have ever been built. A widely accepted approach to doing this is to express the scaling in terms of nondimensional parameters. Geometrical parameters, such as the ratio of the major to the minor radius, the shape of the plasma cross section, and the angle of the magnetic field, can be chosen in current experiments to equal the value desired for a full-scale reactor. The remaining (dimensional) parameters can be taken to be the particle density n, the temperature T, the magnetic field B, and the size (major radius) R. These can be combined into the three dimensionless parameters β (the ratio of plasma pressure to magnetic pressure), ν^* (the product of the collision frequency and the thermal transit time), and ρ^* (the ratio of the Larmor radius to the torus radius). These have the following scalings:

$$\beta \sim nTB^{-2}$$

$$\nu^* \sim nT^{-2}R$$

$$\rho^* \sim T^{1/2}B^{-1}R^{-1}$$

The radius R can be varied while keeping these three parameters constant if n, T, and B are scaled in this way:

$$n \sim R^{-2}$$

$$T \sim R^{-1/2}$$

$$B \sim R^{-5/4}$$

Note that this similarity transformation is distinct from that considered above, which would yield $n \sim R^{-1}$, $T \sim R^{0}$, and $B \sim R^{-1}$. This is because the physical effects to be studied are different.

The scaling of the magnetic field with the minus 5/4 power of the size implies that a 1:3 scale model of a power-producing tokamak with a magnetic field of 10 T at the coils would require a field of about 40 T, which is technologically infeasible.

The next best alternative is to allow ρ^* to vary and to extrapolate according to the dependence found. ρ^* is the parameter considered least likely to harbor surprises, partly for theoretical considerations, but also simply because it is, in contrast to β and ν^*, already much larger than unity. This can be done in a single machine (constant R) by varying the magnetic field and scaling density and temperature as:

$$n \sim B^{4/3}$$

$$T \sim B^{2/3}$$

It should be kept in mind that the assumption has been made that the important turbulent transport processes depend only on the parameters chosen. It is only physical reasoning, not mathematical necessity, that concludes that the ratio of the torus radius to the Larmor radius is important, and not, for example, the ratio to the Debye length. In the same way, it has been assumed that the absolute energy levels of atomic physics do *not* dictate an absolute temperature dependence, or equivalently, that the boundary layer where atomic physics *is* important, is small enough not to determine the overall energy confinement.

References

- Ghoranneviss, M.; S. Meshkani (2016). "Techniques for improving plasma confinement in IR-T1 Tokamak". International Journal of Hydrogen Energy. doi:10.1016/j.ijhydene.2016.03.075.

- Loureiro, J.; C. Silva; J. Horacek; J. Adamek; J. Stockel (2014). "Scrape-off layer width of parallel heat flux on tokamak COMPASS". Plasma Physics&Technology. 1 (3): 121–123. ISSN 2336-2634.

- Block, L. P. (May 1978). "A Double Layer Review". Astrophysics and Space Science. NASA/STI. 55 (1): 59–83. Bibcode:1978Ap&SS..55...59B. doi:10.1007/bf00642580. Retrieved April 16, 2013.

- Adamek, Jiri; Matěj Peterka; Tomaž Gyergyek; Pavel Kudrna; Milan Tichý (2012). "Diagnostics of magnetized low temperature plasma by ball-pen probe". NUKLEONIKA. 57 (2): 297–300.

Plasma Processing Techniques

The techniques used in the processing of plasma are plasma etching, corona treatment, plasma polymerization, plasma cleaning and plasma cutting. This chapter discusses plasma processing techniques in a critical manner providing key analysis to the subject matter.

Plasma Etching

Plasma etching is a form of plasma processing used to fabricate integrated circuits. It involves a high-speed stream of glow discharge (plasma) of an appropriate gas mixture being shot (in pulses) at a sample. The plasma source, known as etch species, can be either charged (ions) or neutral (atoms and radicals). During the process, the plasma will generate volatile etch products at room temperature from the chemical reactions between the elements of the material etched and the reactive species generated by the plasma. Eventually the atoms of the shot element embed themselves at or just below the surface of the target, thus modifying the physical properties of the target.

Mechanisms

Generation of the Plasma

A plasma is a high energetic condition in which a lot of processes can occur. These processes happen because of electrons and atoms. To form the plasma electrons have to be accelerated to gain energy. Highly energetic electrons transfer the energy to atoms by collisions. Three different processes can occur because of this collisions:

- Excitation
- Dissociation
- Ionization

There are different species in the plasma present such as electrons, ions, radicals and neutral particles. Those species are interacting with each other constantly. Plasma etching itself can be divided into two main ways of interactions:

- The generation of the chemical species
- The interaction with the surrounding surfaces

Without a plasma all those processes would occur at a higher temperature. There are different ways to change the plasma chemistry and get different kinds of plasma etching or plasma depositions. One of the excitation techniques to form a plasma is by using RF excitation of a power source of 13.56 MHz.

The mode of operation of the plasma system will change if the operating pressure changes. Also, it is different for different structures of the reaction chamber. In the simple case, the electrode structure is symmetrical, and the sample is placed upon the grounded electrode.

Influences on the Process

The key to developing successful complex etching processes is to find the appropriate gas etch chemistry that will form volatile products with the material to be etched as shown in Table 1. For some difficult materials (such as magnetic materials), the volatility can only be obtained when the wafer temperature is increased. The main factors that influence the plasma process:

- Electron source

- Pressure

- Gas species

- vacuum

Table 1
Halogen-, hydride- and methyl-compounds and their volatility for elements and materials of interest in micro- and nano-technology applications

Elements	Fluorides	Boiling temperature (°C)	Chlorides	Boiling temperature (°C)	Bromides	Boiling temperature (°C)	Hydrides, trimethyls	Boiling temperature (°C)
Al	AlF_3	1297 (subl.)	$AlCl_3$	178 (subl.)	$AlBr_3$	263		
As	AsF_3	−63	$AsCl_3$	130.2	$AsBr_3$	221	AsH_3	−55
	AsF_5	−53			$AsBr_5$			
C	CF_4	−128	CCl_4	77	CBr_4	189	CH_4	−164
Cr	CrF_2	>1300	CrO_2Cl_2	117	$CrBr_2$	842		
Cu	CuF	1100 (subl.)	$CuCl$	1490	$CuBr$	1345		
	CuF_2	950	$CuCl_2$	993			CuH	55–60
Ga	GaF_3	1000	$GaCl_3$	201.3	$GaBr_3$	278.8	$Ga(CH_3)_3$	134
Ge	GeF_4	−37 (subl.)	$GeCl_4$	84	$GeBr_4$	186.5	GeH_4	−88.5
In	InF_3	>1200	$InCl_3$	300 (subl.)			$In(CH_3)_3$	55.7
Mo	MoF_6	213.6	$MoCl_3$	268				
	MoF_8	35	$MoOCl_2$	100 (subl.)				
	MoO_2F_2	270 (subl.)						
	$MoOF_4$	180						
P	PF_3	−101.5	PCl_3	75	PBr_3	172.9	PH_3	−87.7
	PF_5	−75	PCl_5	162 (subl.)	PBr_5	106		
Si	SiF_4	−86	$SiCl_4$	57.6	$SiBr_4$	154	SiH_4	−111.8
Ta	TaF_5	229.5	$TaCl_5$	242	$TaBr_5$	348.8		
Ti	TiF_4	284 (subl.)	$TiCl_4$	136.4	$TiBr_4$	230		
W	WF_6	17.5	WCl_6	346.7				
	WOF_4	187.5	WCl_2	275.6	WBr_2	333		
			$WOCl_4$	227.5	$WOBr_4$	327		

C. Cardinaud et al. / Applied Surface Science 164 (2000) 72–83

Surface Interaction

The reaction of the products depend on the likelihood of dissimilar atoms, photons, or radicals reacting to form chemical compounds. The temperature of the surface also affects the reaction of products. Adsorption happens when a substance is able to gather and reach the surface in a condensed layer, ranging in thickness (usually a thin, oxidized layer.) Volatile products desorb in the plasma phase and help the plasma etching process as the material interacts with the sample's walls. If the products are not volatile, a thin film will form at the surface of the material. Different principles that affect a sample's ability for plasma etching:

- Volatility

- Adsorption

- Chemical Affinity

- Ion-bombarding

- Sputtering

Plasma etching can change the surface contact angles, such as, hydrophilic to hydrophobic or vice versa. The Argon plasma etching has reported to enhance contact angle from 52 deg to 68 deg, and, Oxygen plasma etching to reduce contact angle from 52 deg to 19 deg for CFRP composites for bone plate applications. Similarly, the plasma etching has reported to reduce the surface roughness from hundreds of nanometers to as much lower as 3 nm for metals.

Types of Plasma Etching

The pressure is one factor that influences a plasma etching process a lot. For plasma etching to happen, the chamber has to be under low pressure which means less than 100 Pa. In order to generate low pressure plasma the gas species which is used has to be ionized. The ionization happens by a glow charge. Those excitations happen by an external source, which can deliver up to 30 kW and frequencies from 50 Hz (dc) over 5–10 Hz (pulsed dc) to radio and microwave frequency (MHz-GHz).

Microwave Plasma Etching

Microwave etching happens with an excitation sources in the microwave frequency, so between MHz and GHz. One example of plasma etching is shown here.

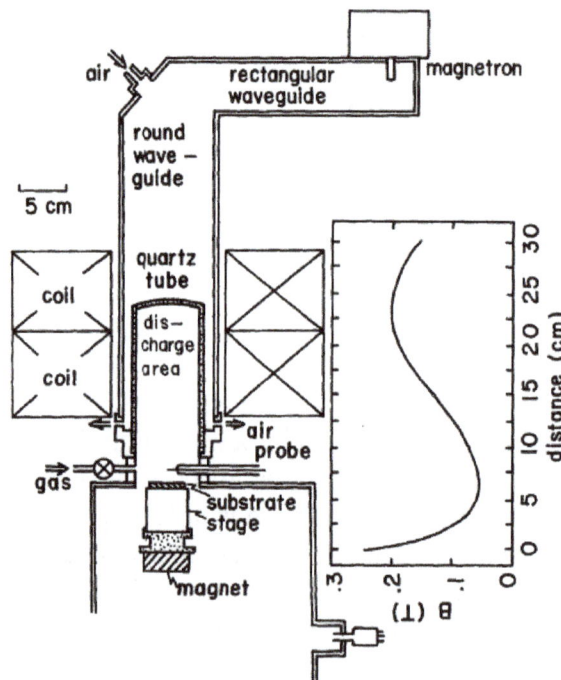

A microwave plasma etching apparatus. The microwave operates at 2.45 GHz. This frequency is generated by a magnetron and discharges through a rectangular and a round waveguide. The discharge area is in a quartz tube with an inner diameter of 66mm. Two coils and a permanent magnet are wrapped around the quartz tube to create a magnetic field which directs the plasma.

Hydrogen Plasma Etching

On form to use gas is the plasma etching is hydrogen plasma etching. Therefore, an experimental apparatus like this can be used:

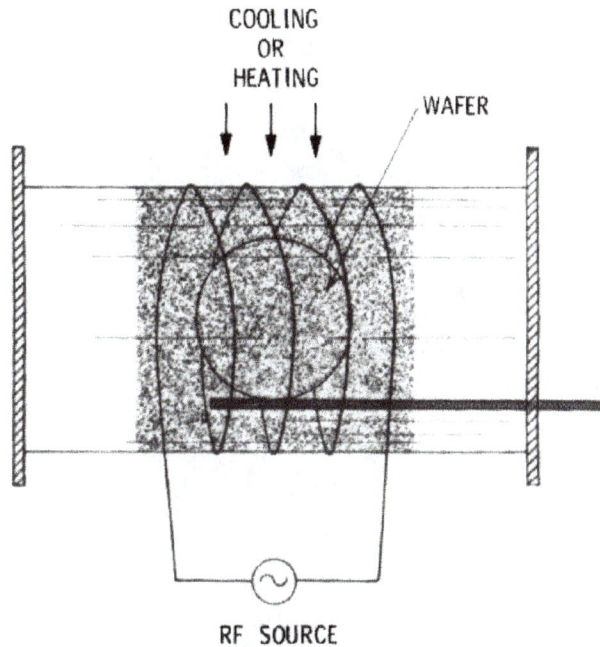

A quartz tube with an rf excitation of 30 MHz is shown. It is coupled with a coil around the tube with a power density of 2-10 W/cm³. The gas species is H_2 gas in the chamber. The range of the gas pressure is 100-300 um.

Applications

Plasma etching is currently being used to process semiconducting materials for their use in the fabrication of electronics. Small features can be etched into the surface of the semiconducting material in order to be more efficient or enhance certain properties when used in electronic devices. For example, plasma etching can be used to create deep trenches on the surface of silicon for uses in Microelectromechanical systems. This application suggests that plasma etching also has the potential to play a major role in the production of microelectronics. Similarly, research is currently being done on how the process can be adjusted to the nanometer scale.

Hydrogen plasma etching, in particular, has other interesting applications. When used in the process of etching semiconductors, hydrogen plasma etching has been shown to be effective in removing portions of native oxides found on the surface. Hydrogen plasma etching also tends to leave a clean and chemically balanced surface, which is ideal for a number of applications.

Corona Treatment

Corona treatment (sometimes referred to as air plasma) is a surface modification technique that uses a low temperature corona discharge plasma to impart changes in the properties of a surface. The corona plasma is generated by the application of high voltage to an electrode that has a sharp tip. The plasma forms at the tip. A linear array of electrodes is often used to create a curtain of corona plasma. Materials such as plastics, cloth, or paper may be passed through the corona plasma curtain in order to change the surface energy of the material. All materials have an inherent surface energy. Surface treatment systems are available for virtually any surface format including dimensional objects, sheets and roll goods that are handled in a web format. Corona treatment is a widely used surface treatment method in the plastic film, extrusion, and converting industries.

Verner Eisby, a Danish engineer, the inventor of corona treatment.

History

The corona treatment was invented by the Danish engineer Verner Eisby in 1951. Verner had been asked by one of his customers if he could find a solution which would make it possible to print on plastic. Verner found that there were already a couple of ways to accomplish this. One was a gas flame method and the other was a spark generating method, both of which were crude and uncontrollable and did not produce a homogeneous product. Verner came up with the theory that a high frequency corona discharge would provide both a more efficient and controllable method to treat the surface. Exhaustive experiments proved him to be correct. Verner's company, Vetaphone, obtained patent rights for the new corona treatment system.

Materials

Many plastics, such as polyethylene and polypropylene, have chemically inert and nonporous surfaces with low surface tensions causing them to be non-receptive to bonding with printing inks, coatings, and adhesives. Although results are invisible to the naked eye, surface treating modifies surfaces to improve adhesion.

Polyethylene, polypropylene, nylon, vinyl, PVC, PET, metalized surfaces, foils, paper, and paperboard stocks are commonly treated by this method. It is safe, economical, and delivers high line speed throughput. Corona treatment is also suitable for the treatment of injection and blow molded parts, and is capable of treating multiple surfaces and difficult parts with a single pass.

Equipment

Corona discharge equipment consists of a high-frequency power generator, a high-voltage transformer, a stationary electrode, and a treater ground roll. Standard utility electrical power is converted into higher frequency power which is then supplied to the treater station. The treater station applies this power through ceramic or metal electrodes over an air gap onto the material's surface.

Two basic corona treater stations are used in extrusion coating applications. Bare Roll and Covered Roll. On a Bare Roll treater station the dielectric encapsulates the electrode. On a Covered Roll station it encapsulates the treater base roll. The treater consists of an electrode and a base roll in both stations. In theory a Covered Roll treater is generally used to treat non-conductive webs, and a Bare Roll treater is used to treat conductive webs. However, manufacturers who treat a variety of substrates on the same production line may choose to use a Bare Roll treater.

Pre-treatment

Many substrates provide a better bonding surface when they are treated at the time they are produced. This is called "pre-treatment." The effects of corona treatment diminish over time. Therefore many surfaces will require a second "bump" treatment at the time they are converted to ensure bonding with printing inks, coatings, and adhesives.

Other Technologies

Other technologies used for surface treatment include in-line atmospheric (air) plasma, flame plasma, and chemical plasma systems.

Atmospheric Plasma Treatment

Atmospheric-pressure plasma treatment is very similar to corona treatment but there are a few differences between them. Both treatments may use one or more high voltage electrodes which charge the surrounding blown gas molecules and ionizes them. However in atmospheric plasma systems, the overall plasma density is much greater which enhances the rate and degree to which the ionized molecules are incorporated onto a materials' surface. An increased rate of ion bombardment occurs which may result in stronger material bonding traits depending on the gas molecules used in the process. Atmospheric plasma treatment technology also eliminates a possibility of treatment on a material's non-treated side; also known as backside treatment.

Flame Plasma

Flame plasma treaters generate more heat than other treating processes, but materials treated through this method tend to have a longer shelf-life. These plasma systems are different from air plasma systems because flame plasma occurs when flammable gas and surrounding air are combusted into an intense blue flame. Objects' surfaces are polarized from the flame plasma affecting the distribution of the surface's electrons in an oxidation form. This treatment requires higher temperatures so many of the materials that are treated with a flame plasma can be damaged.

Chemical Plasma

Chemical plasma is based on the combination of air plasma and flame plasma. Much like air plasma, chemical plasma fields are generated from electrically charged air. But, instead of air, chemical plasma relies on a mixture of other gases depositing various chemical groups onto the treated surface.

Plasma Polymerization

Plasma polymerization (or glow discharge polymerization) uses plasma sources to generate a gas discharge that provides energy to activate or fragment gaseous or liquid monomer, often containing a vinyl group, in order to initiate polymerization. Polymers formed from this technique are generally highly branched and highly cross-linked, and adhere to solid surfaces well. The biggest advantage to this process is that polymers can be directly attached to a desired surface while the chains are growing, which reduces steps necessary for other coating processes such as grafting. This is very useful for pinhole-free coatings of 100 picometers to 1 micrometre thickness with solvent insoluble polymers.

Introduction

In as early as the 1870s "polymers" formed by this process were known, but these polymers were initially thought of as undesirable byproducts associated with electric discharge, with little attention being given to their properties. It was not until the 1960s that the properties of these polymers where found to be useful. It was found that flawless thin polymeric coatings could be formed on metals, although for very thin films (<10mm) this has recently been shown to be an oversimplification. By selecting the monomer type and the energy density per monomer, known as the Yasuda parameter, the chemical composition and structure of the resulting thin film can be varied with a wide range. These films are usually inert, adhesive, and have low dielectric constants. Some common monomers polymerized by this method include styrene, ethylene, methacrylate and pyridine, just to name a few. The 1970s brought about many advances in plasma polymerization, including the polymerization of many different types of monomers. The mechanisms of deposition however were largely ignored until more recently. Since this time most attention devoted to plasma polymerization has been in the fields of coatings, but since it is difficult to control polymer structure, it has limited applications.

Basic Operating Mechanism

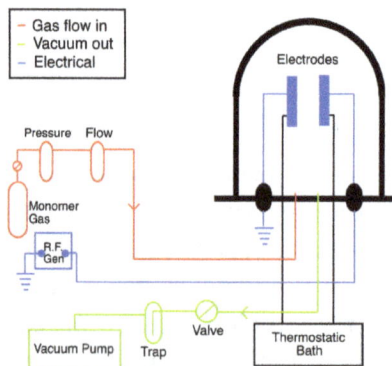

Figure 1. Schematic representation of basic internal electrode glow discharge polymerization apparatus.

Glow Discharge

Plasma consists of a mixture of electrons, ions, radicals, neutrals and photons. Some of these species are in local thermodynamic equilibrium, while others are not. Even for simple gases like argon this mixture can be complex. For plasmas of organic monomers, the complexity can rapidly increase as some components of the plasma fragment, while others interact and form larger species. Glow discharge is a technique in polymerization which forms free electrons which gain energy from an electric field, and then lose energy through collisions with neutral molecules in the gas phase. This leads to many chemically reactive species, which then lead to a plasma polymerization reaction. The electric discharge process for plasma polymerization is the "low-temperature plasma" method, because higher temperatures cause degradation. These plasmas are formed by a direct current, alternating current or radio frequency generator.

Types of Reactors

There are a few designs for apparatus used in plasma polymerization, one of which is the Bell (static type), in which monomer gas is put into the reaction chamber, but does not flow through the chamber. It comes in and polymerizes without removal. This type of reactor is shown in Figure 1. This reactor has internal electrodes, and polymerization commonly takes place on the cathode side. All devices contain the thermostatic bath, which is used to regulate temperature, and a vacuum to regulate pressure.

Operation: The monomer gas comes into the Bell type reactor as a gaseous species, and then is put into the plasma state by the electrodes, in which the plasma may consist of radicals, anions and cations. These monomers are then polymerized on the cathode surface, or some other surface placed in the apparatus by different mechanisms of which details are discussed below. The deposited polymers then propagate off the surface and form growing chains with seemingly uniform consistency.

Another popular reactor type is the flow through reactor (continuous flow reactor), which also has internal electrodes, but this reactor allows monomer gas to flow through the reaction chamber as its name implies, which should give a more even coating for polymer film deposition. It has the advantage that more monomer keeps flowing into the reactor in order to deposit more polymer. It has the disadvantage of forming what is called "tail flame," which is when polymerization extends into the vacuum line.

A third popular type of reactor is the electrodeless. This uses an RF coil wrapped around the glass apparatus, which then uses a radio frequency generator to form the plasma inside of the housing without the use of direct electrodes. The polymer can then be deposited as it is pushed through this RF coil toward the vacuum end of the apparatus. This has the advantage of not having polymer building up on the electrode surface, which is desirable when polymerizing onto other surfaces.

A fourth type of system growing in popularity is the atmospheric-pressure plasma system, which is useful for depositing thin polymer films. This system bypasses the requirements for special hardware involving vacuums, which then makes it favorable for integrated industrial use. It has been shown that polymers formed at atmospheric-pressure can have similar properties for coatings as those found in the low-pressure systems.

Physical Process Characteristics

The formation of a plasma for polymerization depends on many of the following. An electron energy of 1–10 eV is required, with electron densities of 10^9 to 10^{12} per cubic centimeter, in order to form the desired plasma state. The formation of a low-temperature plasma is important; the electron temperatures are not equal to the gas temperatures and have a ratio of T_e/T_g of 10 to 100, so that this process can occur at near ambient temperatures, which is advantageous because polymers degrade at high temperatures, so if a high-temperature plasma was used the polymers would degrade after formation or would never be formed. This entails non-equilibrium plasmas, which means that charged monomer species have more kinetic energy than neutral monomer species, and cause the transfer of energy to a substrate instead of an uncharged monomer.

Kinetics

The kinetic rate of these reactions depends mostly on the monomer gas, which must be either gaseous or vaporized. However, other parameters are also important as well, such as power, pressure, flow rate, frequency, electrode gap and reactor configuration. Low flow rates usually only depend on the amount of reactive species present for polymerization, whereas high flow rates depend on the amount of time that is spent in the reactor. Therefore, the maximum rate of polymerization is somewhere in the middle.

The fastest reactions tend to be in the order of triple-bonded > double-bonded > single bonded molecules, and also lower molecular weight molecules are faster than higher ones. So acetylene is faster than ethylene, and ethylene is faster than propene, etc. The molecular weight factor in polymer deposition is dependent on the monomer flow rate, in which a higher molecular weight monomer typically near 200 g/mol needs a much higher flow rate of 15×10^4 g/cm², whereas lower molecular weights around 50 g/mol require a flow rate of only 5×10^4 g/cm². A heavy monomer therefore needs a faster flow, and would likely lead to increased pressures, decreasing polymerization rates.

Increased pressure tends to decrease polymerization rates reducing uniformity of deposition since uniformity is controlled by constant pressure. This is a reason that high-pressure plasma or atmospheric-pressure plasmas are not usually used in favor of low-pressure systems.At pressures greater than 1 torr, oligomers are formed on the electrode surface, and the monomers also on the surface can dissolve them to get a low degree of polymerization forming an oily substance. At low pressures, the reactive surfaces are low in monomer and facilitate growing high molecular weight polymers.

The rate of polymerization depends on input power, until power saturation occurs and the rate becomes independent of it. A narrower electrode gap also tends to increase polymerization rates because a higher electron density per unit area is formed. Polymerization rates also depend on the type of apparatus used for the process. In general, increasing the frequency of alternating current glow discharge up to about 5 kHz increases the rate due to the formation of more free radicals. After this frequency, inertial effects of colliding monomers inhibit polymerization. This forms the first plateau for polymerization frequencies. A second maximum in frequency occurs at 6 MHz, where side reactions are overcome again and the reaction occurs through free radicals diffused from plasma to the electrodes, at which point a second plateau is obtained. These parameters differ slightly for each monomer and must be optimized in-situ.

Synthetic Routes

Plasma contains many species such as ions, free radicals and electrons, so it is important to look at what contributes to the polymerization process most. The first suggested process by Westwood et al. was that of a cationic polymerization, since in a direct current system polymerization occurs mainly on the cathode. However, more investigation has led to the belief that the mechanism is more of a radical polymerization process, since radicals tend to be trapped in the films, and termination can be overcome by reinitiation of oligomers. Other kinetic studies also appear to support this theory.

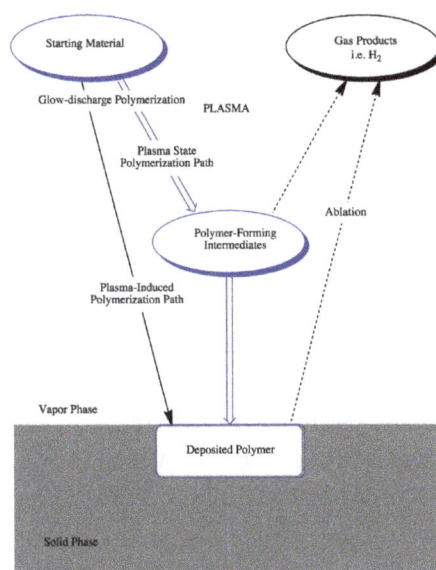

Figure 2. Schematic of plasma polymerization process possibilities, w/blue representing dominant pathway.

However, since the mid 1990s a number of papers focusing on the formation of highly functionalized plasma polymers have postulated a more significant role for cations, particularly where the plasma sheath is collosionless. The assumption that the plasma ion density is low and consequently the ion flux to surfaces is low has been challenged, pointing out that ion flux is determined according to the Bohm sheath criterion i.e. ion flux is proportional to the square root of the electron temperature and not RT.

In polymerization, both gas phase and surface reactions occur, but mechanism differs between high and low frequencies. At high frequencies it occurs in reactive intermediates, whereas at low frequencies polymerization happens mainly on surfaces. As polymerization occurs, the pressure inside the chamber decreases in a closed system, since gas phase monomers go to solid polymers. An example diagram of the ways that polymerization can take place is shown in Figure 2, wherein the most abundant pathway is shown in blue with double arrows, with side pathways shown in black. The ablation occurs by gas formation during polymerization. Polymerization has two pathways, either the plasma state or plasma induced processes, which both lead to deposited polymer.

Polymers can be deposited on many substrates other than the electrode surfaces, such as glass, other organic polymers or metals, when either a surface is placed in front of the electrodes, or placed in the middle between them. The ability for them to build off of electrode surfaces is likely to be an electrostatic interaction, while on other surfaces covalent attachment is possible.

Polymerization is likely to take place through either ionic and/or radical processes which are initiated by plasma formed from the glow discharge. The classic view presented by Yasuda based upon thermal initiation of parylene polymerization is that there are many propagating species present at any given time as shown in Figure 3. This figure shows two different pathways by which the polymerization may take place.

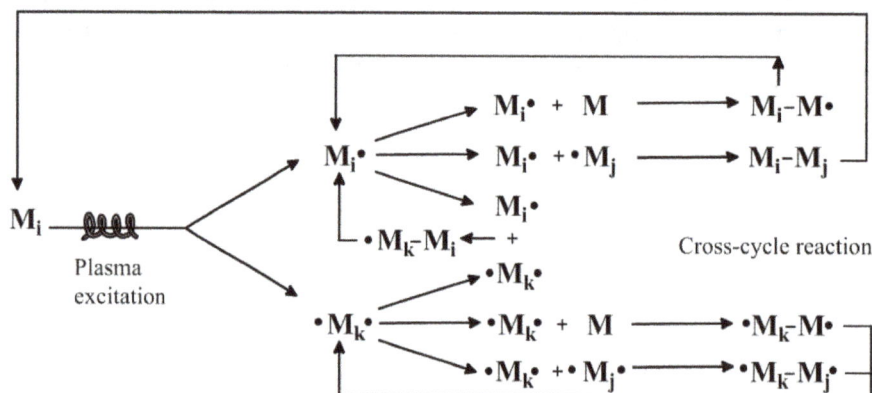

Figure 3. Schematic representation of bicyclic step-growth mechanism of plasma polymerization.

The first pathway is a monofunctionalization process, bears resemblance to a standard free radical polymerization mechanism ($M\bullet$)- although with the caveat that the reactive species may be ionic and not necessarily radical. The second pathway refers to a difunctional mechanism, which by example may contain a cationic and a radical propagating center on the same monomer ($\bullet M\bullet$). A consequence is that 'polymer' can grow in multiple directions by multiple pathways off one species, such as a surface or other monomer. This possibility let Yasuda to term the mechanism as a very rapid step-growth polymerization. In the diagram, M_x refers to the original monomer molecule or any of many dissociation products such as chlorine, fluorine and hydrogen. The $M\bullet$ species refers to those that are activated and capable of participating in reactions to form new covalent bonds. The $\bullet M\bullet$ species refers to an activated difunctional monomer species. The subscripts i, j, and k show the sizes of the different species involved. Even though radicals represent the activated species, any ion or radical could be used in the polymerization. As can be seen here, plasma polymerization is a very complex process, with many parameters effecting everything from rate to chain length.

Selection, or the favouring of one particular pathway can be achieved by altering the plasma parameters. For example, pulsed plasma with selected monomers appears to favour much more regular polymer structures and it has been postulated these grow by a mechanism akin to (radical) chain growth in the plasma off-time.

Common Monomers/Polymers

Common monomers	
Name	Structure
Thiophene	
1,7-Octadiene	

Pyridine	
Acrylonitrile	
Furan	
Styrene	
Acetylene	$H-C{\equiv}C-H$
2-Methyloxazoline	
Tetramethyldisiloxane	$\begin{array}{cc} CH_3 & CH_3 \\ \vert & \vert \\ H-Si-O-Si-H \\ \vert & \vert \\ CH_3 & CH_3 \end{array}$

Monomers

As can be seen in the monomer table, many simple monomers are readily polymerized by this method, but most must be smaller ionizable species because they have to be able to go into the plasma state. Though monomers with multiple bonds polymerize readily, it is not a necessary requirement, as ethane, silicones and many others polymerize also. There are also other stipulations that exist. Yasuda et al. studied 28 monomers and found that those containing aromatic groups, silicon, olefinic group or nitrogen (NH, NH_2, CN) were readily polymerizable, while those containing oxygen, halides, aliphatic hydrocarbons and cyclic hydrocarbons where decomposed more readily. The latter compounds have more ablation or side reactions present, which inhibit stable polymer formation. It is also possible to incorporate N_2, H_2O, and CO into copolymers of styrene.

Plasma polymers can be thought of as a type of graft polymers since they are grown off of a substrate. These polymers are known to form nearly uniform surface deposition, which is one of their desirable properties. Polymers formed from this process often cross-link and form branches due to the multiple propagating species present in the plasma. This often leads to very insoluble polymers, which gives an advantage to this process, since *hyperbranched polymers* can be deposited directly without solvent.

Polymers

Common polymers include: polythiophene, polyhexafluoropropylene, polytetramethyltin, polyhexamethyldisiloxane, polytetramethyldisiloxane, polypyridine, polyfuran, and poly-2-methyloxazoline.

The following are listed in order of decreasing rate of polymerization: polystyrene, polymethyl styrene, polycyclopentadiene, polyacrylate, polyethyl acrylate, polymethyl methacrylate, polyvinyl acetate, polyisoprene, polyisobutene, and polyethylene.

Nearly all polymers created by this method have excellent appearance, are clear, and are significantly cross-linked. Linear polymers are not formed readily by plasma polymerization methods based on propagating species. Many other polymers could be formed by this method.

General Characteristics of Plasma Polymers

The properties of plasma polymers differ greatly from those of conventional polymers. While both types are dependent on the chemical properties of the monomer, the properties of plasma polymers depend more greatly on the design of the reactor and the chemical and physical characteristics of the substrate on which the plasma polymer is deposited. The location within the reactor where the deposition occurs also has an effect on the resultant polymer's properties. In fact by using plasma polymerization with a single monomer and varying the reactor, substrate, etc. a variety of polymers, each having different physical and chemical properties, can be prepared. The large dependence of the polymer features on these factors make it difficult to assign a set of basic characteristics, but a few common properties that set plasma polymers apart from conventional polymers do exist.

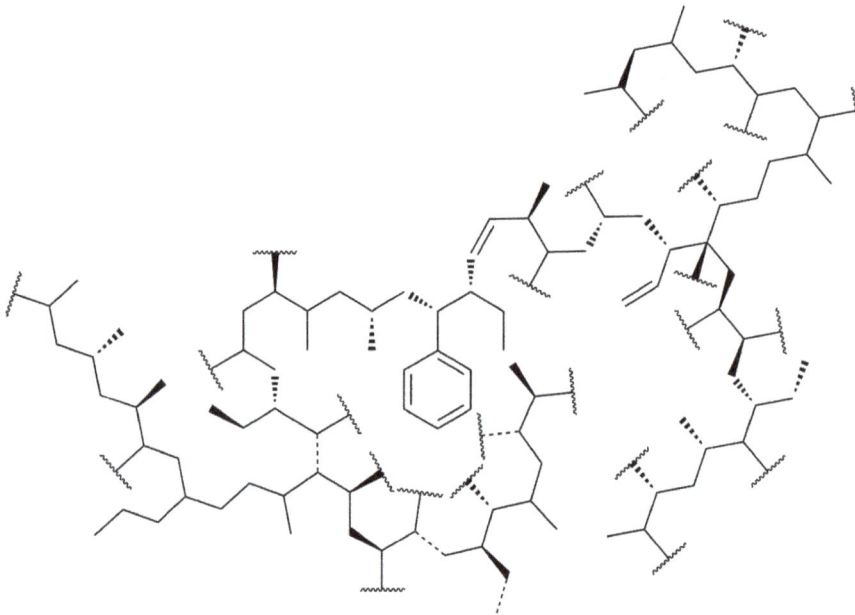

Figure 4. Hypothesized model of plasma-polymerized ethylene film.

The most significant difference between conventional polymers and plasma polymers is that plasma polymers do not contain regular repeating units. Due to the number of different propagating species present at any one time as discussed above, the resultant polymer chains are highly branched and are randomly terminated with a high degree of cross-linking. An example of a proposed structure for plasma polymerized ethylene demonstrating a large extend of cross-linking and branching is shown in Figure 4.

All plasma polymers contain free radicals as well. The amount of free radicals present varies between polymers and is dependent on the chemical structure of the monomer. Because the formation of the trapped free radicals is tied to the growth mechanism of the plasma polymers, the overall properties of the polymers directly correlate to the number of free radicals.

Plasma polymers also contain an internal stress. If a thick layer (e.g. 1 µm) of a plasma polymer is deposited on a glass slide, the plasma polymer will buckle and frequently crack. The curling is attributed to an internal stress formed in the plasma polymer during the polymer deposition. The degree of curling is dependent on the monomer as well as the conditions of the plasma polymerization.

Most plasma polymers are insoluble and infusible. These properties are due to the large amount of cross-linking in the polymers, previously discussed. Consequently, the kinetic path length for these polymers must be sufficiently long, so these properties can be controlled to a point.

The permeabilities of plasma polymers also differ greatly from those of conventional polymers. Because of the absence of large-scale segmental mobility and the high degree of cross-linking within the polymers, the permeation of small molecules does not strictly follow the typical mechanisms of "solution-diffusion" or molecular-level sieve for such small permeants. Really the permeability characteristics of plasma polymers falls between these two ideal cases.

A final common characteristic of plasma polymers is the adhesion ability. The specifics of the adhesion ability for a given plasma polymer, such as thickness and characteristics of the surface layer, again are particular for a given plasma polymer and few generalizations can be made.

Advantages and Disadvantages

Plasma polymerization offers a number of advantages over other polymerization methods and in general. The most significant advantage of plasma polymerization is its ability to produce polymer films of organic compounds that do not polymerize under normal chemical polymerization conditions. Nearly all monomers, even saturated hydrocarbons and organic compounds without a polymerizable structure such as a double bond, can be polymerized with this technique.

A second advantage is the ease of application of the polymers as coatings versus conventional coating processes. While coating a substrate with conventional polymers requires a number of steps, plasma polymerization accomplishes all these in essentially a single step. This leads to a cleaner and 'greener' synthesis and coating process, since no solvent is needed during the polymer preparation and no cleaning of the resultant polymer is needed either. Another 'green' aspect of the synthesis is that no initiator is needed for the polymer preparation since reusable electrodes cause the reaction to proceed. The resultant polymer coatings also have a number of advantages over typical coatings. These advantages include being nearly pinhole free, highly dense, and that the thickness of the coating can easily be varied.

There are also a number of disadvantages relating to plasma polymerization versus conventional methods. The most significant disadvantage is the high cost of the process. A vacuum system is required for the polymerization, significantly increasing the set up price.

Another disadvantage is due to the complexity of plasma processes. Because of the complexity it is not easy to achieve a good control over the chemical composition of the surface after modification. The influence of process parameters on the chemical composition of the resultant polymer means it can take a long time to determine the optimal conditions. The complexity of the process also makes it impossible to theorize what the resultant polymer will look like, unlike conventional polymers which can be easily determined based on the monomer.

Applications

The advantages offered by plasma polymerization have resulted in substantial research on the applications of these polymers. The vastly different chemical and mechanical properties offered by polymers formed with plasma polymerization means they can be applied to countless different systems. Applications ranging from adhesion, composite materials, protective coatings, printing, membranes, biomedical applications, water purification and so on have all been studied.

Of particular interest since the 1980s has been the deposition of functionalized plasma polymer films. For example, functionalized films are used as a means of improving biocompatibility for biological implants6 and for producing super-hydrophobic coatings.They have also been extensively employed in biomaterials for cell attachment, protein binding and as anti-fouling surfaces. Through the use of low power and pressure plasma, high functional retention can be achieved which has led to substantial improvements in the biocompatibility of some products, a simple example being the development of extended wear contact lenses. Due to these successes, the huge potential of functional plasma polymers is slowly being realised by workers in previously unrelated fields such as water treatment and wound management.Emerging technologies such as nanopatterning, 3D scaffolds, micro-channel coating and microencapsulation are now also utilizing functionalized plasma polymers, areas for which traditional polymers are often unsuitable

A significant area of research has been on the use of plasma polymer films as permeation membranes. The permeability characteristics of plasma polymers deposited on porous substrates are different than usual polymer films. The characteristics depend on the deposition and polymerization mechanism. Plasma polymers as membranes for separation of oxygen and nitrogen, ethanol and water, and water vapor permeation have all been studied. The application of plasma polymerized thin films as reverse osmosis membranes has received considerable attention as well. Yasuda et al. have shown membranes prepared with plasma polymerization made from nitrogen containing monomers can yield up to 98% salt rejection with a flux of 6.4 gallons/ft^2 a day. Further research has shown that varying the monomers of the membrane offer other properties as well, such as chlorine resistance.

Plasma-polymerized films have also found electrical applications. Given that plasma polymers frequently contain many polar groups, which form when the radicals react with oxygen in air during the polymerization process, the plasma polymers were expected to be good dielectric materials in thin film form. Studies have shown that the plasma polymers generally do in fact have a higher dielectric property. Some plasma polymers have been applied as chemical sensory devices due to their electrical properties. Plasma polymers have been studied as chemical sensory devices for humidity, propane, and carbon dioxide amongst others. Thus far issues with instability against aging and humidity have limited their commercial applications.

The application of plasma polymers as coatings has also been studied. Plasma polymers formed from tetramethoxysilane have been studied as protective coatings and have shown to increase the hardness of polyethylene and polycarbonate. The use of plasma polymers to coat plastic lenses is increasing in popularity. Plasma depositions are able to easily coat curved materials with a good uniformity, such as those of bifocals. The different plasma polymers used can be not only scratch resistant, but also hydrophobic leading to anti-fogging effects.

Plasma Cleaning

Plasma cleaning involves the removal of impurities and contaminants from surfaces through the use of an energetic plasma or Dielectric barrier discharge (DBD) plasma created from gaseous species. Gases such as argon and oxygen, as well as mixtures such as air and hydrogen/nitrogen are used. The plasma is created by using high frequency voltages (typically kHz to >MHz) to ionise the low pressure gas (typically around 1/1000 atmospheric pressure), although atmospheric pressure plasmas are now also common.

Fig. 1. The surface of a MEMS device is cleaned with bright, blue oxygen plasma in a plasma etcher to rid it of carbon contaminants. (100mTorr, 50W RF)

Methods

In plasma, gas atoms are excited to higher energy states and also ionized. As the atoms and molecules 'relax' to their normal, lower energy states they release a photon of light, this results in the characteristic "glow" or light associated with plasma. Different gases give different colors. For example, oxygen plasma emits a light blue color.

A plasma's activated species include atoms, molecules, ions, electrons, free radicals, metastables, and photons in the short wave ultraviolet (vacuum UV, or VUV for short) range. This 'soup', which incidentally is around room temperature, then interacts with any surface placed in the plasma.

If the gas used is oxygen, the plasma is an effective, economical, environmentally safe method for critical cleaning. The VUV energy is very effective in the breaking of most organic bonds (i.e., C–H, C–C, C=C, C–O, and C–N) of surface contaminants. This helps to break apart high molecular weight contaminants. A second cleaning action is carried out by the oxygen species created in the plasma (O_2^+, O_2^-, O_3, O, O^+, O^-, ionised ozone, metastable excited oxygen, and free electrons). These species react with organic contaminants to form H_2O, CO, CO_2, and lower molecular weight hydrocarbons. These compounds have relatively high vapour pressures (means they escapes to space easily) and are evacuated from the chamber during processing. The resulting surface is ultra-clean. In Fig. 2, a relative content of carbon over material depth is shown before and after cleaning with excited oxygen .

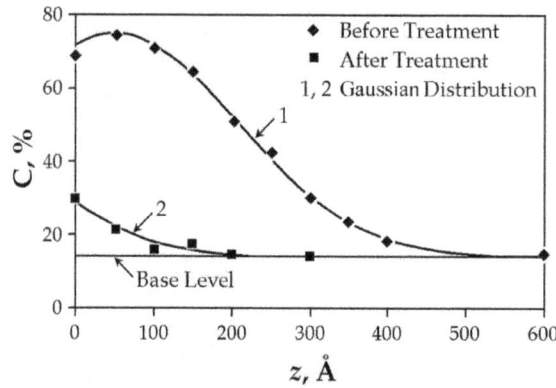

Fig. 2. Content of carbon over material depth z: before a sample treatment - diamond points and after the treatment during 1 s. - square points

If the part to be treated consists of easily oxidized materials such as silver or copper, inert gases such as argon or helium are used instead. The plasma activated atoms and ions behave like a molecular sandblast and can break down organic contaminants. These contaminants are again vapourised and evacuated from the chamber during processing.

Most of these by-products are small quantities of gases such as carbon dioxide and water vapor with trace amounts of carbon monoxide and other hydrocarbons.

Whether or not organic removal is complete can be assessed with contact angle measurements. When an organic contaminant is present, the contact angle of water with the device will be high. After the removal of the contaminant, the contact angle will be reduced to that characteristic of contact with the pure substrate.

If a surface to be treated is coated with a patterned conductive layer (metal, ITO) deposited on it, treatment by a direct contact with plasma (capable for contraction to microarcs) could be destructive. In this case, cleaning by neutral atoms excited in plasma to metastable state can be applied. Results of the same applications to surfaces of glass samples coated with Cr and ITO layers are shown in Fig. 3.

Fig. 3. Contact Angle of Water Droplet of 5 μl on glass coated with different materials.

After treatment, the contact angle of a water droplet is decreased becoming less than its value on the untreated surface. In Fig. 4, the relaxation curve for droplet footprint is shown for glass sample. A photograph of the same droplet on the untreated surface is shown in Fig. 4 inset. Surface relaxation time corresponding to a data shown in Fig. 4 is about 4 hours.

Plasma Ashing is the process in which plasma cleaning is performed for the sole purpose of removing carbon. Plasma Ashing is always done with O_2 gas.

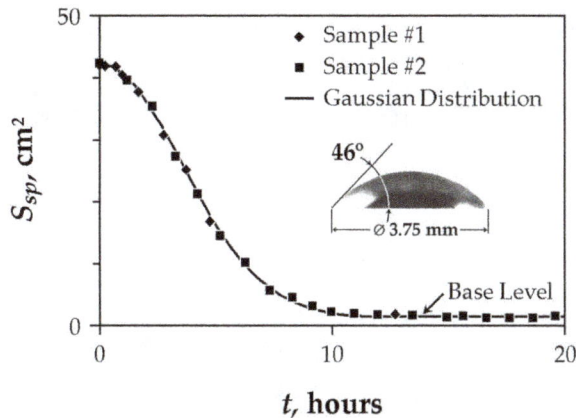

Fig. 4. Surface area of water droplet of 5 µl volume footprint on glass surface versus time t after its treatment. Droplet on untreated glass is shown in inset.

Plasma Electrolytic Oxidation

Plasma electrolytic oxidation (PEO), also known as microarc oxidation (MAO), is an electrochemical surface treatment process for generating oxide coatings on metals. It is similar to anodizing, but it employs higher potentials, so that discharges occur and the resulting plasma modifies the structure of the oxide layer. This process can be used to grow thick (tens or hundreds of micrometers), largely crystalline, oxide coatings on metals such as aluminium, magnesium and titanium. Because they can present high hardness and a continuous barrier, these coatings can offer protection against wear, corrosion or heat as well as electrical insulation.

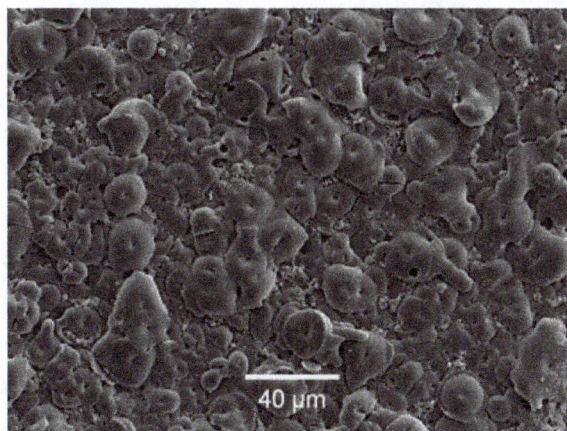

A typical PEO surface on aluminium, as viewed in an SEM.

A yacht winch drum undergoing PEO processing. Below; a finished winch drum installed on a yacht.

The coating is a chemical conversion of the substrate metal into its oxide, and grows both inwards and outwards from the original metal surface. Because it is a conversion coating, rather than a deposited coating (such as a coating formed by plasma spraying), it has excellent adhesion to the substrate metal. A wide range of substrate alloys can be coated, including all wrought aluminium alloys and most cast alloys, although high levels of silicon can reduce coating quality.

Process

Metals such as aluminium naturally form a passivating oxide layer which provides moderate protection against corrosion. The layer is strongly adherent to the metal surface, and it will regrow quickly if scratched off. In conventional anodizing, this layer of oxide is grown on the surface of the metal by the application of electrical potential, while the part is immersed in an acidic electrolyte.

In plasma electrolytic oxidation, higher potentials are applied. For example, in the plasma electrolytic oxidation of aluminium, at least 200 V must be applied. This locally exceeds the dielectric breakdown potential of the growing oxide film, and discharges occur. These discharges result in localised plasma reactions, with conditions of high temperature and pressure which modify the growing oxide. Processes include melting, melt-flow, re-solidification, sintering and densification of the growing oxide. One of the most significant effects, is that the oxide is partially converted from amorphous alumina into crystalline forms such as corundum (α-Al_2O_3) which is much harder. As a result, mechanical properties such as wear resistance and toughness are enhanced.

Equipment Used

The part to be coated is immersed in a bath of electrolyte which usually consists of a dilute alkaline solution such as KOH. It is electrically connected, so as to become one of the electrodes in the electrochemical cell, with the other "counter-electrode" typically being made from an inert material such as stainless steel, and often consisting of the wall of the bath itself.

Potentials of over 200 V are applied between these two electrodes. These may be continuous or pulsed direct current (DC) (in which case the part is simply an anode in DC operation), or alternating pulses (alternating current or "pulsed bi-polar" operation) where the stainless steel counter electrode might just be earthed.

Coating Properties

Plasma electrolytic oxide coatings are generally recognized for high hardness, wear resistance, and corrosion resistance. However, the coating properties are highly dependent on the substrate used, as well as on the composition of the electrolyte and the electrical regime used.

Even on aluminium, the coating properties can vary strongly according to the exact alloy composition. For instance, the hardest coatings can be achieved on 2XXX series aluminium alloys, where the highest proportion of crystalline phase corundum (α-Al_2O_3) is formed, resulting in hardnesses of ~2000 HV, whereas coatings on the 5XXX series have less of this important constituent and are hence softer. Extensive work is being pursued by Prof. T. W. Clyne at the University of Cambridge to investigate the fundamental electrical and plasma physical processes involved in this process, having previously elucidated some of the micromechanical (& pore architectural), mechanical and thermal characteristics of PEO coatings.

Plasma Gasification

Plasma gasification is a process which converts organic matter into synthetic gas, electricity, and slag using plasma. A plasma torch powered by an electric arc, is used to ionize gas and catalyze organic matter into synthetic gas and solid waste (slag). It is used commercially as a form of waste treatment and has been tested for the gasification of biomass and solid hydrocarbons, such as coal, oil sands, and oil shale.

Process

A plasma torch itself typically uses an inert gas such as argon. The electrodes vary from copper or tungsten to hafnium or zirconium, along with various other alloys. A strong electric current under high voltage passes between the two electrodes as an electric arc. Pressurized inert gas is ionized passing through the plasma created by the arc. The torch's temperature ranges from 4,000 to 25,000 °F (2,200 to 13,900 °C). The temperature of the plasma reaction determines the structure of the plasma and forming gas. This can be optimized to minimize ballast contents, composed of the byproducts of oxidation: CO_2, N_2, H_2O, etc.

The waste is heated, melted and finally vaporised. At these conditions molecular dissociation can occur by breaking down molecular bonds. Complex molecules are separated into individual atoms. The resulting elemental components are in a gaseous phase. Molecular dissociation using plasma is referred to as "plasma pyrolysis."

Feedstocks

The feedstock for plasma waste treatment is most often municipal solid waste, organic waste, or both. Feedstocks may also include biomedical waste and hazmat materials. Content and consistency of the waste directly impacts performance of a plasma facility. Pre-sorting and recycling useful material before gasification provides consistency. Too much inorganic material such as metal and construction waste increases slag production, which in turn decreases syngas production. However, a benefit is that the slag itself is chemically inert and safe to handle (certain materials may affect the content of the gas produced, however). Shredding waste before entering the main chamber helps to increase syngas production. This creates an efficient transfer of energy which ensures more materials are broken down.

For better processing, air and/or steam is added into plasma gasificator.

Yields

Pure highly calorific synthetic gas consists predominantly of Carbon monoxide (CO), H_2, CH, among other components. The conversion rate of plasma gasification exceeds 99%. Non-flammable inorganic components in the waste stream are not broken down. This includes various metals. A phase change from solid to liquid adds to the volume of slag.

Plasma processing of waste is ecologically clean. The lack of oxygen prevents the formation of many toxic materials. The high temperatures in a reactor also prevent the main components of the gas from forming toxic compounds such as furans, dioxins, nitrogen oxides, or sulfur dioxide. Water filtration removes ash and gaseous pollutants.

The production of ecologically clean synthetic gas is the standard goal. The gas product contains no phenols or complex hydrocarbons however circulating water from filtering systems is toxic. The water removes toxins (poisons) and the hazardous substances which must be cleaned.

Metals resulting from plasma pyrolysis can be recovered from the slag and eventually sold as a commodity. Inert slag is granulated. This slag grain is used in construction. A portion of the syngas produced feeds on-site turbines, which power the plasma torches and thus support the feed system. This is self-sustaining electric power.

Equipment

Gasification reactors operate at negative pressure and recover both gaseous and solid resources.

Advantages

The main advantages of plasma technologies for waste treatment are:

- Clean destruction of hazardous waste

- Preventing hazardous waste from reaching landfills

- Some processes are designed to recover fly ash, bottom ash, and most other particulates, for 95% or better diversion from landfills, and no harmful emissions of toxic waste

- Production of vitrified slag which could be used as construction material

- Processing of organic waste into combustible syngas for electric power and thermal energy and

- Production of value-added products (metals) from slag

- Plasma arc gasification is a safe means to destroy both medical and other hazardous waste.

- Gasification with starved combustion and rapid quenching of syngas from elevated temperatures can avoid the production of dioxins and furans that are common to incinerators

- Air emissions can be cleaner than landfills and incinerators. Plasma gasification plants using IC engines to produce electricity do not require stacks, and can be integrated in small spaces inside the urban envelope

Disadvantages

Main disadvantages of plasma technologies for waste treatment are:

- Large initial investment costs relative to that of alternatives, including landfill and incineration. As with all major waste treatment processes, plasma gasification is an infrastructure process.

- Operational costs are high relative to that of incineration.

- The industry has a history of exaggerated claims about performance in order to sell equipment and licenses before processes were proven. This has resulted in a number of very costly failures from gasifiers that were not properly integrated with feed systems, gas cleanup, and power production, or from gasifiers that were scaled up without operational data to support the scaleups.

- Little or even negative net energy production. Processes that use plasma for catalysis and refinement of syngas to operate IC engines and/or gas turbines can theoretically be more efficient than incinerators of equal size, but this is yet to be demonstrated in practice.

- Frequent maintenance and limited plant availability.

- For some early technologies, the plasma flame reduces the diameter of the sampler orifice over time, necessitating frequent maintenance.

Commercialization

Plasma gasification is used commercially for waste disposal in five locations representing to a total design capacity of 250 tonnes waste per day worldwide.

In the Northeast of England in the United Kingdom plasma gasification technology was attempted implemented within the Northeast of England Process Industry Cluster(NEPIC) on Teesside by Air Products. Two large units were erected, designed to gasify societal waste to produce power with the synthesis gas produced. By late 2015, Air Products halted construction on the second phase until it fixed issues encountered during commissioning of the first phase. On April 4, 2016, Air Products announced it was leaving the waste-to-energy business, and was taking a write-down of $0.9-$1.0B.

Military Use

The US Navy is employing Plasma Arc Waste Destruction System (PAWDS) on its latest generation Gerald R. Ford-class aircraft carrier. The compact system being used will treat all combustible solid waste generated on board the ship. After having completed factory acceptance testing in Montreal, the system is scheduled to be shipped to the Huntington Ingalls shipyard for installation on the carrier.

Plasma Arc Welding

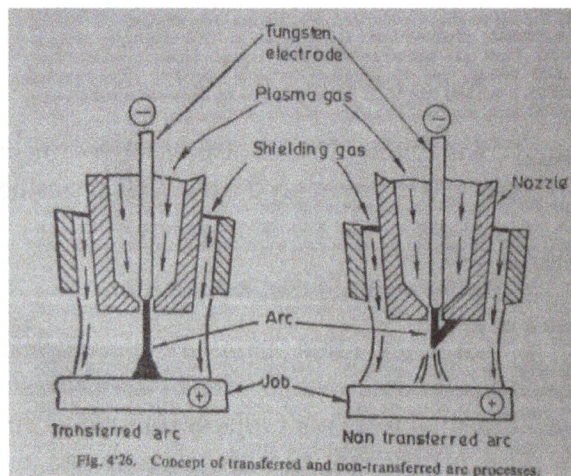

Plasma arc types.

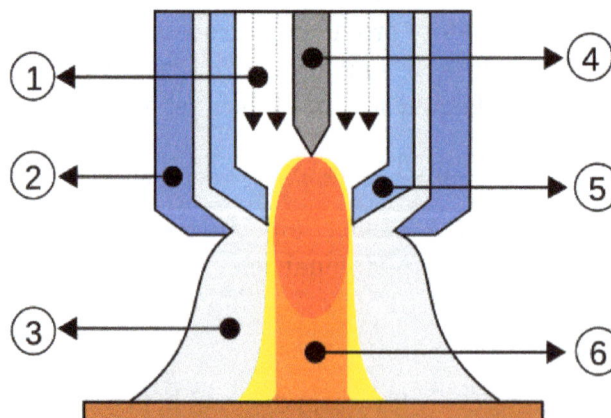

1. Gas plasma, 2. Nozzle protection, 3. Shield Gas, 4. Electrode, 5. Nozzle constriction, 6. Electric arc

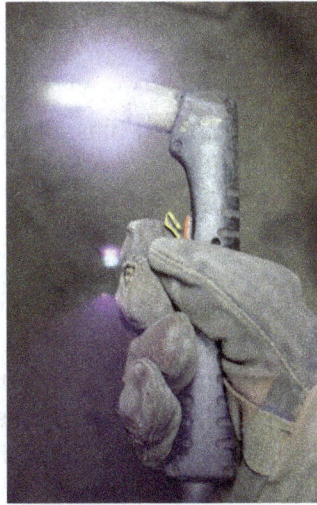

Plasma cutting torch.

Metal cutting by plasma arc.

Plasma arc welding (PAW) is an arc welding process similar to gas tungsten arc welding (GTAW). The electric arc is formed between an electrode (which is usually but not always made of sintered tungsten) and the workpiece. The key difference from GTAW is that in PAW, by positioning the electrode within the body of the torch, the plasma arc can be separated from the shielding gas envelope. The plasma is then forced through a fine-bore copper nozzle which constricts the arc and the plasma exits the orifice at high velocities (approaching the speed of sound) and a temperature approaching 28,000 °C (50,000 °F) or higher.

Just as oxy-fuel torches can be used for either welding or cutting, so too can plasma torches, which can achieve plasma arc welding or plasma cutting.

Arc plasma is the temporary state of a gas. The gas gets ionized after passage of electric current through it and it becomes a conductor of electricity. In ionized state atoms break into electrons (−) and cations (+) and the system contains a mixture of ions, electrons and highly excited atoms. The degree of ionization may be between 1% and greater than 100% i.e.; double and triple degrees of ionization. Such states exist as more electrons are pulled from their orbits.

The energy of the plasma jet and thus the temperature is dependent upon the electrical power employed to create arc plasma. A typical value of temperature obtained in a plasma jet torch may be of the order of 28000 °C(50000 °F) against about 5500 °C (10000 °F) in ordinary electric welding arc. Actually all welding arcs are (partially ionized) plasmas, but the one in plasma arc welding is a constricted arc plasma.

Concept

Plasma arc welding is an arc welding process wherein coalescence is produced by the heat obtained from a constricted arc setup between a tungsten/alloy tungsten electrode and the water-cooled (constricting) nozzle (non-transferred arc) or between a tungsten/alloy tungsten electrode and the job (transferred arc). The process employs two inert gases, one forms the arc plasma and the second shields the arc plasma. Filler metal may or may not be added.

History

The plasma arc welding and cutting process was invented by Robert M. Gage in 1953 and patented in 1957. The process was unique in that it could achieve precision cutting and welding on both thin and thick metals. It was also capable of spray coating hardening metals onto other metals. One example was the spray coating of the turbine blades of the moon bound Saturn rocket.

Principle of Operation

Plasma arc welding is a constricted arc process. The arc is constricted with the help of a water-cooled small diameter nozzle which squeezes the arc, increases its pressure, temperature and heat intensely and thus improves arc stability, arc shape and heat transfer characteristics. Plasma arc welding processes can be divided into two basic types:

Non-transferred arc process

> The arc is formed between the electrode(-) and the water cooled constricting nozzle(+). Arc plasma comes out of the nozzle as a flame. The arc is independent of the work piece and the work piece does not form a part of the electrical circuit. Just like an arc flame (as in atomic hydrogen welding), it can be moved from one place to another and can be better controlled. The non transferred plasma arc possesses comparatively less energy density as compared to a transferred arc plasma and it is employed for welding and in applications involving ceramics or metal plating (spraying). High density metal coatings can be produced by this process. A non-transferred arc is initiated by using a high frequency unit in the circuit.

Transferred arc process

> The arc is formed between the electrode(-) and the work piece(+). In other words, arc is transferred from the electrode to the work piece. A transferred arc possesses high energy density and plasma jet velocity. For this reason it is employed to cut and melt metals. Besides carbon steels this process can cut stainless steel and nonferrous metals where an oxyacetylene torch does not succeed. Transferred arc can also be used for welding at high arc travel speeds. For initiating a transferred arc, a current limiting resistor is put in the circuit, which permits a flow of about 50 amps, between the nozzle and electrode and a pilot arc is

established between the electrode and the nozzle. As the pilot arc touches the job main current starts flowing between electrode and job, thus igniting the transferred arc. The pilot arc initiating unit gets disconnected and pilot arc extinguishes as soon as the arc between the electrode and the job is started. The temperature of a constricted plasma arc may be of the order of 8000 - 25000°C.

Equipment

The equipment needed in plasma arc welding along with their functions are as follows:

Power Supply

A direct current power source (generator or rectifier) having drooping characteristics and open circuit voltage of 70 volts or above is suitable for plasma arc welding. Rectifiers are generally preferred over DC generators. Working with helium as an inert gas needs open circuit voltage above 70 volts. This higher voltage can be obtained by series operation of two power sources; or the arc can be initiated with argon at normal open circuit voltage and then helium can be switched on.

Typical welding parameters for plasma arc welding are as follows:

Current 50 to 350 amps, voltage 27 to 31 volts, gas flow rates 2 to 40 liters/minute (lower range for orifice gas and higher range for outer shielding gas), direct current electrode negative (DCEN) is normally employed for plasma arc welding except for the welding of aluminum in which cases water cooled electrode is preferable for reverse polarity welding, i.e. direct current electrode positive (DCEP).

High Frequency Generator and Current Limiting Resistors

A high frequency generator and current limiting resistors are used for arc ignition. The arc starting system may be separate or built into the system.

Plasma Torch

It is either transferred arc or non transferred arc typed. It is hand operated or mechanized. At present, almost all applications require automated system. The torch is water cooled to increase the life of the nozzle and the electrode. The size and the type of nozzle tip are selected depending upon the metal to be welded, weld shapes and desired penetration depth.

Shielding Gases

Two inert gases or gas mixtures are employed. The orifice gas at lower pressure and flow rate forms the plasma arc. The pressure of the orifice gas is intentionally kept low to avoid weld metal turbulence, but this low pressure is not able to provide proper shielding of the weld pool. To have suitable shielding protection same or another inert gas is sent through the outer shielding ring of the torch at comparatively higher flow rates. Most of the materials can be welded with argon, helium, argon+hydrogen and argon+helium, as inert gases or gas mixtures. Argon is very commonly used. Helium is preferred where a broad heat input pattern and flatter cover pass is desired without key hole mode weld. A mixture of argon and hydrogen supplies heat energy higher than when only argon is used and thus permits keyhole mode welds in nickel base alloys, copper base alloys and stainless steels.

For cutting purposes a mixture of argon and hydrogen (10-30%) or that of nitrogen may be used. Hydrogen, because of its dissociation into atomic form and thereafter recombination generates temperatures above those attained by using argon or helium alone. In addition, hydrogen provides a reducing atmosphere, which helps in preventing oxidation of the weld and its vicinity. (Care must be taken, as hydrogen diffusing into the metal can lead to embrittlement in some metals and steels.)

Voltage Control

Voltage control is required in contour welding. In normal key hole welding a variation in arc length up to 1.5 mm does not affect weld bead penetration or bead shape to any significant extent and thus a voltage control is not considered essential.

Current and Gas Decay Control

It is necessary to close the key hole properly while terminating the weld in the structure.

Fixture

It is required to avoid atmospheric contamination of the molten metal under bead.

Process Description

Technique of work piece cleaning and filler metal addition is similar to that in TIG welding. Filler metal is added at the leading edge of the weld pool. Filler metal is not required in making root pass weld.

Type of Joints: For welding work piece up to 25 mm thick, joints like square butt, J or V are employed. Plasma welding is used to make both key hole and non-key hole types of welds.

Making a non-key hole weld: The process can make non key hole welds on work pieces having thickness 2.4 mm and under.

Making a keyhole welds: An outstanding characteristics of plasma arc welding, owing to exceptional penetrating power of plasma jet, is its ability to produce keyhole welds in work piece having thickness from 2.5 mm to 25 mm. A keyhole effect is achieved through right selection of current, nozzle orifice diameter and travel speed, which create a forceful plasma jet to penetrate completely through the work piece. Plasma jet in no case should expel the molten metal from the joint. The major advantages of keyhole technique are the ability to penetrate rapidly through relatively thick root sections and to produce a uniform under bead without mechanical backing. Also, the ratio of the depth of penetration to the width of the weld is much higher, resulting narrower weld and heat-affected zone. As the weld progresses, base metal ahead the keyhole melts, flow around the same solidifies and forms the weld bead. Key holing aids deep penetration at faster speeds and produces high quality bead. While welding thicker pieces, in laying others than root run, and using filler metal, the force of plasma jet is reduced by suitably controlling the amount of orifice gas.

Plasma arc welding is an advancement over the GTAW process. This process uses a non-consumable tungsten electrode and an arc constricted through a fine-bore copper nozzle. PAW can be used to join all metals that are weldable with GTAW (i.e., most commercial metals and alloys). Difficult-to-weld in metals by PAW include bronze, cast iron, lead and magnesium. Several basic PAW process variations are possible by varying the current, plasma gas flow rate, and the orifice diameter, including:

- Micro-plasma (< 15 Amperes)

- Melt-in mode (15–100 Amperes)

- Keyhole mode (>100 Amperes)

- Plasma arc welding has a greater energy concentration as compared to GTAW.

- A deep, narrow penetration is achievable, with a maximum depth of 12 to 18 mm (0.47 to 0.71 in) depending on the material.

- Greater arc stability allows a much longer arc length (stand-off), and much greater tolerance to arc length changes.

- PAW requires relatively expensive and complex equipment as compared to GTAW; proper torch maintenance is critical

- Welding procedures tend to be more complex and less tolerant to variations in fit-up, etc.

- Operator skill required is slightly greater than for GTAW.

- Orifice replacement is necessary.

Process Variables

Gases

At least two separate (and possibly three) flows of gas are used in PAW:

- Plasma gas – flows through the orifice and becomes ionized.

- Shielding gas – flows through the outer nozzle and shields the molten weld from the atmosphere

- Back-purge and trailing gas – required for certain materials and applications.

These gases can all be same, or of differing composition.

Key Process Variables

- Current Type and Polarity

- DCEN from a CC source is standard

- AC square-wave is common on aluminum and magnesium

- Welding current and pulsing - Current can vary from 0.5 A to 1200 A; Current can be constant or pulsed at frequencies up to 20 kHz

- Gas flow rate (This critical variable must be carefully controlled based upon the current, orifice diameter and shape, gas mixture, and the base material and thickness.)

Other Plasma Arc Processes

Depending upon the design of the torch (e.g., orifice diameter), electrode design, gas type and velocities, and the current levels, several variations of the plasma process are achievable, including:

- Plasma arc cutting (PAC)
- Plasma arc gouging
- Plasma arc surfacing
- Plasma arc spraying

Plasma Arc Cutting

When used for cutting, the plasma gas flow is increased so that the deeply penetrating plasma jet cuts through the material and molten material is removed as cutting dross. PAC differs from oxy-fuel cutting in that the plasma process operates by using the arc to melt the metal whereas in the oxy-fuel process, the oxygen oxidizes the metal and the heat from the exothermic reaction melts the metal. Unlike oxy-fuel cutting, the PAC process can be applied to cutting metals which form refractory oxides such as stainless steel, cast iron, aluminum, and other non-ferrous alloys. Since PAC was introduced by Praxair Inc. at the American Welding Society show in 1954, many process refinements, gas developments, and equipment improvements have occurred.

Plasma Cutting

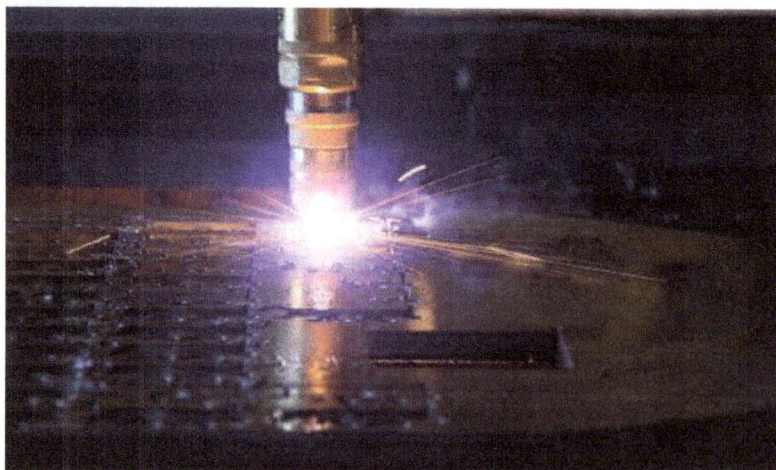

CNC Plasma Cutting

Plasma cutting is a process that cuts through electrically conductive materials by means of an accelerated jet of hot plasma. Typical materials cut with a plasma torch include steel, aluminum, brass and copper, although other conductive metals may be cut as well. Plasma cutting is often used in fabrication shops, automotive repair and restoration, industrial construction, and salvage and scrapping operations. Due to the high speed and precision cuts combined with low cost, plasma cutting sees widespread use from large-scale industrial CNC applications down to small hobbyist shops.

Plasma cutting performed by an industrial robot

Process

The basic plasma cutting process involves creating an electrical channel of superheated, electrically ionized gas i.e. plasma from the plasma cutter itself, through the work piece to be cut, thus forming a completed electric circuit back to the plasma cutter via a grounding clamp. This is accomplished by a compressed gas (oxygen, air, inert and others depending on material being cut) which is blown through a focused nozzle at high speed toward the work piece. An electrical arc is then formed within the gas, between an electrode near or integrated into the gas nozzle and the work piece itself. The electrical arc ionizes some of the gas, thereby creating an electrically conductive channel of plasma. As electricity from the cutter torch travels down this plasma it delivers sufficient heat to melt through the work piece. At the same time, much of the high velocity plasma and compressed gas blow the hot molten metal away, thereby separating i.e. cutting through the work piece.

Freehand cut of a thick steel plate

Plasma cutting is an effective means of cutting thin and thick materials alike. Hand-held torches can usually cut up to 38mm thick steel plate, and stronger computer-controlled torches can cut steel up to 150 mm thick. Since plasma cutters produce a very hot and very localized "cone" to cut with, they are extremely useful for cutting sheet metal in curved or angled shapes.

History

Plasma cutting grew out of plasma welding in the 1960s, and emerged as a very productive way to cut sheet metal and plate in the 1980s. It had the advantages over traditional "metal against metal" cutting of producing no metal chips, giving accurate cuts, and producing a cleaner edge than oxy-fuel cutting. Early plasma cutters were large, somewhat slow and expensive and, therefore, tended to be dedicated to repeating cutting patterns in a "mass production" mode.

Plasma cutting with a tilting head

As with other machine tools, CNC (computer numerical control) technology was applied to plasma cutting machines in the late 1980s into the 1990s, giving plasma cutting machines greater flexibility to cut diverse shapes "on demand" based on a set of instructions that were programmed into the machine's numerical control. These CNC plasma cutting machines were, however, generally limited to cutting patterns and parts in flat sheets of steel, using only two axes of motion (referred to as X Y cutting).

Safety

Proper eye protection (but not gas welding goggles as these do not give UV protection) and face shields are needed to prevent eye damage called arc eye as well as damage from debris, as per arc welding. It is recommended to use green lens shade #8 or #9 safety glasses for cutting to prevent the retinas from being "flashed" or burned. OSHA recommends a shade 8 for arc current less than 300, but notes that "*These values apply where the actual arc is clearly seen. Experience has shown that lighter filters may be used when the arc is hidden by the workpiece.*" Lincoln Electric, a manufacturer of plasma cutting equipment, says, "*Typically a darkness shade of #7 to #9 is acceptable.*" Longevity Global, Inc., another manufacturer, offers this more specific table for Eye Protection for Plasma Arc Cutting at lower amperages :

Current Level in Amps	Minimum Shade Number
Below 20	#4
20-40	#5
40-60	#6
60-80	#8

Leather gloves, apron and jacket are also recommended to prevent burns from sparks and debris.

Starting Methods

Plasma cutters use a number of methods to start the arc. In some units, the arc is created by putting the torch in contact with the work piece. Some cutters use a high voltage, high frequency circuit to start the arc. This method has a number of disadvantages, including risk of electrocution, difficulty of repair, spark gap maintenance, and the large amount of radio frequency emissions. Plasma cutters working near sensitive electronics, such as CNC hardware or computers, start the pilot arc by other means. The nozzle and electrode are in contact. The nozzle is the cathode, and the electrode is the anode. When the plasma gas begins to flow, the nozzle is blown forward. A third, less common method is capacitive discharge into the primary circuit via a silicon controlled rectifier.

Inverter Plasma Cutters

Analog plasma cutters, typically requiring more than 2 kilowatts, use a heavy mains-frequency transformer. Inverter plasma cutters rectify the mains supply to DC, which is fed into a high-frequency transistor inverter between 10 kHz to about 200 kHz. Higher switching frequencies allow smaller transformer resulting in overall size and weight reduction.

Plasma cutting

The transistors used were initially MOSFETs, but are now increasingly using IGBTs. With paralleled MOSFETs, if one of the transistors activates prematurely it can lead to a cascading failure of one quarter of the inverter. A later invention, IGBTs, are not as subject to this failure mode. IGBTs can be generally found in high current machines where it is not possible to parallel sufficient MOSFET transistors.

The switch mode topology is referred to as a dual transistor off-line forward converter. Although lighter and more powerful, some inverter plasma cutters, especially those without power factor correction, cannot be run from a generator (that means manufacturer of the inverter unit forbids doing so; it is only valid for small, light portable generators). However newer models have internal circuitry that allow units without power factor correction to run on light power generators.

CNC Cutting Methods

Some plasma cutter manufacturers build CNC cutting tables, and some have the cutter built into the table. CNC tables allow a computer to control the torch head producing clean sharp cuts. Modern CNC plasma equipment is capable of multi-axis cutting of thick material, allowing opportunities for complex welding seams that are not possible otherwise. For thinner material, plasma cutting is being progressively replaced by laser cutting, due mainly to the laser cutter's superior hole-cutting abilities.

A specialized use of CNC Plasma Cutters has been in the HVAC industry. Software processes information on ductwork and creates flat patterns to be cut on the cutting table by the plasma torch. This technology has enormously increased productivity within the industry since its introduction in the early 1980s.

CNC Plasma Cutters are also used in many workshops to create decorative metalwork. For instance, commercial and residential signage, wall art, address signs, and outdoor garden art.

In recent years there has been even more development. Traditionally the machines' cutting tables were horizontal, but now vertical CNC plasma cutting machines are available, providing for a smaller footprint, increased flexibility, optimum safety and faster operation.

CNC Plasma Cutting Configurations

There are 3 main configurations of CNC Plasma Cutting, and they are largely differentiated by the forms of materials before processing, and the flexibility of the cutting head.

2 Dimensional / 2-Axis Plasma Cutting

This is the most common and conventional form of CNC Plasma Cutting. Producing flat profiles, where the cut edges are at 90 Degrees to the material surface. High powered cnc plasma cutting beds are configured in this way, able to cut profiles from metal plate up to 150mm thick.

3 Dimensional / 3+ Axis Plasma Cutting

Once again, a process for producing flat profiles from sheet or plate metal, however with the introduction of an additional axis of rotation, the cutting head of a CNC Plasma Cutting machine can tilt whilst being taken through a conventional 2 dimensional cutting path. The result of this is cut edges at an angle other than 90 Degrees to the material surface, for example 30-45 Degree angles. This angle is continuous throughout the thickness of the material. This is typically applied in situations where the profile being cut is to be used as part of a welded fabrication as the angled edge forms part of the weld preparation. When the weld preparation is applied during the cnc plasma cutting process, secondary operations such as grinding or machining can be avoided, reducing cost. The angular cutting capability of 3 Dimensional plasma cutting can also be used to create countersunk holes and chamfer edges of profiled holes.

Tube & Section Plasma Cutting

Used in the processing of tube, pipe or any form of long section. The plasma cutting head usually remains stationary whilst the workpiece is fed through, and rotated around its longitudinal axis. There are some configurations where, as with 3 Dimensional Plasma Cutting, the cutting head can tilt and rotate. This allows angled cuts to be made through the thickness of the tube or section, commonly taken advantage of in the fabrication of process pipework where cut pipe can be provided with a weld preparation in place of a straight edge.

New Technology

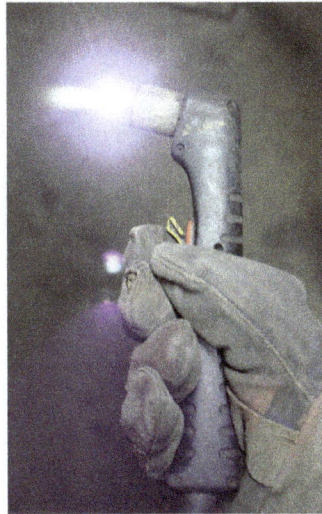

Hand held plasma cutter

High performance cut

In the past decade plasma torch manufacturers have engineered new models with a smaller nozzle and a thinner plasma arc. This allows near-laser precision on plasma cut edges. Several manufacturers have combined precision CNC control with these torches to allow fabricators to produce parts that require little or no finishing.

Costs

Plasma torches were once quite expensive. For this reason they were usually only found in professional welding shops and very well-stocked private garages and shops. However, modern plasma torches are becoming cheaper, and now are within the price range of many hobbyists. Older units may be very heavy, but still portable, while some newer ones with inverter technology weigh only a little, yet equal or exceed the capacities of older ones.

References

- Shen, Mitchel; Alexis T. Bell (1979). Plasma Polymerization. Washington D.C.: American Chemical Society. ISBN 0-8412-0510-8.

- Inagaki, N. (1996). Plasma surface modification and plasma polymerization. Lancaster, Pa.: Technomic Pub. Co. ISBN 1-56676-337-1.

- A. Pizzi; K. L. Mittal (2003). Handbook of Adhesive Technology, Revised and Expanded (2, illustrated, revised ed.). CRC Press. p. 1036. ISBN 978-0824709860.

- Sacks, Raymond; Bohnart, E. (2005). "17". Welding Principles and Practices (Third ed.). New York: McGraw_Hill. p. 597. ISBN 978-0-07-825060-6.

- Macgregor-Ramiasa, Melanie N.; Cavallaro, Alex A.; Vasilev, Krasimir (2015). "Properties and reactivity of polyoxazoline plasma polymer films". J. Mater. Chem. B. 3 (30): 6327–6337. doi:10.1039/C5TB00901D.

- Pourali, M. "Application of Plasma Gasification Technology in Waste to Energy #x2014;Challenges and Opportunities". IEEE Transactions on Sustainable Energy. 1 (3): 125–130. doi:10.1109/TSTE.2010.2061242. ISSN 1949-3029.

- Jimbo, Hajime (1996). "Plasma Melting and Useful Application of Molten Slag". Waste management. 16 (5): 417–422. doi:10.1016/S0956-053X(96)00087-6. Retrieved 2013-03-20.

- Leal-Quirós, Edbertho (2004). "Plasma Processing of Municipal Solid Waste". Brazilian Journal of Physics. 34 (4B): 1587–1593. Bibcode:2004BrJPh..34.1587L. doi:10.1590/S0103-97332004000800015. Retrieved 2013-03-20.

- Huang, Haitao; Lan Tang (2007). "Treatment of Organic Waste Using Thermal Plasma Pyrolysis Technology". Energy Conversion and Management. 48 (4): 1331–1337. doi:10.1016/j.enconman.2006.08.013. Retrieved 2013-03-12.

- "Alter NRG Announces Commissioning of Biomass Gasifier at Waste To Liquids Facility in China" (Press release). Alter NRG. Retrieved 2013-01-29.

- Messenger, Ben (12 April 2013). "Second Plasma Gasification Plant for Teesside Following Government Deal". Waste Management News.

- Bratsev, A. N.; V. E. Popov; A. F. Rutberg; S. V. Shtengel' (2006). "A Facility for Plasma Gasification of Waste of Various Types" (PDF). High temperature. 44 (6): 823–828. Retrieved 2013-03-12.

Electric Discharge in Gases: An Overview

Electric discharge in gases usually arises when current flows through a gaseous medium because of ionization of the gas. Electric arc, glow discharge, Paschen's law and Townsend discharge are the topics elucidates in the chapter. This section on electric discharge in gases offers an insightful focus, keeping in mind the complex subject matter.

Electric Discharge in Gases

Electric discharge in gases occurs when electric current flows through a gaseous medium due to ionization of the gas. Depending on several factors, the discharge may radiate visible light. The properties of electric discharges in gases are studied in connection with design of lighting sources and in the design of high voltage electrical equipment.

Discharge Types

In cold cathode tubes, the electric discharge in gas has three regions, with distinct current-voltage characteristics:

- I: Townsend discharge, below the breakdown voltage. At low voltages, the only current is that due to the generation of charge carriers in the gas by cosmic rays or other sources of ionizing radiation. As the applied voltage is increased, the free electrons carrying the current gain enough energy to cause further ionization, causing an electron avalanche. In this regime, the current increases from femtoamperes to microamperes, i.e. by nine orders of magnitude, for very little further increase in voltage. The voltage-current characteristics begins tapering off near the breakdown voltage and the glow becomes visible.

- II: glow discharge, which occurs once the breakdown voltage is reached. The voltage across the electrodes suddenly drops and the current increases to milliampere range. At lower currents, the voltage across the tube is almost current-independent; this is used in glow discharge voltage stabilizers. At lower currents, the area of the electrodes covered by the glow discharge is proportional to the current. At higher currents the normal glow turns into abnormal glow, the voltage across the tube gradually increases, and the glow discharge covers more and more of the surface of the electrodes. Low-power switching (glow-discharge thyratrons), voltage stabilization, and lighting applications (e.g. Nixie tubes, decatrons, neon lamps) operate in this region.

- III: arc discharge, which occurs in the ampere range of the current; the voltage across the tube drops with increasing current. High-current switching tubes, e.g. triggered spark gap, ignitron, thyratron and krytron (and its vacuum tube derivate, sprytron, using vacuum arc), high-power mercury-arc valves and high-power light sources, e.g. mercury-vapor lamps and metal halide lamps, operate in this range.

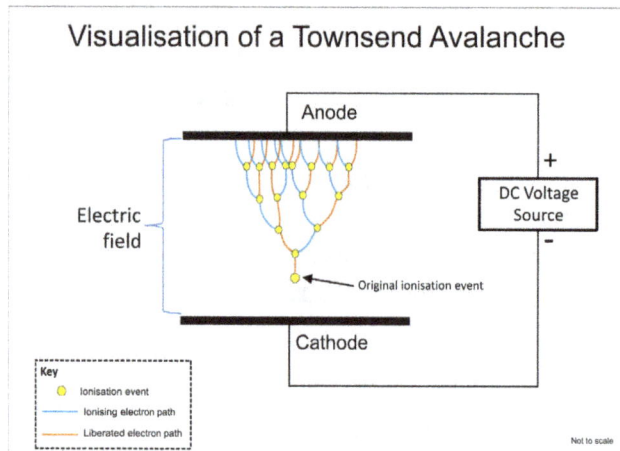

Visualisation of a Townsend Avalanche

Avalanche effect between two electrodes. The original ionisation event liberates one electron, and each subsequent collision liberates a further electron, so two electrons emerge from each collision: the ionising electron and the liberated electron.

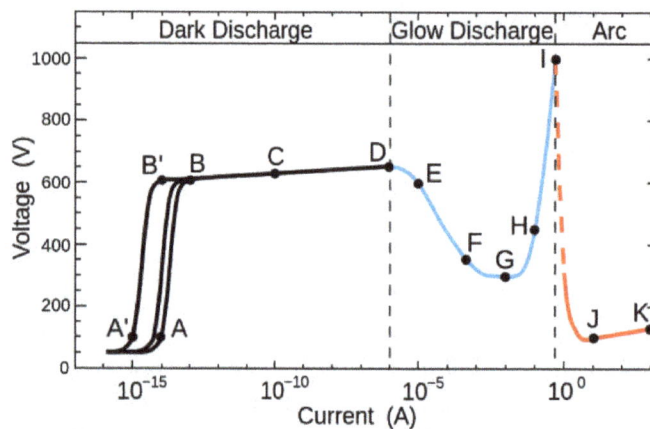

Voltage-current characteristics of electrical discharge in neon at 1 torr, with two planar electrodes separated by 50 cm.
A: random pulses by cosmic radiation
B: saturation current
C: avalanche Townsend discharge
D: self-sustained Townsend discharge
E: unstable region: corona discharge
F: sub-normal glow discharge
G: normal glow discharge
H: abnormal glow discharge
I: unstable region: glow-arc transition
J: electric arc
K: electric arc
The A-D region is called a dark discharge; there is some ionization, but the current is below 10 microamperes and there is no significant amount of radiation produced.
The F-H region is a region of glow discharge; the plasma emits a faint glow that occupies almost all the volume of the tube; most of the light is emitted by excited neutral atoms.
The I-K region is a region of arc discharge; the plasma is concentrated in a narrow channel along the center of the tube; a great amount of radiation is produced.

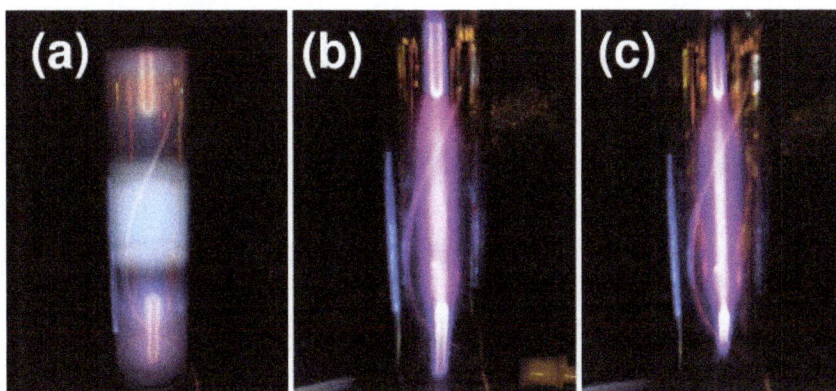

Transition from glow to arc discharge in argon, by increasing the gas pressure.

Glow discharge is facilitated by electrons striking the gas atoms and ionizing them. For formation of glow discharge, the mean free path of the electrons has to be reasonably long but shorter than the distance between the electrodes; glow discharges therefore do not readily occur at both too low and too high gas pressures.

The breakdown voltage for the glow discharge depends nonlinearly on the product of gas pressure and electrode distance according to Paschen's law. For a certain pressure × distance value, there is a lowest breakdown voltage. The increase of strike voltage for shorter electrode distances is related to too long mean free path of the electrons in comparison with the electrode distance.

A small amount of a radioactive element may be added into the tube, either as a separate piece of material (e.g. nickel-63 in krytrons) or as addition to the alloy of the electrodes (e.g. thorium), to preionize the gas and increase the reliability of electrical breakdown and glow or arc discharge ignition. A gaseous radioactive isotope, e.g. krypton-85, can also be used. Ignition electrodes and keepalive discharge electrodes can also be employed.

The E/N ratio between the electric field E and the concentration of neutral particles N is often used, because the mean energy of electrons (and therefore many other properties of discharge) is a function of E/N. Increasing the electric intensity E by some factor q has the same consequences as lowering gas density N by factor q.

Its SI unit is V·cm^2, but the Townsend unit (Td) is frequently used.

Application in Analog Computation

The use of a glow discharge for solution of certain mapping problems was described in 2002. According to a Nature news article describing the work, researchers at Imperial College London demonstrated how they built a mini-map that gives tourists luminous route indicators. To make the one-inch London chip, the team etched a plan of the city centre on a glass slide. Fitting a flat lid over the top turned the streets into hollow, connected tubes. They filled these with helium gas, and inserted electrodes at key tourist hubs. When a voltage is applied between two points, electricity naturally runs through the streets along the shortest route from A to B – and the gas glows like a tiny glowing strip light. The approach itself provides a novel visible analog computing approach for solving a wide class of maze searching problems based on the properties of lighting up of a glow discharge in a microfluidic chip.

Electric Arc

An electric arc or arc discharge is an electrical breakdown of a gas that produces an ongoing electrical discharge. The current through a normally nonconductive medium such as air produces a plasma; the plasma may produce visible light. An arc discharge is characterized by a lower voltage than a glow discharge, and it relies on thermionic emission of electrons from the electrodes supporting the arc. An archaic term is voltaic arc, as used in the phrase "voltaic arc lamp".

An electric arc between two nails

History

Lightning is a natural electric arc.

The phenomenon is believed to be first described by Sir Humphry Davy in an 1801 paper published in William Nicholson's Journal of Natural Philosophy, Chemistry and the Arts. However, Davy's description was not an electric arc as this phenomenon is considered by the modern science: "This is evidently the description, not of an arc, but of a spark. For the essence of an arc is that it should be continuous, and that the poles should not be in contact after it has once started. The spark produced by Sir Humphry Davy was plainly not continuous; and although the carbons remained red hot for some time after contact, there can have been no arc joining them, or so close an observer would have mentioned it". In the same year Davy publicly demonstrated the effect, before the Royal Society, by transmitting an electric current through two touching carbon rods and then pulling them a short distance apart. The demonstration produced a "feeble" arc, not readily distinguished from a sustained spark, between charcoal points. The Society subscribed for a more powerful battery of 1000 plates and in 1808 he demonstrated the large-scale arc. He is credited with naming the arc. He called it an arc because it assumes the shape of an upward bow when the distance between the electrodes is not small. This is due to the buoyant force on the hot gas. Independently the phenomenon was discovered in 1802 and described in 1803 as a "special fluid with electrical properties", by Vasily V. Petrov, a Russian scientist experimenting with a copper-zinc battery consisting of 4200 discs.

In the late nineteenth century, electric arc lighting was in wide use for public lighting. The tendency of electric arcs to flicker and hiss was a major problem. In 1895, Hertha Marks Ayrton wrote a series of articles for the Electrician, explaining that these phenomena were the result of oxygen coming into contact with the carbon rods used to create the arc. In 1899, she was the first woman ever to read her own paper before the Institution of Electrical Engineers (IEE). Her paper was entitled "The Hissing of the Electric Arc". Shortly thereafter, Ayrton was elected the first female member of the IEE; the next woman to be admitted to the IEE was in 1958. She petitioned to present a paper before the Royal Society but was not allowed because of her sex, and "The Mechanism of the Electric Arc" was read by John Perry in her stead in 1901.

Overview

An electric arc is the form of electric discharge with the highest current density. The maximum current through an arc is limited only by the external circuit, not by the arc itself.

An arc between two electrodes can be initiated by ionization and glow discharge, as the current through the electrodes is increased. The breakdown voltage of the electrode gap is a function of the pressure and type of gas surrounding the electrodes. When an arc starts, its terminal voltage is much less than a glow discharge, and current is higher. An arc in gases near atmospheric pressure is characterized by visible light emission, high current density, and high temperature. An arc is distinguished from a glow discharge partly by the approximately equal effective temperatures of both electrons and positive ions; in a glow discharge, ions have much less thermal energy than the electrons.

Electric arcs between the power line and pantographs of an electric train after catenary icing

Electricity arcs between the power rail and electrical pickup "shoe" on a London Underground train

A drawn arc can be initiated by two electrodes initially in contact and drawn apart; this can initiate an arc without the high-voltage glow discharge. This is the way a welder starts to weld a joint, momentarily touching the welding electrode against the workpiece then withdrawing it till a stable arc is formed. Another example is separation of electrical contacts in switches, relays and circuit breakers; in high-energy circuits arc suppression may be required to prevent damage to contacts.

Electrical resistance along the continuous electric arc creates heat, which ionizes more gas molecules (where degree of ionization is determined by temperature), and as per this sequence: solid-liquid-gas-plasma; the gas is gradually turned into a thermal plasma. A thermal plasma is in thermal equilibrium; the temperature is relatively homogeneous throughout the atoms, molecules, ions and electrons. The energy given to electrons is dispersed rapidly to the heavier particles by elastic collisions, due to their great mobility and large numbers.

Current in the arc is sustained by thermionic emission and field emission of electrons at the cathode. The current may be concentrated in a very small hot spot on the cathode; current densities on the order of one million amperes per square centimetre can be found. Unlike a glow discharge, an arc has little discernible structure, since the positive column is quite bright and extends nearly to the electrodes on both ends. The cathode fall and anode fall of a few volts occurs within a fraction of a millimetre of each electrode. The positive column has a lower voltage gradient and may be absent in very short arcs.

A low-frequency (less than 100 Hz) alternating current arc resembles a direct current arc; on each cycle, the arc is initiated by breakdown, and the electrodes interchange roles as anode and cathode as current reverses. As the frequency of the current increases, there is not enough time for all ionization to disperse on each half cycle and the breakdown is no longer needed to sustain the arc; the voltage vs. current characteristic becomes more nearly ohmic.

The various shapes of electric arcs are emergent properties of non-linear patterns of current and electric field. The arc occurs in the gas-filled space between two conductive electrodes (often made of tungsten or carbon) and it results in a very high temperature, capable of melting or vaporizing most materials. An electric arc is a continuous discharge, while the similar electric spark discharge is momentary. An electric arc may occur either in direct current circuits or in alternating current circuits. In the latter case, the arc may re-strike on each half cycle of the current. An electric arc differs from a glow discharge in that the current density is quite high, and the voltage drop within the arc is low; at the cathode, the current density can be as high as one megaampere per square centimeter.

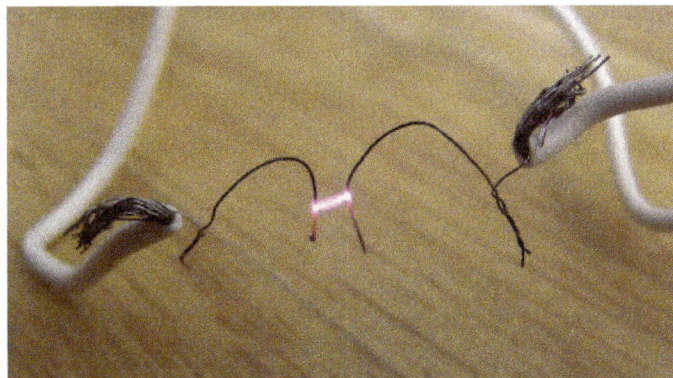

Electric arc between strands of wire.

An electric arc has a non-linear relationship between current and voltage. Once the arc is established (either by progression from a glow discharge or by momentarily touching the electrodes then separating them), increased current results in a lower voltage between the arc terminals. This negative resistance effect requires that some positive form of impedance—an electrical ballast—be placed in the circuit to maintain a stable arc. This property is the reason uncontrolled electrical arcs in apparatus become so destructive, since once initiated, an arc will draw more and more current from a fixed-voltage supply until the apparatus is destroyed.

Uses

Industrially, electric arcs are used for welding, plasma cutting, for electrical discharge machining, as an arc lamp in movie projectors and followspots in stage lighting. Electric arc furnaces are used to produce steel and other substances. Calcium carbide is made in this way as it requires a large amount of energy to promote an endothermic reaction (at temperatures of 2500 °C).

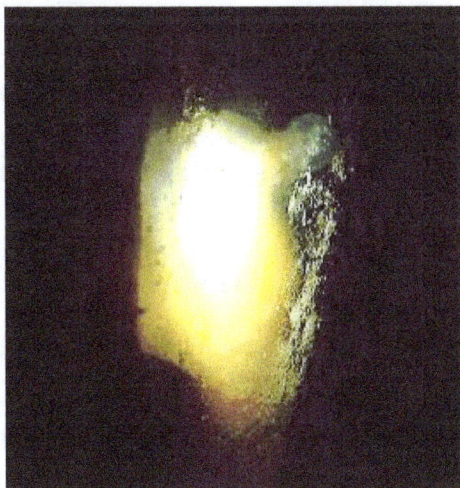

An electric arc can melt calcium oxide

Spark plugs are used in internal combustion engines of vehicles to initiate the combustion of the fuel in a timed fashion.

Spark gaps are also used in electric stove lighters (both external and built-in).

Carbon arc lights were the first electric lights. They were used for street lights in the 19th century and for specialized applications such as searchlights until World War 2. Today, low-pressure electric arcs are used in many applications. For example, fluorescent tubes, mercury, sodium, and metal halide lamps are used for lighting; xenon arc lamps are used for movie projectors.

Formation of an intense electric arc, similar to a small-scale arc flash, is the foundation of exploding-bridgewire detonators.

A major remaining application is in high voltage switchgear for high voltage transmission networks. Modern devices use Sulphur Hexafluoride at high pressure in a nozzle flow between separated electrodes within a pressure vessel. The AC fault current is interrupted at current zero by the highly electronegative SF6 ions absorbing free electrons from the decaying plasma. A similar air based technology has largely been replaced as many noisy units in series were required to prevent the current re-igniting under similar supergrid conditions.

Electric arcs have been studied for electric propulsion of spacecraft.

Guiding the Arc

Scientists have discovered a method to control the path of an arc between two electrodes by firing laser beams at the gas between the electrodes. The gas becomes a plasma and guides the arc. By constructing the plasma path between the electrodes with different laser beams, the arc can be formed into curved and S-shaped paths. The arc could also hit an obstacle and reform on the other side of the obstacle. The laser guided arc technology could be useful in applications to deliver a spark of electricity to a precise spot.

Undesired Arcing

Undesired or unintended electric arcing can have detrimental effects on electric power transmission, distribution systems and electronic equipment. Devices which may cause arcing include switches, circuit breakers, relay contacts, fuses and poor cable terminations. When an inductive circuit is switched off, the current cannot instantaneously jump to zero: a transient arc will be formed across the separating contacts. Switching devices susceptible to arcing are normally designed to contain and extinguish an arc, and snubber circuits can supply a path for transient currents, preventing arcing. If a circuit has enough current and voltage to sustain an arc formed outside of a switching device, the arc can cause damage to equipment such as melting of conductors, destruction of insulation, and fire. An arc flash describes an explosive electrical event that presents a hazard to people and equipment.

Undesired arcing in electrical contacts of contactors, relays and switches can be reduced by devices such as contact arc suppressors and RC Snubbers or through techniques including:

- immersion in transformer oil, dielectric gas or vacuum
- arc chutes

- magnetic blowouts

- pneumatic blowouts

- sacrificial ("arcing") contacts

- damping materials to absorb arc energy, either thermally or through chemical decomposition

Arcing can also occur when a low resistance channel (foreign object, conductive dust, moisture…) forms between places with different voltage. The conductive channel then can facilitate formation of an electric arc. The ionized air has high electrical conductivity approaching that of metals, and it can conduct extremely high currents, causing a short circuit and tripping protective devices (fuses and circuit breakers). A similar situation may occur when a lightbulb burns out and the fragments of the filament pull an electric arc between the leads inside the bulb, leading to overcurrent that trips the breakers.

Electric arc over the surface of plastics causes their degradation. A conductive carbon-rich track tends to form in the arc path, negatively influencing their insulation properties. The arc susceptibility is tested according to ASTM D495, by point electrodes and continuous and intermittent arcs; it is measured in seconds required to form a track that is conductive under high-voltage low-current conditions. Some materials are less susceptible to degradation than others. For example, polytetrafluoroethylene has arc resistance of about 200 seconds. From thermosetting plastics, alkyds and melamine resins are better than phenolic resins. Polyethylenes have arc resistance of about 150 seconds; polystyrenes and polyvinyl chlorides have relatively low resistance of about 70 seconds. Plastics can be formulated to emit gases with arc-extinguishing properties; these are known as *arc-extinguishing plastics*.

Arcing over some types of printed circuit boards, possibly due to cracks of the traces or the failure of a solder, renders the affected insulating layer conductive as the dielectric is combusted due to the high temperatures involved. This conductivity prolongs the arcing due to cascading failure of the surface.

Arc Suppression

Arc suppression is a method of attempting to reduce or eliminate the electrical arc. There are several possible areas of use of arc suppression methods, among them metal film deposition and sputtering, arc flash protection, electrostatic processes where electrical arcs are not desired (such as powder painting, air purification, PVDF film poling) and contact current arc suppression. In industrial, military and consumer electronic design, the latter method generally applies to devices such as electromechanical power switches, relays and contactors. In this context, arc suppression refers to the concept of contact protection.

Part of the energy of an electrical arc forms new chemical compounds from the air surrounding the arc: these include oxides of nitrogen and ozone, the second of which can be detected by its distinctive sharp smell. These chemicals can be produced by high-power contacts in relays and motor commutators, and they are corrosive to nearby metal surfaces. Arcing also erodes the surfaces of the contacts, wearing them down and creating high contact resistance when closed.

Glow Discharge

A glow discharge is a plasma formed by the passage of electric current through a low-pressure gas. It is created by applying a voltage between two metal electrodes in a glass tube containing gas. When the voltage exceeds a certain value called the striking voltage, the gas in the tube ionizes, becoming a plasma, and begins conducting electricity, causing it to glow with a colored light. The color depends on the gas used. Glow discharge is widely used as a source of light in devices such as neon lights, fluorescent lamps, and plasma-screen televisions. Analyzing the light produced by spectroscopy can reveal much about the atomic interactions in the gas, so glow discharge is used in plasma physics and analytical chemistry. It is also used in the surface treatment technique called sputtering.

NE-2 type neon lamp powered by alternating current

Basic Operating Mechanism

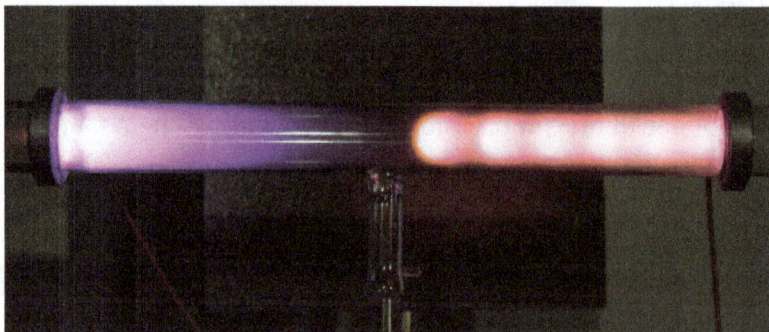

The simplest type of glow discharge is a direct-current glow discharge. In its simplest form, it consists of two electrodes in a cell held at low pressure (0.1–10 torr; about 1/10000th to 1/100th of atmospheric pressure). The cell is typically filled with neon, but other gases can also be used. An electric potential of several hundred volts is applied between the two electrodes. A small fraction of the population of atoms within the cell is initially ionized through random processes (thermal collisions between atoms or with gamma rays, for example). The ions (which are positively charged) are driven towards the cathode by the electric potential, and the electrons are driven towards the anode by the same potential. The initial population of ions and electrons collides with other atoms, ionizing them. As long as the potential is maintained, a population of ions and electrons remains.

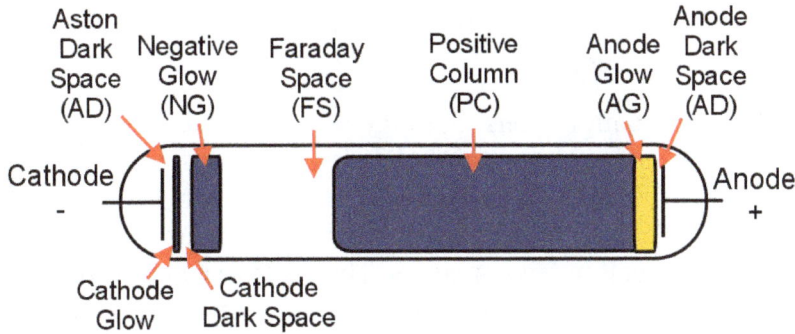

A Crookes tube illustrating the different glowing regions that make up a glow discharge and a diagram giving their names.

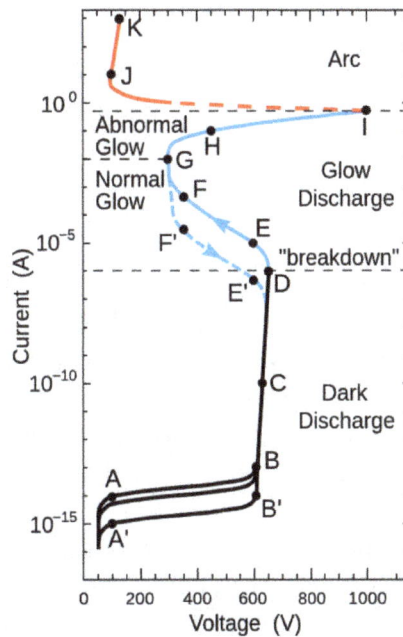

Current-voltage curve (characteristic curve) of a discharge in neon gas at a pressure of 1 torr, showing the different regions. The vertical (current) scale is logarithmic.

Some of the ions' kinetic energy is transferred to the cathode. This happens partially through the ions striking the cathode directly. The primary mechanism, however, is less direct. Ions strike the more numerous neutral gas atoms, transferring a portion of their energy to them. These neutral atoms then strike the cathode. Whichever species (ions or atoms) strike the cathode, collisions within the cathode redistribute this energy until a portion of the cathode is ejected, typically in the form of free atoms. This process is known as sputtering. Once free of the cathode, atoms move into the bulk of the glow discharge through drift and due to the energy they gained from sputtering. The atoms can then be excited by collisions with ions, electrons, or other atoms that have been previously excited by collisions. Once excited, atoms will lose their energy fairly quickly. Of the various ways that this energy can be lost, the most important is radiatively, meaning that a photon is released to carry the energy away. In optical atomic spectroscopy, the wavelength of this photon can be used to determine the identity of the atom (that is, which chemical element it is) and the number of photons is directly proportional to the concentration of that element in the sample. Some collisions (those of high enough energy) will cause ionization. In atomic mass spectrometry, these ions are detected. Their mass identifies the type of atoms and their quantity reveals the amount of that element in the sample.

The figure above shows the main regions that may be present in a glow discharge. Regions described as "glows" emit significant light; regions labeled as "dark spaces" do not. As the discharge becomes more extended (i.e., stretched horizontally in the geometry of the figure), the positive column may become striated. That is, alternating dark and bright regions may form. Compressing the discharge horizontally will result in fewer regions. The positive column will be compressed while the negative glow will remain the same size, and, with small enough gaps, the positive column will disappear altogether. In an analytical glow discharge, the discharge is primarily a negative glow with dark region above and below it.

Below the ionization voltage or breakdown voltage there is no glow, but as the voltage increases to the ionization point the Townsend discharge happens just as glow discharge becomes visible; this is the start of the normal glow range. As the voltage is increased above the normal glow range, abnormal glow begins. If the voltage is increased to the point the cathode glow covers the entire cathode arc discharge begins.

Use in Analytical Chemistry

Glow discharges can be used to analyze the elemental, and sometimes molecular, composition of solids, liquids, and gases, but elemental analysis of solids is the most common. In this arrangement, the sample is used as the cathode. As mentioned earlier, gas ions and atoms striking the sample surface knock atoms off of it, a process known as sputtering. The sputtered atoms, now in the gas phase, can be detected by atomic absorption, but this is a comparatively rare strategy. Instead, atomic emission and mass spectrometry are usually used. Collisions between the gas-phase sample atoms and the plasma gas pass energy to the sample atoms. This energy can excite the atoms, after which they can lose their energy through atomic emission. By observing the wavelength of the emitted light, the atom's identity can be determined. By observing the intensity of the emission, the concentration of atoms of that type can be determined. Energy gained through collisions can also ionize the sample atoms. The ions can then be detected by mass spectrometry. In this case, it is the mass of the ions that identify the element and the number of ions that reflect the concentration. This method is referred to as glow discharge mass spectrometry (GDMS) and it has detection limits down to the sub-ppb range for most elements that are nearly matrix-independent.

Both bulk and depth analysis of solids may be performed with glow discharge. Bulk analysis assumes that the sample is fairly homogeneous and averages the emission or mass spectrometric signal over time. Depth analysis relies on tracking the signal in time, therefore, is the same as tracking the elemental composition in depth. Depth analysis requires greater control over operational parameters. For example, conditions (current, potential, pressure) need to be adjusted so that the crater produced by sputtering is flat bottom (that is, so that the depth analyzed over the crater area is uniform). In bulk measurement, a rough or rounded crater bottom would not adversely impact analysis. Under the best conditions, depth resolution in the single nanometer range has been achieved (in fact, within-molecule resolution has been demonstrated).

The chemistry of ions and neutrals in vacuum is called gas phase ion chemistry and is part of the analytical study that includes glow discharge.

Powering Modes

In analytical chemistry, glow discharges are most often operated in direct-current mode. For this mode, the cathode (which is the sample in solids analysis) must be conductive. The potential, pressure, and current are interrelated. Only two can be directly controlled at once, while the third must be allowed to vary. The pressure is most typically held constant, but other schemes may be used. The pressure and current may be held constant, while potential is allowed to vary. The pressure and voltage may be held constant while the current is allowed to vary. The power (product of voltage and current) may be held constant while the pressure is allowed to vary.

DC powered neon lamp, showing glow discharge surrounding only the cathode

Glow discharges may also be operated in radio-frequency. The use of this frequency will establish a negative DC-bias voltage on the sample surface. The DC-bias is the result of an alternating current waveform that is centered about negative potential; as such it more or less represent the average potential residing on the sample surface. Radio-frequency has ability to appear to flow through insulators (non-conductive materials).

Both radio-frequency and direct-current glow discharges can be operated in pulsed mode, where the potential is turned on and off. This allows higher instantaneous powers to be applied without excessively heating the cathode. These higher instantaneous powers produce higher instantaneous signals, aiding detection. Combining time-resolved detection with pulsed powering results in additional benefits. In atomic emission, analyte atoms emit during different portions of the pulse than background atoms, allowing the two to be discriminated. Analogously, in mass spectrometry, sample and background ions are created at different times.

Types

There are various types of glow discharge examples include: high pressure glow discharge, hollow cathode discharge, spray discharge.

Application to Analog Computing

An interesting application for using glow discharge was described in a 2002 scientific paper by Ryes, Ghanem *et al*. According to a Nature news article describing the work, researchers at Imperial College London demonstrated how they built a mini-map that glows along the shortest route between two points. The Nature news article describes the system as follows:

> To make the one-inch London chip, the team etched a plan of the city centre on a glass slide. Fitting a flat lid over the top turned the streets into hollow, connected tubes. They filled these with helium gas, and inserted electrodes at key tourist hubs. When a voltage is applied between two points, electricity naturally runs through the streets along the shortest route from A to B - and the gas glows like a tiny neon strip light.

The approach itself provides a novel visible analog computing approach for solving a wide class of maze searching problems based on the properties of lighting up of a glow discharge in a microfluidic chip.

Application to Voltage Regulation

In the mid-20th century, prior to the development of solid state components such as Zener diodes, voltage regulation in circuits was often accomplished with voltage-regulator tubes, which used glow discharge.

A 5651 voltage-regulator tube in operation

Paschen's Law

Paschen's Law is an equation that gives the breakdown voltage, that is the voltage necessary to start a discharge or electric arc, between two electrodes in a gas as a function of pressure and gap length. It is named after Friedrich Paschen who discovered it empirically in 1889.

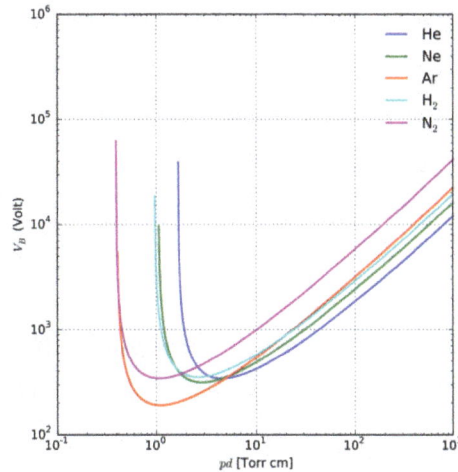

Paschen curves obtained for helium, neon, argon, hydrogen and nitrogen, using the expression for the breakdown voltage as a function of the parameters A,B that interpolate the first Townsend coefficient.

Paschen studied the breakdown voltage of various gases between parallel metal plates as the gas pressure and gap distance were varied. The voltage necessary to arc across the gap decreased as the pressure was reduced and then increased gradually, exceeding its original value. He also found that at normal pressure, the voltage needed to cause an arc reduced as the gap size was reduced but only to a point. As the gap was reduced further, the voltage required to cause an arc began to rise and again exceeded its original value. For a given gas, the voltage is a function only of the product of the pressure and gap length. The curve he found of voltage versus the pressure-gap length product *(right)* is called Paschen's curve. He found an equation that fit these curves, which is now called Paschen's law.

At higher pressures and gap lengths, the breakdown voltage is approximately *proportional* to the product of pressure and gap length, and the term Paschen's law is sometimes used to refer to this simpler relation. However this is only roughly true, over a limited range of the curve.

Paschen Curve

Early vacuum experimenters found a rather surprising behavior. An arc would sometimes take place in a long irregular path rather than at the minimum distance between the electrodes. For example, in air, at a pressure of 10^{-3} atmospheres, the distance for minimum breakdown voltage is about 7.5 μm. The voltage required to arc this distance is 327 V which is insufficient to ignite the arcs for gaps that are either wider or narrower. For a 3.5 μm gap, the required voltage is 533 V, nearly twice as much. If 500 V were applied, it would not be sufficient to arc at the 2.85 μm distance, but would arc at a 7.5 μm distance.

It was found that breakdown voltage was described by the equation:

$$V_B = \frac{Bpd}{\ln(Apd) - ln[ln(1+\dfrac{1}{\gamma_{se}})]}$$

Where V is the breakdown voltage in Volts, γ_{se} is the pressure in Pascals, d is the gap distance in meters, γ_{se} is the secondary electron emission coefficient at the cathode, A is the saturation ionization in the gas at a particular E/p (electric field/pressure), and B is related to the excitation and ionization energies. The constants A and B are determined experimentally and found to be roughly constant over a restricted range of E/p for any given gas. For example, air with an E/p in the range of 450 to 7500 V·(kPa·cm)⁻¹, A = 112.50 (kPa·cm)⁻¹ and B = 2737.50 V·(kPa·cm)⁻¹. The graph of this equation is the Paschen curve. By differentiating it with respect to pd and setting the derivative to zero, the minimum voltage can be found. This yields

$$pd = e^{1-b}$$

and predicts the occurrence of a minimum breakdown voltage for pd = 7.5×10⁻⁶ m·atm. This is 327 V in air at standard atmospheric pressure at a distance of 7.5 µm. The composition of the gas determines both the minimum arc voltage and the distance at which it occurs. For argon, the minimum arc voltage is 137 V at a larger 12 µm. For sulfur dioxide, the minimum arc voltage is 457 V at only 4.4 µm.

For air at STP, the voltage needed to arc a 1-meter gap is about 3.4 MV. The intensity of the electric field for this gap is therefore 3.4 MV/m. The electric field needed to arc across the minimum voltage gap is much greater than what is necessary to arc a gap of one meter. For a 7.5 µm gap the arc voltage is 327 V which is 43 MV/m. This is about 13 times greater than the field strength for the 1 meter gap. The phenomenon is well verified experimentally and is referred to as the Paschen minimum. The equation loses accuracy for gaps under about 10 µm in air at one atmosphere and incorrectly predicts an infinite arc voltage at a gap of about 2.7 micrometers. Breakdown voltage can also differ from the Paschen curve prediction for very small electrode gaps when field emission from the cathode surface becomes important.

Physical Mechanism

The mean free path of a molecule in a gas is the average distance between its collision with other molecules. This is inversely proportional to the pressure of the gas. In air the mean free path of molecules is about 96 nm. Since electrons are much faster, their average distance between colliding with molecules is about 5.6 times longer or about 0.5 µm. This is a substantial fraction of the 7.5 µm spacing between the electrodes for minimum arc voltage. If the electron is in an electric field of 43 MV/m, it will be accelerated and acquire 21.5 electron volts of energy in 0.5 µm of travel in the direction of the field. The first ionization energy needed to dislodge an electron from nitrogen is about 15 eV. The accelerated electron will acquire more than enough energy to ionize a nitrogen atom. This liberated electron will in turn be accelerated which will lead to another collision. A chain reaction then leads to avalanche breakdown and an arc takes place from the cascade of released electrons.

More collisions will take place in the electron path between the electrodes in a higher pressure gas. When the pressure-gap product pd is high, an electron will collide with many different gas molecules as it travels from the cathode to the anode. Each of the collisions randomizes the electron direction, so the electron is not always being accelerated by the electric field—sometimes it travels back towards the cathode and is decelerated by the field.

Collisions reduce the electron's energy and make it more difficult for it to ionize a molecule. Energy losses from a greater number of collisions require larger voltages for the electrons to accumulate sufficient energy to ionize many gas molecules, which is required to produce an avalanche breakdown.

On the left side of the Paschen minimum, the pd product is small. The electron mean free path can become long compared to the gap between the electrodes. In this case, the electrons might gain lots of energy, but have fewer ionizing collisions. A greater voltage is therefore required to assure ionization of enough gas molecules to start an avalanche.

Derivation

Basics

To calculate the breakthrough voltage a homogeneous electrical field is assumed. This is the case in a parallel plate capacitor setup. The electrodes may have the distance d. The cathode is located at the point $x = 0$.

To get impact ionization the electron energy E_e must become greater than the ionization energy E_I of the gas atoms between the plates. Per length of path x a number of α ionizations will occur. α is known as the first Townsend coefficient as it was introduced by Townsend in, section 17. The increase of the electron current Γ_e can be described for the assumed setup as

$$\Gamma_e(x = d) = \Gamma_e(x = 0)e^{\alpha d} \qquad (1)$$

(So the number of free electrons at the anode is equal to the number of free electrons at the cathode that were multiplied by impact ionization. The larger d and/or α the more free electrons are created.)

The number of created electrons is

$$\Gamma_e(d) - \Gamma_e(0) = \Gamma_e(0)\left(e^{\alpha d} - 1\right) \qquad (2)$$

Neglecting possible multiple ionizations of the same atom, the number of created ions is the same as the number of created electrons:

$$\Gamma_i(0) - \Gamma_i(d) = \Gamma_e(0)\left(e^{\alpha d} - 1\right) \qquad (3)$$

Γ_i is the ion current. To keep the discharge going on, free electrons must be created at the cathode surface. This is possible because the ions hitting the cathode release secondary electrons at the impact. (For very large applied voltages also field electron emission can occur.) Without field emission, we can write

$$\Gamma_e(0) = \gamma \Gamma_i(0) \qquad (4)$$

where   is the mean number of generated secondary electrons per ion. This is also known as the second Townsend coefficient. Assuming that $\Gamma_i(d) = 0$ one gets the relation between the Townsend coefficients by putting (4) into (3) and transforming:

$$\alpha d = \ln\left(1 + \frac{1}{\gamma}\right) \qquad (5)$$

Impact Ionization

What is the amount of α? The number of ionization depends upon the probability that an electron hits an ion. This probability P is the relation of the cross-sectional area of a collision between electron and ion σ in relation to the overall area A that is available for the electron to fly through:

$$P = \frac{N\sigma}{A} = \frac{x}{\lambda} \qquad (6)$$

As expressed by the second part of the equation, it is also possible to express the probability as relation of the path traveled by the electron x to the mean free path λ (distance at which another collision occurs).

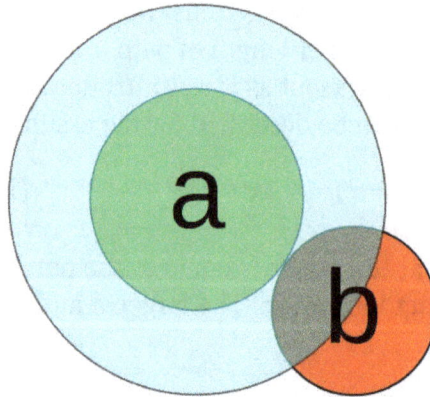

Visualization of the cross-section σ: If the center of particle b penetrates the blue circle, a collision occurs with particle a. So the area of the circle is the cross-section and its radius r is the sum of the radii of the particles.

N is the number of molecules which electrons can hit. It can be calculated using the equation of state of the ideal gas

$$pV = Nk_BT \qquad (7)$$

(p : pressure, V : volume, k_B : Boltzmann constant, T : temperature)

The adjoining sketch illustrates that $\sigma = \pi(r_a + r_b)^2$. As the radius of an electron can be neglected compared to the radius of an ion r_I it simplifies to $\sigma = \pi r_I^2$. Using this relation, putting (7) into (6) and transforming to λ one gets

$$\lambda = \frac{k_BT}{p\pi r_I^2} = \frac{1}{L \cdot p} \qquad (8)$$

where the factor L was only introduced for a better overview.

The alteration of the current of not yet collided electrons at every point in the path x can be expressed as

$$d\Gamma_e(x) = -\Gamma_e(x)\frac{dx}{\lambda_e} \qquad (9)$$

This differential equation can easily be solved:

$$\Gamma_e(x) = \Gamma_e(0)\exp\left(-\frac{x}{\lambda_e}\right) \qquad (10)$$

The probability that $\lambda > x$ (that there was not yet a collision at the point) is

$$P(\lambda > x) = \frac{\Gamma_e(x)}{\Gamma_e(0)} = \exp\left(-\frac{x}{\lambda_e}\right) \qquad (11)$$

According to its definition α is the number of ionizations per length of path and thus the relation of the probability that there was no collision in the mean free path of the ions, and the mean free path of the electrons:

$$\alpha = \frac{P(\lambda > \lambda_I)}{\lambda_e} = \frac{1}{\lambda_e}\exp\left(0\frac{\lambda_I}{\lambda_e}\right) = \frac{1}{\lambda_e}\exp\left(0\frac{E_I}{E_e}\right) \qquad (12)$$

It was hereby considered that the energy E that a charged particle can get between a collision depends on the electric field strength $\mathcal{E}$ and the charge Q:

$$E = \lambda Q\mathcal{E} \qquad (13)$$

Breakdown Voltage

For the parallel-plate capacitor we have $\mathcal{E} = \frac{U}{d}$, where U is the applied voltage. As a single ionization was assumed Q is the elementary charge e. We can now put (13) and (8) into (12) and get

$$\alpha = L\cdot p\exp\left(-\frac{L\cdot p\cdot d\cdot E_I}{eU}\right) \qquad (14)$$

Putting this into (5) and transforming to U we get the Paschen law for the breakdown voltage $U_{breakdown}$ that was first investigated by Paschen in and whose formula was first derived by Townsend in, section 227:

$$U_{breakdown} = \frac{L\cdot p\cdot d\cdot E_I}{e\left(\ln(L\cdot p\cdot d) - \ln\left(\ln\left(1+\gamma^{-1}\right)\right)\right)} \qquad (15)$$

$$\text{with } L = \frac{\pi r_I^2}{k_B T}$$

Plasma Ignition

Plasma ignition in definition of Townsend (Townsend discharge) is a self-sustaining discharge, independent of an external source of free electrons. This means that electrons from the cathode can reach the anode in the distance d and ionize at least one atom on its way. So according to the definition of α this relation must be fulfilled:

$$\alpha d \geq 1 \qquad (16)$$

If $\alpha d = 1$ is used instead of (5) one gets for the breakdown voltage

$$U_{\text{breakdown Townsend}} = \frac{L \cdot p \cdot d \cdot E_I}{e \cdot \ln(L \cdot p \cdot d)} = \frac{d \cdot E_I}{e \cdot \lambda_e \ln\left(\frac{d}{\lambda_e}\right)} \qquad (17)$$

Conclusions / Validity

Paschen's law requires that

- there are already free electrons at the cathode ($\tilde{A}_e(x=0) \neq 0$) which can be accelerated to trigger impact ionization. Such so-called *seed electrons* can be created by ionization by cosmic x-ray background.

- the creation of further free electrons is only achieved by impact ionization. Thus Paschen's law is not valid if there are external electron sources. This can for example be a light source creating secondary electrons via the photoelectric effect. This has to be considered in experiments.

- each ionized atom leads to only one free electron. But multiple ionizations occur always in practice.

- free electrons at the cathode surface are created by the impacting ions. The problem is that the number of thereby created electrons strongly depends on the material of the cathode, its surface (roughness, impurities) and the environmental conditions (temperature, humidity etc.). The experimental, reproducible determination of the factor is therefore nearly impossible.

- the electrical field is homogeneous.

Effects with Different Gases

Different gases will have different mean free paths for molecules and electrons. This is because different molecules have different diameters. Noble gases like helium and argon are monatomic and tend to have smaller diameters. This gives them a greater mean free path length.

Ionization potentials differ between molecules as well as the speed that they recapture electrons after they have been knocked out of orbit. All three effects change the number of collisions needed to cause an exponential growth in free electrons. These free electrons are necessary to cause an arc.

Townsend Discharge

The Townsend discharge or Townsend avalanche is a gas ionisation process where free electrons are accelerated by an electric field, collide with gas molecules, and consequently free additional electrons. Those electrons are in turn accelerated and free additional electrons. The result is an avalanche multiplication that permits electrical conduction through the gas. The discharge requires a source of free electrons and a significant electric field; without both, the phenomenon does not occur.

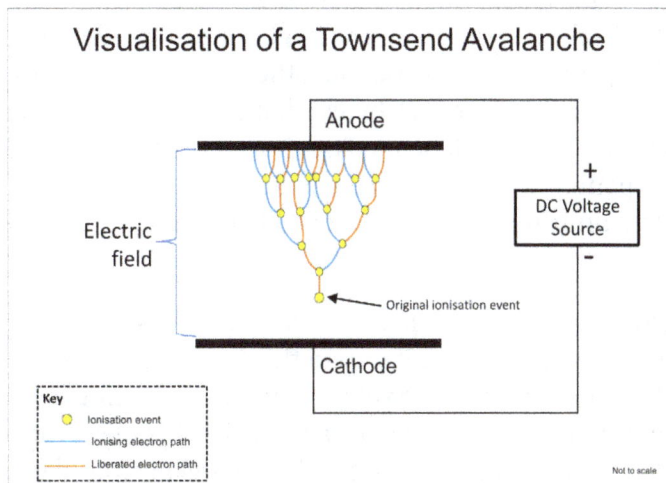

Avalanche effect in gas subject to ionising radiation between two plate electrodes. The original ionisation event liberates one electron, and each subsequent collision liberates a further electron, so two electrons emerge from each collision to sustain the avalanche.

The Townsend discharge is named after John Sealy Townsend, who discovered the fundamental ionisation mechanism by his work between 1897 and 1901.

General Description of the Phenomenon

The avalanche occurs in a gaseous medium that can be ionised (such as air). The electric field and the mean free path of the electron must allow free electrons to acquire an energy level (velocity) that can cause impact ionisation. If the electric field is too small, then the electrons do not acquire enough energy. If the mean free path is too short, the electron gives up its acquired energy in a series of non-ionising collisions. If the mean free path is too long, then the electron reaches the anode before colliding with another molecule.

The avalanche mechanism is shown in the accompanying diagram. The electric field is applied across a gaseous medium; initial ions are created with ionising radiation (for example, cosmic rays). An original ionisation event produces an ion pair; the positive ion accelerates towards the cathode while the free electron accelerates towards the anode. If the electric field is strong enough, the free electron can gain sufficient velocity (energy) to liberate another electron when it next collides with a molecule. The two free electrons then travel towards the anode and gain sufficient energy from the electric field to cause further impact ionisations, and so on. This process is effectively a chain reaction that generates free electrons. The total number of electrons reaching the anode is equal to the number of collisions, plus the single initiating free electron. Initially, the number of collisions grows exponentially. The limit to the multiplication in an electron avalanche is known as the Raether limit.

The Townsend avalanche can have a large range of current densities. In common gas-filled tubes, such as those used as gaseous ionisation detectors, magnitudes of currents flowing during this process can range from about 10^{-18} amperes to about 10^{-5} amperes.

Quantitative Description of the Phenomenon

Townsend's early experimental apparatus consisted of planar parallel plates forming two sides of a chamber filled with a gas. A direct current high-voltage source was connected between the plates; the lower voltage plate being the cathode while the other was the anode. He forced the cathode to emit electrons using the photoelectric effect by irradiating it with X-rays, and he found that the current I flowing through the chamber depended on the electric field between the plates. However, this current showed an exponential increase as the plate gaps became small, leading to the conclusion that the gas ions were multiplying as they moved between the plates due to the high electric field.

Townsend observed currents varying exponentially over ten or more orders of magnitude with a constant applied voltage when the distance between the plates was varied. He also discovered that gas pressure influenced conduction: he was able to generate ions in gases at low pressure with a much lower voltage than that required to generate a spark. This observation overturned conventional thinking about the amount of current that an irradiated gas could conduct.

The experimental data obtained from his experiments are described by the following formula

$$\frac{I}{I_0} = e^{\alpha_n d},$$

where

- I is the current flowing in the device,

- I_0 is the photoelectric current generated at the cathode surface,

- e is Euler's number

- α_n is the *first Townsend ionisation coefficient*, expressing the number of ion pairs generated per unit length (e.g. meter) by a negative ion (anion) moving from cathode to anode,

- d is the distance between the plates of the device.

The almost constant voltage between the plates is equal to the breakdown voltage needed to create a self-sustaining avalanche: it *decreases* when the current reaches the glow discharge regime.

Subsequent experiments revealed that the current I rises faster than predicted by the above formula as the distance d increases: two different effects were considered in order to better model the discharge: positive ions and cathode emission.

Gas Ionisation Caused by Motion of Positive Ions

Townsend put forward the hypothesis that positive ions also produce ion pairs, introducing a coefficient α_p expressing the number of ion pairs generated per unit length by a positive ion (cation) moving from anode to cathode. The following formula was found

$$\frac{I}{I_0} = \frac{(\alpha_n - \alpha_p)e^{(\alpha_n - \alpha_p)d}}{\alpha_n - \alpha_p e^{(\alpha_n - \alpha_p)d}} \quad \Rightarrow \quad \frac{I}{I_0} \cong \frac{e^{\alpha_n d}}{1 - (\alpha_p / \alpha_n)e^{\alpha_n d}}$$

since $\alpha_p \ll \alpha_n$, in very good agreement with experiments.

The *first Townsend coefficient* (α), also known as *first Townsend avalanche coefficient* is a term used where secondary ionisation occurs because the primary ionisation electrons gain sufficient energy from the accelerating electric field, or from the original ionising particle. The coefficient gives the number of secondary electrons produced by primary electron per unit path length.

Cathode Emission Caused by Impact of Ions

Townsend, Holst and Oosterhuis also put forward an alternative hypothesis, considering the augmented emission of electrons by the cathode caused by impact of positive ions. This introduced *Townsend's second ionisation coefficient* $\dot{o}_i$; the average number of electrons released from a surface by an incident positive ion, according to the following formula:

$$\frac{I}{I_0} = \frac{e^{\alpha_n d}}{1 - \epsilon_i \left(e^{\alpha_n d} - 1\right)}.$$

These two formulas may be thought as describing limiting cases of the effective behavior of the process: either can be used to describe the same experimental results. Other formulas describing various intermediate behaviors are found in the literature, particularly in reference 1 and citations therein.

Conditions

A Townsend discharge can be sustained only over a limited range of gas pressure and electric field intensity. The accompanying plot shows the variation of voltage drop and the different operating regions for a gas-filled tube with a constant pressure, but a varying current between its electrodes. The Townsend avalanche phenomena occurs on the sloping plateau B-D. Beyond D the ionisation is sustained.

At higher pressures, discharges occur more rapidly than the calculated time for ions to traverse the gap between electrodes, and the streamer theory of spark discharge of Raether, Meek and Loeb is applicable. In highly non-uniform electric fields, the corona discharge process is applicable. See Electron avalanche for further description of these mechanisms.

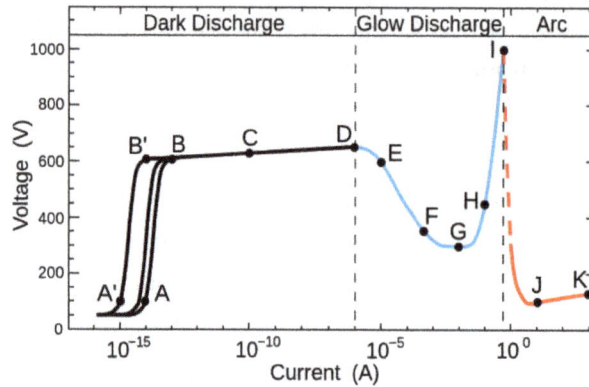

Voltage-current characteristics of electrical discharge in neon at 1 torr, with two planar electrodes separated by 50 cm.
A: random pulses by cosmic radiation
B: saturation current
C: avalanche Townsend discharge
D: self-sustained Townsend discharge
E: unstable region: corona discharge
F: sub-normal glow discharge
G: normal glow discharge
H: abnormal glow discharge
I: unstable region: glow-arc transition
J: electric arc
K: electric arc
A-D region: dark discharge; ionisation occurs, current below 10 microamps.
F-H region: glow discharge; the plasma emits a faint glow.
I-K region: arc discharge; larges amounts of radiation produced.

Discharges in vacuum require vaporization and ionisation of electrode atoms. An arc can be initiated without a preliminary Townsend discharge; for example when electrodes touch and are then separated.

Applications

Gas-discharge Tubes

The starting of Townsend discharge sets the upper limit to the blocking voltage a glow discharge gas-filled tube can withstand. This limit is the Townsend discharge breakdown voltage, also called ignition voltage of the tube.

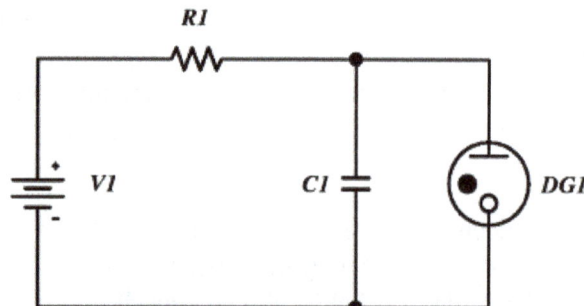

Neon lamp/cold-cathode gas diode relaxation oscillator

The occurrence of Townsend discharge, leading to glow discharge breakdown shapes the current-voltage characteristic of a gas discharge tube such as a neon lamp in a way such that it has a negative differential resistance region of the S-type. The negative resistance can be used to generate electrical oscillations and waveforms, as in the relaxation oscillator whose schematic is shown in the picture on the right. The sawtooth shaped oscillation generated has frequency

$$f \cong \frac{1}{R_1 C_1 \ln \dfrac{V_1 - V_{\text{GLOW}}}{V_1 - V_{\text{TWN}}}},$$

where

- V_{GLOW} is the glow discharge breakdown voltage,

- V_{TWN} is the Townsend discharge breakdown voltage,

- C_1, R_1 and V_1 are respectively the capacitance, the resistance and the supply voltage of the circuit.

Since temperature and time stability of the characteristics of gas diodes and neon lamps is low, and also the statistical dispersion of breakdown voltages is high, the above formula can only give a qualitative indication of what the real frequency of oscillation is.

Gas Phototubes

Avalanche multiplication during Townsend discharge is naturally used in gas phototubes, to amplify the photoelectric charge generated by incident radiation (visible light or not) on the cathode: achievable current is typically 10~20 times greater respect to that generated by vacuum phototubes.

Ionising Radiation Detectors

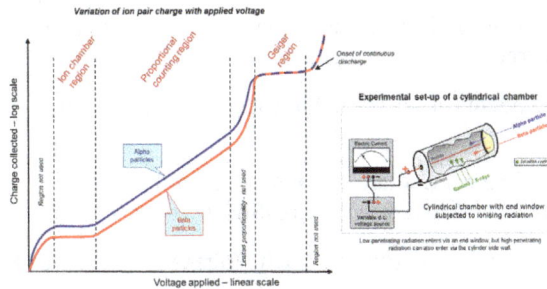

Plot of variation of ionisation current against applied voltage for a co-axial wire cylinder gaseous radiation detector.

Townsend avalanche discharges are fundamental to the operation of gaseous ionisation detectors such as the Geiger–Müller tube and the proportional counter in either detecting ionising radiation or measuring its energy. The incident radiation will ionise atoms or molecules in the gaseous medium to produce ion pairs, but different use is made by each detector type of the resultant avalanche effects.

In the case of a GM tube the high electric field strength is sufficient to cause complete ionisation of the fill gas surrounding the anode from the initial creation of just one ion pair. The GM tube output carries information that the event has occurred, but no information about the energy of the incident radiation.

In the case of proportional counters, multiple creation of ion pairs occurs in the "ion drift" region near the cathode. The electric field and chamber geometries are selected so that an "avalanche region" is created in the immediate proximity of the anode. A negative ion drifting towards the anode enters this region and creates a localised avalanche that is independent of those from other ion pairs, but which can still provide a multiplication effect. In this way spectroscopic information on the energy of the incident radiation is available by the magnitude of the output pulse from each initiating event.

The accompanying plot shows the variation of ionisation current for a co-axial cylinder system. In the ion chamber region, there are no avalanches and the applied voltage only serves to move the ions towards the electrodes to prevent re-combination. In the proportional region, localised avalanches occur in the gas space immediately round the anode which are numerically proportional to the number of original ionising events. Increasing the voltage further increases the number of avalanches until the Geiger region is reached where the full volume of the fill gas around the anodes ionised, and all proportional energy information is lost. Beyond the Geiger region the gas is in continuous discharge owing to the high electric field strength.

References

- Reference Data for Engineers: Radio, Electronics, Computers and Communications By Wendy Middleton, Mac E. Van Valkenburg, p. 16-42, Newnes, 2002 ISBN 0-7506-7291-9

- Handbook of optoelectronics, Volume 1 by John Dakin, Robert G. W. Brown, p. 52, CRC Press, 2006 ISBN 0-7503-0646-7

- Kartsev, V. P. (1983). Shea, William R, ed. Nature Mathematized. Boston, MA: Kluwer Academic. p. 279. ISBN 90-277-1402-9.

- Howatson, A.M. (1965). An Introduction to Gas Discharges. Oxford: Pergamon Press. pp. 47–101. ISBN 0-08-020575-5.

- Mehta, V.K. (2005). Principles of Electronics: for Diploma, AMIE, Degree & Other Engineering Examinations (9th ed., multicolour illustrative ed.). New Delhi: S. Chand. pp. 101–107. ISBN 81-219-2450-2.

- Davy, Humphry (1812). Elements of Chemical Philosophy. p. 85. ISBN 0-217-88947-6. This is the likely origin of the term arc.

- Lieberman, Michael A.; Lichtenberg, Allan J. (2005). Principles of plasma discharges and materials processing (2nd ed.). Hoboken, N.J.: Wiley-Interscience. 546. ISBN 978-0471005773. OCLC 59760348.

- Kuffel, E.; Zaengl, W. S.; Kuffel, J. (2004). High Voltage Engineering Fundamentals (2nd ed.). Butterworth-Heinemann. ISBN 0-7506-3634-3.

- "Lab Note #106 Environmental Impact of Arc Suppression". Arc Suppression Technologies. April 2011. Retrieved October 10, 2011.

Fusion Power: An Essential Aspect

This section focuses on all the themes concerned with fusion power. Fusion power is the energy that is produced by nuclear fusion. Pinch and types of pinch have also been explicated in the text. This chapter has been carefully written to provide an easy understanding of the varied facets of fusion power.

Fusion Power

Fusion power is energy generated by nuclear fusion. Fusion reactions fuse two lighter atomic nuclei to form a heavier nucleus. It is a major area of plasma physics research that attempts to harness such reactions as a source of large scale sustainable energy. Fusion reactions are how stars transmute matter into energy.

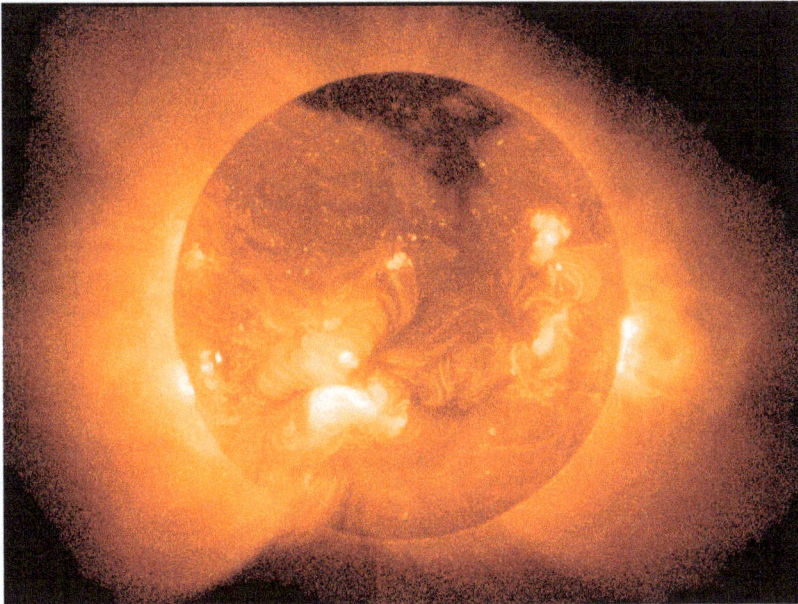

The Sun is a natural fusion reactor.

In most large scale commercial programs, heat from neutron scattering in a controlled reaction is used to operate a steam turbine that drives electric generators. Many fusion concepts are under investigation. The current leading designs are the tokamak and inertial confinement by laser. As of January 2016, these technologies were not viable, as they cannot produce more energy than is required to initiate and sustain a fusion reaction.

Alternative approaches rely on other means of energy transfer, mostly that capture energy without relying on neutron capture.

Background

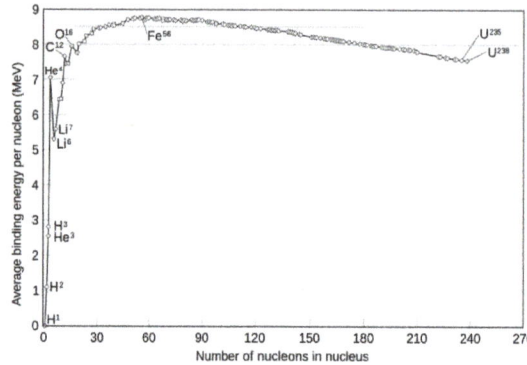

Binding energy for different atoms. Iron-56 has the highest, making it the most stable. Atoms to the left are likely to fuse; atoms to the right are likely to split.

Mechanism

Fusion reactions occur when two (or more) atomic nuclei come close enough for long enough that the strong nuclear force pulling them together exceeds the electrostatic force pushing them apart, fusing them into heavier nuclei. For nuclei lighter than iron-56, the reaction is exothermic, releasing energy. For nuclei heavier than iron-56, the reaction is endothermic, requiring an external source of energy. Hence, nuclei smaller than iron-56 are more likely to fuse while those heavier than iron-56 are more likely to break apart.

The strong force acts only over short distances. The repulsive electrostatic force acts over longer distances, so kinetic energy is needed to overcome this "Coulomb barrier" before the reaction can take place. Way of doing this include speeding up atoms in a particle accelerator, or heating them to high temperatures.

Once an atom is heated above its ionization energy, its electrons are stripped away (it is ionized), leaving just the bare nucleus (the ion). The result is a hot cloud of ions and the electrons formerly attached to them. This cloud is known as a plasma. Because the charges are separated, plasmas are electrically conductive and magnetically controllable. Many fusion devices take advantage of this to control the particles as they are heated.

Cross Section

A reaction's cross section, denoted σ, is the measure of the probability that a fusion reaction will happen. This depends on the relative velocity of the two nuclei. Higher relative velocities increase the probability. Cross sections for many fusion reactions were measured (mainly in the 1970s) using particle beams.

In a plasma, particle velocity can be characterized using a probability distribution. If the plasma is thermalized, the distribution looks like a bell curve, or maxwellian distribution. In this case, it is useful to take the average cross section over the velocity distribution. This is entered into the volumetric fusion rate:

$$P_{\text{fusion}} = n_A n_B \langle \sigma v_{A,B} \rangle E_{\text{fusion}}$$

where:

- P_{fusion} is the energy made by fusion, per time and volume

- n is the number density of species A or B, the particles in the volume

- $\langle \sigma v_{A,B} \rangle$ is the cross section of that reaction, average over all the velocities of the two species v

- E_{fusion} is the energy released by that fusion reaction.

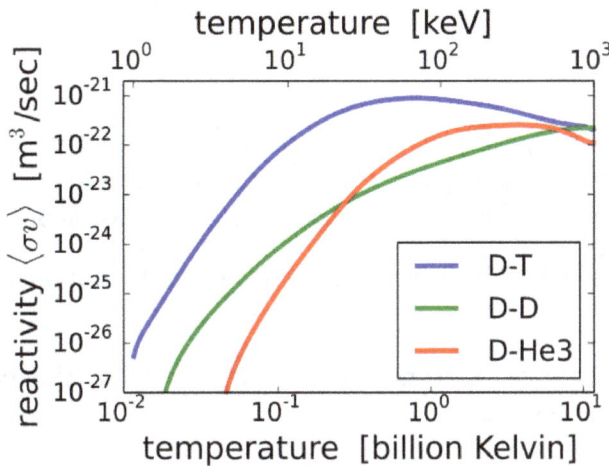

The fusion reaction rate increases rapidly with temperature until it maximizes and then gradually drops off. The deuterium-tritium fusion rate peaks at a lower temperature (about 70 keV, or 800 million kelvin) and at a higher value than other reactions commonly considered for fusion energy.

Lawson Criterion

The Lawson Criterion shows how energy varies with temperature, density, speed of collision and fuel. This equation was central to John Lawson's analysis of fusion working with a hot plasma. Lawson assumed an energy balance.

*Net Power = Efficiency * (Fusion - Radiation Loss - Conduction Loss)*

- *Net Power* is the net power for any fusion power station.

- *Efficiency* how much energy is needed to drive the device and how well it collects power.

- *Fusion* is rate of energy generated by the fusion reactions.

- *Radiation* is the energy lost as light, leaving the plasma.

- *Conduction* is the energy lost, as momentum leaves the plasma.

Plasma clouds lose energy through conduction and radiation. Conduction occurs when ions, electrons or neutrals impact a surface and transfer a portion of their kinetic energy to the atoms of the surface. Radiation is energy that leaves the cloud as light in the visible, UV, IR, or X-ray spectra. Radiation increases with temperature. Fusion power technologies must overcome these losses.

Triple Product: Density, Temperature, Time

The Lawson criterion argues that a machine holding a thermalized (hot) and quasi-neutral plasma has to meet basic criteria to overcome radiation losses, conduction losses and reach efficiency of 30 percent. This became known as the "triple product": the plasma density, temperature and confinement time. Attempts to increase the triple product led to targeting larger plants. Larger plants move structural materials further away from the centre of the plasma, which reduces conduction and radiation losses since more of the radiation is internally reflected. This emphasis on $(nT\tau)$ as a metric of success has impacted other considerations such as cost, size, complexity and efficiency. This has led to larger, more complicated and more expensive machines such as ITER and NIF.

Plasma Behavior

Plasma is an ionized gas that conducts electricity. In bulk, it is modeled using hydrodynamics, which is a combination of the Navier-Stokes equations governing fluids and Maxwell's equations governing how magnetic and electric fields behave. Fusion exploits several plasma properties, including:

Self-organizing plasma conducts electric and magnetic fields. Its motions can generate fields that can in turn contain it.

Diamagnetic plasma can generate its own internal magnetic field. This can reject an externally applied magnetic field, making it diamagnetic.

Magnetic mirrors can reflect plasma when it moves from a low to high density field.

Energy Capture

Multiple approaches have been proposed for energy capture. The simplest is to heat a fluid. The neutrons generated by fusion can re-generate a spent fission fuel. Direct energy conversion was developed (at LLNL in the 1980s) as a method to maintain a voltage using the fusion reaction products. This has demonstrated energy capture efficiency of 48 percent.

Approaches

Magnetic Confinement

Tokamak: the most well-developed and well-funded approach to fusion energy. This method races hot plasma around in a magnetically confined, donut-shaped ring, with an internal current. When completed, ITER will be the world's largest tokamak. As of April 2012 an estimated 215 experimental tokamaks were either planned, decommissioned or currently operating (35) worldwide.

Spherical tokamak: A variation on the tokamak with a spherical shape.

Stellarator: Twisted rings of hot plasma. The stellarator attempts to create a natural twist plasma path, using external magnets, while tokamaks create those magnetic fields using an internal current. Stellarators were developed by Lyman Spitzer in 1950 and have four designs: Torsatron, Heliotron, Heliac and Helias. One example is Wendelstein 7-X, a German fusion device that produced its first plasma on December 10, 2015. It is the world's largest stellarator, designed to investigate the suitability of this type of device for a power station.

Levitated Dipole Experiment (LDX): These use a solid superconducting torus. This is magnetically levitated inside the reactor chamber. The superconductor forms an axisymmetric magnetic field that contains the plasma. The LDX was developed by MIT and Columbia University after 2000 by Jay Kesner and Michael E. Mauel.

Magnetic mirror: Developed by Richard F. Post and teams at LLNL in the 1960s. Magnetic mirrors reflected hot plasma back and forth in a line. Variations included the magnetic bottle and the biconic cusp. A series of well-funded, large, mirror machines were built by the US government in the 1970s and 1980s.

Field-reversed configuration: This device traps plasma in a self-organized quasi-stable structure; where the particle motion makes an internal magnetic field which then traps itself.

Reversed field pinch: Here the plasma moves inside a ring. It has an internal magnetic field. Moving out from the center of this ring, the magnetic field reverses direction.

Inertial Confinement

Direct drive: In this technique, lasers directly blast a pellet of fuel. The goal is to ignite a fusion chain reaction. Ignition was first suggested by John Nuckolls, in 1972. Notable direct drive experiments have been conducted at the Laboratory for Laser Energetics, Laser Mégajoule and the GEKKO XII facilities. Good implosions require fuel pellets with close to a perfect shape in order to generate a symmetrical inward shock wave that produces the high-density plasma.

Fast ignition: This method uses two laser blasts. The first blast compresses the fusion fuel, while the second high energy pulse ignites it. Experiments have been conducted at the Laboratory for Laser Energetics using the Omega and Omega EP systems and at the GEKKO XII laser at the Institute for Laser Engineering in Osaka Japan.

Indirect drive: In this technique, lasers blasts a structure around the pellet of fuel. This structure is known as a Hohlraum. As it disintegrates the pellet is bathed in a more uniform x-ray light, creating better compression. The largest system using this method is the National Ignition Facility.

Magneto-inertial fusion or *Magnetized Liner Inertial Fusion:* This combines a laser pulse with a magnetic pinch. The pinch community refers to it as magnetized liner Inertial fusion while the ICF community refers to it as magneto-inertial fusion.

Heavy Ion Beams There are also proposals to do inertial confinement fusion with ion beams instead of laser beams. The main difference is the mass of the beam has momentum, whereas lasers do not.

Magnetic or Electric Pinches

Z-Pinch: This method sends a strong current (in the z-direction) through the plasma. The current generates a magnetic field that squeezes the plasma to fusion conditions. Pinches were the first method for man-made controlled fusion. Some examples include the Dense plasma focus and the Z machine at Sandia National Laboratories.

Theta-Pinch: This method sends a current inside a plasma, in the theta direction.

Screw Pinch: This method combines a theta and z-pinch for improved stabilization.

Inertial Electrostatic Confinement

Fusor: This method uses an electric field to heat ions to fusion conditions. The machine typically uses two spherical cages, a cathode inside the anode, inside a vacuum. These machines are not considered a viable approach to net power because of their high conduction and radiation losses. They are simple enough to build that amateurs have fused atoms using them.

Polywell: This designs attempts to combine magnetic confinement with electrostatic fields, to avoid the conduction losses generated by the cage.

Other

Magnetized target fusion: This method confines hot plasma using a magnetic field and squeezes it using inertia. Examples include LANL FRX-L machine, General Fusion and the plasma liner experiment.

Uncontrolled: Fusion has been initiated by man, using uncontrolled fission explosions to ignite so-called Hydrogen Bombs. Early proposals for fusion power included using bombs to initiate reactions.

Beam fusion: A beam of high energy particles can be fired at another beam or target and fusion will occur. This was used in the 1970s and 1980s to study the cross sections of high energy fusion reactions.

Bubble fusion: This was a fusion reaction that was supposed to occur inside extraordinarily large collapsing gas bubbles, created during acoustic liquid cavitation. This approach was discredited.

Cold fusion: This is a hypothetical type of nuclear reaction that would occur at, or near, room temperature. Cold fusion is discredited and gained a reputation as pathological science.

Muon-catalyzed fusion: Muons allow atoms to get much closer and thus reduce the kinetic energy required to initiate fusion. Muons require more energy to produce than can be obtained from muon-catalysed fusion, making this approach impractical for power generation.

Gravitational-confinement fusion (GCF) Direct Photo-Electric Conversion: Also known as Space-Based Solar Power argues that a majority of available fusion fuels exists within the sphere of the Sun where it is gravitationally confined, and that a tractable way to accomplish large-scale fusion power is to build very large space-borne platforms that capture energy via photons rather than via a carnot cycle. The theoretical limit of power via this means is a type-2 civilization via a Dyson Sphere.

Common Tools

Heating

Gas must be first heated to form a plasma. This then needs to be hot enough to start fusion reactions. A number of heating schemes have been explored:

Radiofrequency Heating A radio wave is applied to the plasma, causing it to oscillate. This is basically the same concept as a microwave oven. This is also known as electron cyclotron resonance heating or Dielectric heating.

Electrostatic Heating An electric field can do work on charged ions or electrons, heating them.

Neutral Beam Injection An external source of hydrogen is ionized and accelerated by an electric field to form a charged beam which is shone through a source of neutral hydrogen gas towards the plasma which itself is ionized and contained in the reactor by a magnetic field. Some of the intermediate hydrogen gas is accelerated towards the plasma by collisions with the charged beam while remaining neutral: this neutral beam is thus unaffected by the magnetic field and so shines through it into the plasma. Once inside the plasma the neutral beam transmits energy to the plasma by collisions as a result of which it becomes ionized and thus contained by the magnetic field thereby both heating and refuelling the reactor in one operation. The remainder of the charged beam is diverted by magnetic fields onto cooled beam dumps.

Measurement

Thomson Scattering Light scatters from plasma. This light can be detected and used to reconstruct the plasmas' behavior. This technique can be used to find its density and temperature. It is common in Inertial confinement fusion, Tokamaks and fusors. In ICF systems, this can be done by firing a second beam into a gold foil adjacent to the target. This makes x-rays that scatter or traverse the plasma. In Tokamaks, this can be done using mirrors and detectors to reflect light across a plane (two dimensions) or in a line (one dimension).

Langmuir probe This is a metal object placed in a plasma. A potential is applied to it, giving it a positive or negative voltage against the surrounding plasma. The metal collects charged particles, drawing a current. As the voltage changes, the current changes. This makes a IV Curve. The IV-curve can be used to determine the local plasma density, potential and temperature.

Geiger counter Deuterium or tritium fusion produces neutrons. Geiger counters record the rate of neutron production, so they are an essential tool for demonstrating success.

Flux loop A loop of wire is inserted into the magnetic field. As the field passes through the loop, a current is made. The current is measured and used to find the total magnetic flux through that loop. This has been used on the National Compact Stellarator Experiment, the polywell and the LDX machines.

X-ray detector All plasma loses energy by emitting light. This covers the whole spectrum: visible, IR, UV, and X-rays. This occurs anytime a particle changes speed, for any reason. If the reason is deflection by a magnetic field, the radiation is Cyclotron radiation at low speeds and Synchrotron radiation at high speeds. If the reason is deflection by another particle, plasma radiates X-rays, known as Bremsstrahlung radiation. X-rays are termed in both hard and soft, based on their energy.

Power Production

Steam turbines It has been proposed that steam turbines be used to convert the heat from the fusion chamber into electricity. The heat is transferred into a working fluid that turns into steam, driving electric generators.

Neutron blankets Deuterium and tritium fusion generates neutrons. This varies by technique (NIF has a record of 3E14 neutrons per second while a typical fusor produces 1E5–1E9 neutrons per second). It has been proposed to use these neutrons as a way to regenerate spent fission fuel or as a way to breed tritium from a liquid lithium blanket.

Direct conversion This is a method where the kinetic energy of a particle is converted into voltage. It was first suggested by Richard F. Post in conjunction with magnetic mirrors, in the late sixties. It has also been suggested for Field-Reversed Configurations. The process takes the plasma, expands it, and converts a large fraction of the random energy of the fusion products into directed motion. The particles are then collected on electrodes at various large electrical potentials. This method has demonstrated an experimental efficiency of 48 percent.

Confinement

To produce self-sustaining fusion, the energy released by the reaction (or at least a fraction of it) must be used to heat new reactant nuclei and keep them hot long enough that they also undergo fusion reactions.

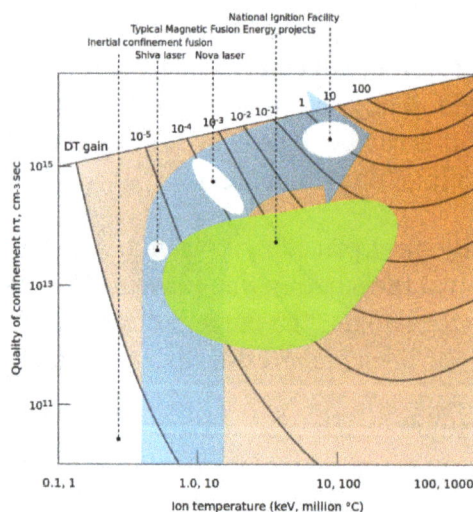

Parameter space occupied by inertial fusion energy and magnetic fusion energy devices as of the mid 1990s. The regime allowing thermonuclear ignition with high gain lies near the upper right corner of the plot.

Confinement refers to all the conditions necessary to keep a plasma dense and hot long enough to undergo fusion. Here are some general principles.

- Equilibrium: The forces acting on the plasma must be balanced for containment. One exception is inertial confinement, where the relevant physics must occur faster than the disassembly time.

- Stability: The plasma must be so constructed so that disturbances will not lead to the plasma disassembling.

- Transport or conduction: The loss of material must be sufficiently slow. The plasma carries off energy with it, so rapid loss of material will disrupt any machines power balance. Material can be lost by transport into different regions or conduction through a solid or liquid.

Unconfined

The first human-made, large-scale fusion reaction was the test of the hydrogen bomb, Ivy Mike, in 1952. As part of the PACER project, it was once proposed to use hydrogen bombs as a source of power by detonating them in underground caverns and then generating electricity from the heat produced, but such a power station is unlikely ever to be constructed.

Magnetic Confinement

At the temperatures required for fusion, the fuel is heated to a plasma state. In this state it has a very good electrical conductivity. This opens the possibility of confining the plasma with magnetic fields. This is the case of magnetized plasma, where the magnetic fields and plasma intermix. This is generally known as magnetic confinement. The field lines put a Lorentz force on the plasma. The force works perpendicular to the magnetic fields, so one problem in magnetic confinement is preventing the plasma from leaking out the ends of the field lines. A general measure of magnetic trapping in fusion is the beta ratio:

$$\beta = \frac{p}{p_{mag}} = \frac{nk_B T}{(B^2 / 2\mu_0)}$$

This is the ratio of the externally applied field to the internal pressure of the plasma. A value of 1 is ideal trapping. Some examples of beta vales include:

1. The START machine: 0.32

2. The Levitated dipole experiment: 0.26

3. Spheromaks: $\approx$ 0.1, Maximum 0.2 based on Mercier limit.

4. The DIII-D machine: 0.126

5. The Gas Dynamic Trap a magnetic mirror: 0.6 for 5E-3 seconds.

Magnetic Mirror One example of magnetic confinement is with the magnetic mirror effect. If a particle follows the field line and enters a region of higher field strength, the particles can be reflected. There are several devices that try to use this effect. The most famous was the magnetic mirror machines, which was a series of large, expensive devices built at the Lawrence Livermore National Laboratory from the 1960s to mid 1980s. Some other examples include the magnetic bottles and Biconic cusp. Because the mirror machines were straight, they had some advantages over a ring shape. First, mirrors were easier to construct and maintain and second direct conversion energy capture, was easier to implement. As the confinement achieved in experiments was poor, this approach was abandoned.

Magnetic Loops Another example of magnetic confinement is to bend the field lines back on themselves, either in circles or more commonly in nested toroidal surfaces. The most highly developed system of this type is the *tokamak*, with the *stellarator* being next most advanced, followed by the Reversed field pinch. Compact toroids, especially the *Field-Reversed Configuration* and the spheromak, attempt to combine the advantages of toroidal magnetic surfaces with those of a simply connected (non-toroidal) machine, resulting in a mechanically simpler and smaller confinement area.

Inertial Confinement

Inertial confinement is the use of rapidly imploding shell to heat and confine plasma. The shell is imploded using a direct laser blast (direct drive) or a secondary x-ray blast (indirect drive) or heavy ion beams. Theoretically, fusion using lasers would be done using tiny pellets of fuel that explode several times a second. To induce the explosion, the pellet must be compressed to about 30 times solid density with energetic beams. If direct drive is used—the beams are focused directly on the pellet—it can in principle be very efficient, but in practice is difficult to obtain the needed uniformity. The alternative approach, indirect drive, uses beams to heat a shell, and then the shell radiates x-rays, which then implode the pellet. The beams are commonly laser beams, but heavy and light ion beams and electron beams have all been investigated.

Electrostatic Confinement

There are also electrostatic confinement fusion devices. These devices confine ions using electrostatic fields. The best known is the Fusor. This device has a cathode inside an anode wire cage. Positive ions fly towards the negative inner cage, and are heated by the electric field in the process. If they miss the inner cage they can collide and fuse. Ions typically hit the cathode, however, creating prohibitory high conduction losses. Also, fusion rates in fusors are very low because of competing physical effects, such as energy loss in the form of light radiation. Designs have been proposed to avoid the problems associated with the cage, by generating the field using a non-neutral cloud. These include a plasma oscillating device, a magnetically-shielded-grid a penning trap and the polywell. The technology is relatively immature, however, and many scientific and engineering questions remain.

History of Research

1920s

Research into nuclear fusion started in the early part of the 20th century. In 1920 the British physicist Francis William Aston discovered that the total mass equivalent of four hydrogen atoms (two protons and two neutrons) are heavier than the total mass of one helium atom (He-4), which implied that net energy can be released by combining hydrogen atoms together to form helium, and provided the first hints of a mechanism by which stars could produce energy in the quantities being measured. Through the 1920s, Arthur Stanley Eddington became a major proponent of the proton–proton chain reaction (PP reaction) as the primary system running the Sun.

1930s

A theory was verified by Hans Bethe in 1939 showing that beta decay and quantum tunneling in the Sun's core might convert one of the protons into a neutron and thereby producing deuterium rather than a diproton. The deuterium would then fuse through other reactions to further increase the energy output. For this work, Bethe won the Nobel Prize in Physics.

1940s

In 1942, nuclear fusion research was subsumed into the Manhattan Project when the secrecy surrounding the field obscured by the science. The first patent related to a fusion reactor was registered in 1946 by the United Kingdom Atomic Energy Authority. The inventors were Sir George Paget Thomson and Moses Blackman. This was the first detailed examination of the Z-pinch concept.

Z-pinch is based on the fact that plasmas are electrically conducting. Running a current through the plasma, will generate a magnetic field around the plasma. This field will, according to Lenz's law, create an inward directed force that causes the plasma to collapse inward, raising its density. Denser plasmas generate denser magnetic fields, increasing the inward force, leading to a chain reaction. If the conditions are correct, this can lead to the densities and temperatures needed for fusion. The difficulty is getting the current into the plasma, which would normally melt any sort of mechanical electrode. A solution emerges again because of the conducting nature of the plasma; by placing the plasma in the middle of an electromagnet, induction can be used to generate the current.

Starting in 1947, two UK teams carried out small experiments and began building a series of ever-larger experiments. When the Huemul results hit the news, James L. Tuck, a UK physicist working at Los Alamos, introduced the pinch concept in the US and produced a series of machines known as the Perhapsatron. The Soviet Union, unbeknownst to the West, was also building a series of similar machines. All of these devices quickly demonstrated a series of instabilities when the pinch was applied. This broke up the plasma column long before it reached the densities and temperatures required for fusion.

1950s

The first successful man-made fusion device was the boosted fission weapon tested in 1951 in the Greenhouse Item test. This was followed by true fusion weapons in 1952's Ivy Mike, and the first practical examples in 1954's Castle Bravo. This was uncontrolled fusion. In these devices, the energy released by the fission explosion is used to compress and heat fusion fuel, starting a fusion reaction. Fusion releases neutrons. These neutrons hit the surrounding fission fuel, causing the atoms to split apart much faster than normal fission processes—almost instantly by comparison. This increases the effectiveness of bombs: normal fission weapons blow themselves apart before all their fuel is used; fusion/fission weapons do not have this practical upper limit.

The first man-made device to achieve ignition was the detonation of this fusion device, codenamed Ivy Mike.

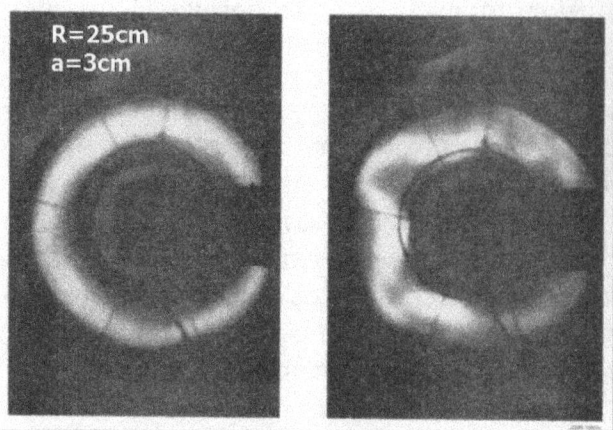

Early photo of plasma inside a pinch machine (imperial college 1950/1951)

In 1949 an expatriate German, Ronald Richter, proposed the Huemul Project in Argentina, announcing positive results in 1951. These turned out to be fake, but it prompted considerable interest in the concept as a whole. In particular, it prompted Lyman Spitzer to begin considering ways to solve some of the more obvious problems involved in confining a hot plasma, and, unaware of the z-pinch efforts, he developed a new solution to the problem known as the stellarator. Spitzer applied to the US Atomic Energy Commission for funding to build a test device. During this period, Jim Tuck who had worked with the UK teams had been introducing the z-pinch concept to his coworkers at his new job at Los Alamos National Laboratory (LANL). When he heard of Spitzer's pitch for funding, he applied to build a machine of his own, the Perhapsatron.

Spitzer's idea won funding and he began work on the stellarator under the code name Project Matterhorn. His work led to the creation of the Princeton Plasma Physics Laboratory. Tuck returned to LANL and arranged local funding to build his machine. By this time, however, it was clear that all of the pinch machines were suffering from the same issues involving stability, and progress stalled. In 1953, Tuck and others suggested a number of solutions to the stability problems. This led to the design of a second series of pinch machines, led by the UK ZETA and Sceptre devices.

Spitzer had planned an aggressive development project of four machines, A, B, C, and D. A and B were small research devices, C would be the prototype of a power-producing machine, and D would be the prototype of a commercial device. A worked without issue, but even by the time B was being used it was clear the stellarator was also suffering from instabilities and plasma leakage. Progress on C slowed as attempts were made to correct for these problems.

By the mid-1950s it was clear that the simple theoretical tools being used to calculate the performance of all fusion machines were simply not predicting their actual behavior. Machines invariably leaked their plasma from their confinement area at rates far higher than predicted. In 1954, Edward Teller held a gathering of fusion researchers at the Princeton Gun Club, near the Project Matterhorn (now known as Project Sherwood) grounds. Teller started by pointing out the problems that everyone was having, and suggested that any system where the plasma was confined within concave fields was doomed to fail. Attendees remember him saying something to the effect that the fields were like rubber bands, and they would attempt to snap back to a straight configuration whenever the power was increased, ejecting the plasma. He went on to say that it appeared the only way to confine the plasma in a stable configuration would be to use convex fields, a "cusp" configuration.

When the meeting concluded, most of the researchers quickly turned out papers saying why Teller's concerns did not apply to their particular device. The pinch machines did not use magnetic fields in this way at all, while the mirror and stellarator seemed to have various ways out. This was soon followed by a paper by Martin David Kruskal and Martin Schwarzschild discussing pinch machines, however, which demonstrated instabilities in those devices were inherent to the design.

The largest "classic" pinch device was the ZETA, including all of these suggested upgrades, starting operations in the UK in 1957. In early 1958, John Cockcroft announced that fusion had been achieved in the ZETA, an announcement that made headlines around the world. When physicists in the US expressed concerns about the claims they were initially dismissed. US experiments soon demonstrated the same neutrons, although temperature measurements suggested these could not be from fusion reactions. The neutrons seen in the UK were later demonstrated to be from different versions of the same instability processes that plagued earlier machines. Cockcroft was forced to retract the fusion claims, and the entire field was tainted for years. ZETA ended its experiments in 1968.

The first controlled fusion experiment was accomplished using Scylla I at the Los Alamos National Laboratory in 1958. This was a pinch machine, with a cylinder full of deuterium. Electric current shot down the sides of the cylinder. The current made magnetic fields that compressed the plasma to 15 million degrees Celsius, squeezed the gas, fused it and produced neutrons.

In 1950–1951 I.E. Tamm and A.D. Sakharov in the Soviet Union, first discussed a tokamak-like approach. Experimental research on those designs began in 1956 at the Kurchatov Institute in Moscow by a group of Soviet scientists led by Lev Artsimovich. The tokamak essentially combined a low-power pinch device with a low-power simple stellarator. The key was to combine the fields in such a way that the particles orbited within the reactor a particular number of times, today known as the "safety factor". The combination of these fields dramatically improved confinement times and densities, resulting in huge improvements over existing devices.

1960s

A key plasma physics text was published by Lyman Spitzer at Princeton in 1963. Spitzer took the ideal gas laws and adopted them to an ionized plasma, developing many of the fundamental equations used to model a plasma.

Laser fusion was suggested in 1962 by scientists at Lawrence Livermore National Laboratory, shortly after the invention of the laser itself in 1960. At the time, Lasers were low power machines, but low-level research began as early as 1965. Laser fusion, formally known as inertial confinement fusion, involves imploding a target by using laser beams. There are two ways to do this: indirect drive and direct drive. In direct drive, the laser blasts a pellet of fuel. In indirect drive, the lasers blast a structure around the fuel. This makes x-rays that squeeze the fuel. Both methods compress the fuel so that fusion can take place.

At the 1964 World's Fair, the public was given its first demonstration of nuclear fusion. The device was a θ-pinch from General Electric. This was similar to the Scylla machine developed earlier at Los Alamos.

The magnetic mirror was first published in 1967 by Richard F. Post and many others at the Lawrence Livermore National Laboratory. The mirror consisted of two large magnets arranged so they had strong fields within them, and a weaker, but connected, field between them. Plasma introduced in the area between the two magnets would "bounce back" from the stronger fields in the middle.

The A.D. Sakharov group constructed the first tokamaks, the most successful being the T-3 and its larger version T-4. T-4 was tested in 1968 in Novosibirsk, producing the world's first quasistationary fusion reaction. When this were first announced, the international community was highly skeptical. A British team was invited to see T-3, however, and after measuring it in depth they released their results that confirmed the Soviet claims. A burst of activity followed as many planned devices were abandoned and new tokamaks were introduced in their place — the C model stellarator, then under construction after many redesigns, was quickly converted to the Symmetrical Tokamak.

In his work with vacuum tubes, Philo Farnsworth observed that electric charge would accumulate in regions of the tube. Today, this effect is known as the Multipactor effect. Farnsworth reasoned that if ions were concentrated high enough they could collide and fuse. In 1962, he filed a patent on a design using a positive inner cage to concentrate plasma, in order to achieve nuclear fusion. During this time, Robert L. Hirsch joined the Farnsworth Television labs and began work on what became the fusor. Hirsch patented the design in 1966 and published the design in 1967.

1970s

In 1972, John Nuckolls outlined the idea of ignition. This is a fusion chain reaction. Hot helium made during fusion reheats the fuel and starts more reactions. John argued that ignition would require lasers of about 1 kJ. This turned out to be wrong. Nuckolls's paper started a major development effort. Several laser systems were built at LLNL. These included the argus, the Cyclops, the Janus, the long path, the Shiva laser and the Nova in 1984. This prompted the UK to build the Central Laser Facility in 1976.

Shiva laser, 1977, the largest ICF laser system built in the seventies

During this time, great strides in understanding the tokamak system were made. A number of improvements to the design are now part of the "advanced tokamak" concept, which includes non-circular plasma, internal diverters and limiters, often superconducting magnets, and operate in the so-called "H-mode" island of increased stability. Two other designs have also become fairly well studied; the compact tokamak is wired with the magnets on the inside of the vacuum chamber, while the spherical tokamak reduces its cross section as much as possible.

The Tandem Mirror Experiment (TMX) in 1979

In 1974 a study of the ZETA results demonstrated an interesting side-effect; after an experimental run ended, the plasma would enter a short period of stability. This led to the reversed field pinch concept, which has seen some level of development since. On May 1, 1974, the KMS fusion company (founded by Kip Siegel) achieves the world's first laser induced fusion in a deuterium-tritium pellet.

In the mid-1970s, Project PACER, carried out at Los Alamos National Laboratory (LANL) explored the possibility of a fusion power system that would involve exploding small hydrogen bombs (fusion bombs) inside an underground cavity. As an energy source, the system is the only fusion power system that could be demonstrated to work using existing technology. It would also require a large, continuous supply of nuclear bombs, however, making the economics of such a system rather questionable.

In 1976, the two beam Argus laser becomes operational at livermore. In 1977, The 20 beam Shiva laser at Livermore is completed, capable of delivering 10.2 kilojoules of infrared energy on target. At a price of $25 million and a size approaching that of a football field, Shiva is the first of the megalasers. That same year, the JET project is approved by the European Commission and a site is selected.

1980s

Magnetic mirrors suffered from end losses, requiring high power, complex magnetic designs, such as the baseball coil pictured here.

The magnetic mirror test facility during construction

The Novette target chamber (metal sphere with diagnostic devices protruding radially), which was reused from the Shiva project and two newly built laser chains visible in background.

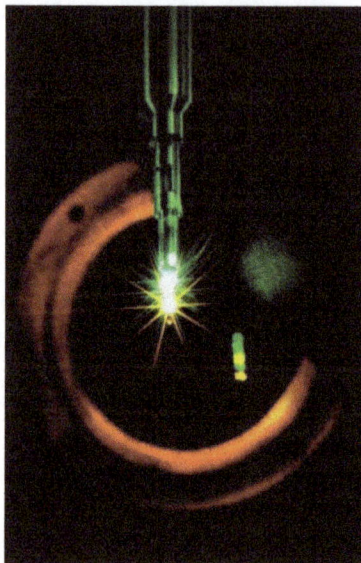

Inertial confinement fusion implosion on the Nova laser during the 1980s was a key driver of fusion development.

As a result of advocacy, the cold war, and the 1970s energy crisis a massive magnetic mirror program was funded by the US federal government in the late 1970s and early 1980s. This program resulted in a series of large magnetic mirror devices including: 2X, Baseball I, Baseball II, the Tandem Mirror Experiment, the Tandem mirror experiment upgrade, the Mirror Fusion Test Facility and the MFTF-B. These machines were built and tested at Livermore from the late 1960s to the mid 1980s. A number of institutions collaborated on these machines, conducting experiments. These included the Institute for Advanced Study and the University of Wisconsin–Madison. The last machine, the Mirror Fusion Test Facility cost 372 million dollars and was, at that time, the most expensive project in Livermore history. It opened on February 21, 1986 and was promptly shut down. The reason given was to balance the United States federal budget. This program was supported from within the Carter and early Reagan administrations by Edwin E. Kintner, a US Navy captain, under Alvin Trivelpiece.

In Laser fusion progressed: in 1983, the NOVETTE laser was completed. The following December 1984, the ten beam NOVA laser was finished. Five years later, NOVA would produce a maximum of 120 kilojoules of infrared light, during a nanosecond pulse. Meanwhile, efforts focused on either fast delivery or beam smoothness. Both tried to deliver the energy uniformly to implode the target. One early problem was that the light in the infrared wavelength, lost lots of energy before hitting the fuel. Breakthroughs were made at the Laboratory for Laser Energetics at the University of Rochester. Rochester scientists used frequency-tripling crystals to transform the infrared laser beams into ultraviolet beams. In 1985, Donna Strickland and Gérard Mourou invented a method to amplify lasers pulses by "chirping". This method changes a single wavelength into a full spectrum. The system then amplifies the laser at each wavelength and then reconstitutes the beam into one color. Chirp pulsed amplification became instrumental in building the National Ignition Facility and the Omega EP system. Most research into ICF was towards weapons research, because the implosion is relevant to nuclear weapons.

During this time Los Alamos National Laboratory constructed a series of laser facilities. This included Gemini (a two beam system), Helios (eight beams), Antares (24 beams) and Aurora (96 beams). The program ended in the early nineties with a cost on the order of one billion dollars.

In 1987, Akira Hasegawa noticed that in a dipolar magnetic field, fluctuations tended compress the plasma without energy loss. This effect was noticed in data taken by Voyager 2, when it encountered Uranus. This observation would become the basis for a fusion approach known as the Levitated dipole.

In Tokamaks, the Tore Supra was under construction over the middle of the eighties (1983 to 1988). This was a Tokamak built in Cadarache, France. In 1983, the JET was completed and first plasmas achieved. In 1985, the Japanese tokamak, JT-60 was completed. In 1988, the T-15 a Soviet tokamak was completed. It was the first industrial fusion reactor to use superconducting magnets to control the plasma. These were Helium cooled.

In 1989, Pons and Fleischmann submitted papers to the *Journal of Electroanalytical Chemistry* claiming that they had observed fusion in a room temperature device and disclosing their work in a press release. Some scientists reported excess heat, neutrons, tritium, helium and other nuclear effects in so-called cold fusion systems, which for a time gained interest as showing promise. Hopes fell when replication failures were weighed in view of several reasons cold fusion is not likely to occur, the discovery of possible sources of experimental error, and finally the discovery that Fleischmann and Pons had not actually detected nuclear reaction byproducts. By late 1989, most scientists considered cold fusion claims dead, and cold fusion subsequently gained a reputation as pathological science. However, a small community of researchers continues to investigate cold fusion claiming to replicate Fleishmann and Pons' results including nuclear reaction byproducts. Claims related to cold fusion are largely disbelieved in the mainstream scientific community. In 1989, the majority of a review panel organized by the US Department of Energy (DOE) found that the evidence for the discovery of a new nuclear process was not persuasive. A second DOE review, convened in 2004 to look at new research, reached conclusions similar to the first.

In 1984, Martin Peng of ORNL proposed an alternate arrangement of the magnet coils that would greatly reduce the aspect ratio while avoiding the erosion issues of the compact tokamak: a Spherical tokamak. Instead of wiring each magnet coil separately, he proposed using a single large conductor in the center, and wiring the magnets as half-rings off of this conductor. What was once a series of individual rings passing through the hole in the center of the reactor was reduced to a single post, allowing for aspect ratios as low as 1.2. The ST concept appeared to represent an enormous advance in tokamak design. However, it was being proposed during a period when US fusion research budgets were being dramatically scaled back. ORNL was provided with funds to develop a suitable central column built out of a high-strength copper alloy called "Glidcop". However, they were unable to secure funding to build a demonstration machine, "STX". Failing to build an ST at ORNL, Peng began a worldwide effort to interest other teams in the ST concept and get a test machine built. One way to do this quickly would be to convert a spheromak machine to the Spherical tokamak layout. Peng's advocacy also caught the interest of Derek Robinson, of the United Kingdom Atomic Energy Authority fusion center at Culham. Robinson was able to gather together a team and secure funding on the order of 100,000 pounds to build an experimental machine, the Small Tight Aspect Ratio Tokamak, or START. Several parts of the machine were recycled from earlier projects, while others were loaned from other labs, including a 40 keV neutral beam injector from ORNL. Construction of START began in 1990, it was assembled rapidly and started operation in January 1991.

1990s

In 1991 the Preliminary Tritium Experiment at the Joint European Torus in England achieved the world's first controlled release of fusion power.

Z Machine (a pinch at SNL) went through a number of upgrades during the mid to late nineties

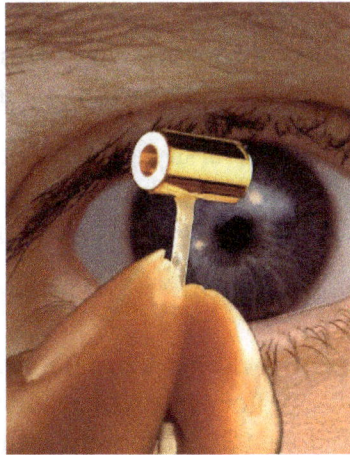

Mockup of a gold-plated hohlraum designed for use in the National Ignition Facility

In 1992, a major article was published in Physics Today by Robert McCory at the Laboratory for laser energetics outlying the current state of ICF and advocating for a national ignition facility. This was followed up by a major review article, from John Lindl in 1995, advocating for NIF. During this time a number of ICF subsystems were developing, including target manufacturing, cryogenic handling systems, new laser designs (notably the NIKE laser at NRL) and improved diagnostics like time of flight analyzers and Thomson scattering. This work was done at the NOVA laser system, General Atomics, Laser Mégajoule and the GEKKO XII system in Japan. Through this work and lobbying by groups like the fusion power associates and John Sethian at NRL, a vote was made in congress, authorizing funding for the NIF project in the late nineties.

In the early nineties, theory and experimental work regarding fusors and polywells was published. In response, Todd Rider at MIT developed general models of these devices. Rider argued that all plasma systems at thermodynamic equilibrium were fundamentally limited. In 1995, William Nevins published a criticism arguing that the particles inside fusors and polywells would build up angular momentum, causing the dense core to degrade.

In 1995, the University of Wisconsin–Madison built a large fusor, known as HOMER, which is still in operation. Meanwhile, Dr George H. Miley at Illinois, built a small fusor that has produced neutrons using deuterium gas and discovered the "star mode" of fusor operation. The following year, the first "US-Japan Workshop on IEC Fusion", was conducted. At this time in Europe, an IEC device was developed as a commercial neutron source by Daimler-Chrysler and NSD Fusion.

In 1996, the Z-machine was upgraded and opened to the public by the US Army in August 1998 in Scientific American. The key attributes of Sandia's Z machine are its 18 million amperes and a discharge time of less than 100 nanoseconds. This generates a magnetic pulse, inside a large oil tank, this strikes an array of tungsten wires called a *liner*. Firing the Z-machine has become a way to test very high energy, high temperature (2 billion degrees) conditions. In 1996, the Tore Supra creates a plasma for two minutes with a current of almost 1 million amperes driven non-inductively by 2.3 MW of lower hybrid frequency waves. This is 280 MJ of injected and extracted energy. This result was possible because of the actively cooled plasma-facing components

In 1997, JET produced a peak of 16.1MW of fusion power (65% of heat to plasma), with fusion power of over 10MW sustained for over 0.5 sec. Its successor, the International Thermonuclear

Experimental Reactor (ITER), was officially announced as part of a seven-party consortium (six countries and the EU). ITER is designed to produce ten times more fusion power than the power put into the plasma. ITER is currently under construction in Cadarache, France.

In the late nineties, a team at Columbia University and MIT developed the Levitated dipole a fusion device which consisted of a superconducting electromagnet, floating in a saucer shaped vacuum chamber. Plasma swirled around this donut and fused along the center axis.

2000s

In the March 8, 2002 issue of the peer-reviewed journal *Science*, Rusi P. Taleyarkhan and colleagues at the Oak Ridge National Laboratory (ORNL) reported that acoustic cavitation experiments conducted with deuterated acetone (C_{3D60}) showed measurements of tritium and neutron output consistent with the occurrence of fusion. Taleyarkhan was later found guilty of misconduct, the Office of Naval Research debarred him for 28 months from receiving Federal Funding, and his name was listed in the 'Excluded Parties List'.

Starting in 1999, a growing number of amateurs have been able to fuse atoms using homemade fusors, shown here.

The Mega Ampere Spherical Tokamak became operational in the UK in 1999

"Fast ignition" was developed in the late nineties, and was part of a push by the Laboratory for Laser Energetics for building the Omega EP system. This system was finished in 2008. Fast ignition showed such dramatic power savings that ICF appears to be a useful technique for energy production. There are even proposals to build an experimental facility dedicated to the fast ignition approach, known as HiPER.

In April 2005, a team from UCLA announced it had devised a way of producing fusion using a machine that "fits on a lab bench", using lithium tantalate to generate enough voltage to smash deuterium atoms together. The process, however, does not generate net power. Such a device would be useful in the same sort of roles as the fusor. In 2006, China's EAST test reactor is completed. This was the first tokamak to use superconducting magnets to generate both the toroidal and poloidal fields.

In the early 2000s, Researchers at LANL reasoned that a plasma oscillating could be at local thermodynamic equilibrium. This prompted the POPS and Penning trap designs. At this time, researchers at MIT became interested in fusors for space propulsion and powering space vehicles. Specifically, researchers developed fusors with multiple inner cages. Greg Piefer graduated from Madison and founded Phoenix Nuclear Labs, a company that developed the fusor into a neutron source for the mass production of medical isotopes. Robert Bussard began speaking openly about the Polywell in 2006. He attempted to generate interest in the research, before his death. In 2008, Taylor Wilson achieved notoriety for achieving nuclear fusion at 14, with a homemade fusor.

In 2009, a high-energy laser system, the National Ignition Facility (NIF), was finished in the US, which can heat hydrogen atoms to temperatures only existing in nature in the cores of stars. The new laser is expected to have the ability to produce, for the first time, more energy from controlled, inertially confined nuclear fusion than was required to initiate the reaction.

2010s

In 2010, NIF researchers were conducting a series of "tuning" shots to determine the optimal target design and laser parameters for high-energy ignition experiments with fusion fuel in the following months. Two firing tests were performed on October 31, 2010 and November 2, 2010. In early 2012, NIF director Mike Dunne expected the laser system to generate fusion with net energy gain by the end of 2012. However, it was delayed and not achieved by that date.

The preamplifiers of the National Ignition Facility. In 2012, the NIF achieved a 500-terawatt shot.

Inertial (laser) confinement is being developed at the United States National Ignition Facility (NIF) based at Lawrence Livermore National Laboratory in California, the French Laser Mégajoule, and the planned European Union High Power laser Energy Research (HiPER) facility. NIF reached initial operational status in 2010 and has been in the process of increasing the power and energy of its "shots", with fusion ignition tests to follow. A three-year goal announced in 2009 to produce net energy from fusion by 2012 was missed; in September 2013, however, the facility announced a significant milestone from an August 2013 test that produced more energy from the fusion reaction than had been provided to the fuel pellet. This was reported as the first time this had been accomplished in fusion power research. The facility reported that their next step involved improving the system to prevent the hohlraum from either breaking up asymmetrically or too soon.

The Wendelstein7X under construction

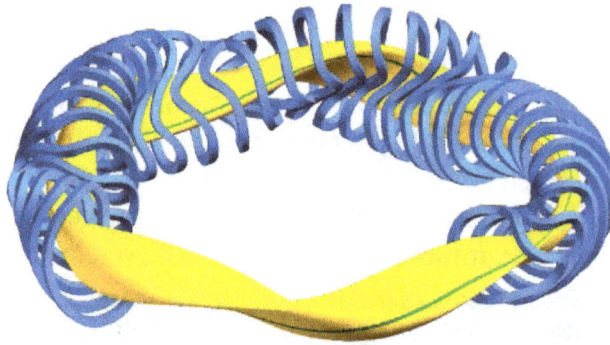

Example of a stellarator design: A coil system (blue) surrounds plasma (yellow). A magnetic field line is highlighted in green on the yellow plasma surface.

A 2012 paper demonstrated that a dense plasma focus had achieved temperatures of 1.8 billion degrees Celsius, sufficient for boron fusion, and that fusion reactions were occurring primarily within the contained plasmoid, a necessary condition for net power. The focus consists of two coaxial cylindrical electrodes made from copper or beryllium and housed in a vacuum chamber containing a low-pressure fusible gas. An electrical pulse is applied across the electrodes, heating the gas into a plasma. The current forms into a minuscule vortex along the axis of the machine, which then kinks into a cage of current with an associated magnetic field. The cage of current and magnetic-field-entrapped plasma is called a plasmoid. The acceleration of the electrons about the magnetic field lines heats the nuclei within the plasmoid to fusion temperatures.

In April 2014, Lawrence Livermore National Laboratory ended the Laser Inertial Fusion Energy (LIFE) program and redirected their efforts towards NIF. In August 2014, Phoenix Nuclear Labs announced the sale of a high-yield neutron generator that could sustain 5×10^{11} deuterium fusion reactions per second over a 24-hour period. In October 2014, Lockheed Martin's Skunk Works announced the development of a high-beta fusion reactor that they hope to yield a functioning 100-megawatt prototype by 2017 and to be ready for regular operation by 2022.

Deep-space exploration, as well as higher-velocity lower-cost space transport services in general would be enabled by this compact fusion reactor technology.

In January 2015, the polywell was presented at Microsoft Research.

In August, 2015, MIT announced a tokamak it named ARC fusion reactor design using rare-earth barium-copper oxide (REBCO) superconducting tapes to produce high-magnetic field coils that it claimed produce comparable magnetic field strength in a smaller configuration than other designs.

In October 2015, researchers at the Max Planck Institute of Plasma Physics completed building the largest stellarator to date, named Wendelstein 7-X. On December 10, they successfully produced the first helium plasma, and on February 3, 2016 produced the device's first Hydrogen plasma. With plasma discharges lasting up to 30 minutes, Wendelstein 7-X will try to demonstrate the essential stellarator attribute: continuous operation of a high-temperature hydrogen plasma.

Fuels

By firing particle beams at targets, many fusion reactions have been tested, while the fuels considered for power have all been light elements like the isotopes of hydrogen—deuterium and tritium. Other reactions like the deuterium and Helium[3] reaction or the Helium[3] and Helium[3] reactions, would require a supply of Helium[3]. This can either come from other nuclear reactions or from extraterrestrial sources. Finally, researchers hope to do the p-11 B
reaction, because it does not directly produce neutrons, though side reactions can.

Deuterium, Tritium

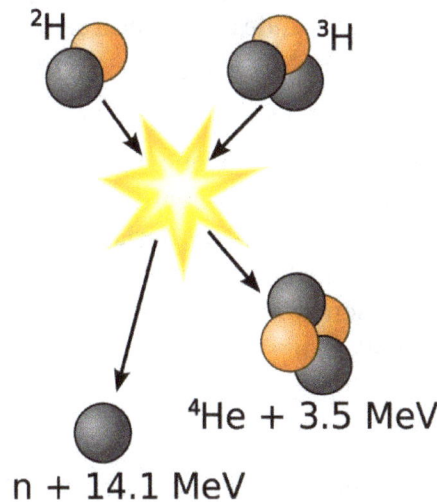
Diagram of the D-T reaction

The easiest nuclear reaction, at the lowest energy, is:

$$^2_1D + {}^3_1T \rightarrow {}^4_2He + {}^1_0n$$

This reaction is common in research, industrial and military applications, usually as a convenient source of neutrons. Deuterium is a naturally occurring isotope of hydrogen and is commonly available. The large mass ratio of the hydrogen isotopes makes their separation easy compared to the difficult uranium enrichment process. Tritium is a natural isotope of hydrogen, but because it has a short half-life of 12.32 years, it is hard to find, store, produce, and is expensive. Consequently, the deuterium-tritium fuel cycle requires the breeding of tritium from lithium using one of the following reactions:

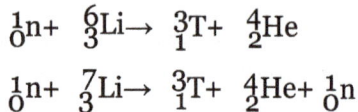

$$_0^1n + {}_3^6Li \rightarrow {}_1^3T + {}_2^4He$$

$$_0^1n + {}_3^7Li \rightarrow {}_1^3T + {}_2^4He + {}_0^1n$$

The reactant neutron is supplied by the D-T fusion reaction shown above, and the one that has the greatest yield of energy. The reaction with ^{6}Li is exothermic, providing a small energy gain for the reactor. The reaction with ^{7}Li is endothermic but does not consume the neutron. At least some ^{7}Li reactions are required to replace the neutrons lost to absorption by other elements. Most reactor designs use the naturally occurring mix of lithium isotopes.

Several drawbacks are commonly attributed to D-T fusion power:

1. It produces substantial amounts of neutrons that result in the neutron activation of the reactor materials.

2. Only about 20% of the fusion energy yield appears in the form of charged particles with the remainder carried off by neutrons, which limits the extent to which direct energy conversion techniques might be applied.

3. It requires the handling of the radioisotope tritium. Similar to hydrogen, tritium is difficult to contain and may leak from reactors in some quantity. Some estimates suggest that this would represent a fairly large environmental release of radioactivity.

The neutron flux expected in a commercial D-T fusion reactor is about 100 times that of current fission power reactors, posing problems for material design. After a series of D-T tests at JET, the vacuum vessel was sufficiently radioactive that remote handling was required for the year following the tests.

In a production setting, the neutrons would be used to react with lithium in order to create more tritium. This also deposits the energy of the neutrons in the lithium, which would then be transferred to drive electrical production. The lithium neutron absorption reaction protects the outer portions of the reactor from the neutron flux. Newer designs, the advanced tokamak in particular, also use lithium inside the reactor core as a key element of the design. The plasma interacts directly with the lithium, preventing a problem known as "recycling". The advantage of this design was demonstrated in the Lithium Tokamak Experiment.

Deuterium

This is the second easiest fusion reaction, fusing deuterium with itself. The reaction has two branches that occur with nearly equal probability:

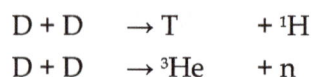

$$D + D \rightarrow T + {}^1H$$
$$D + D \rightarrow {}^3He + n$$

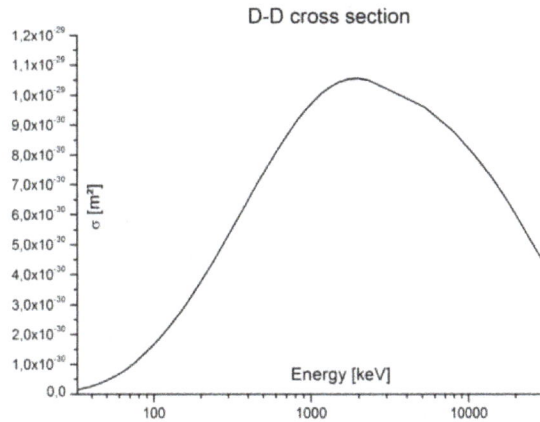

Deuterium fusion cross section (in square meters) at different ion collision energies.

This reaction is also common in research. The optimum energy to initiate this reaction is 15 keV, only slightly higher than the optimum for the D-T reaction. The first branch does not produce neutrons, but it does produce tritium, so that a D-D reactor will not be completely tritium-free, even though it does not require an input of tritium or lithium. Unless the tritons can be quickly removed, most of the tritium produced would be burned before leaving the reactor, which would reduce the handling of tritium, but would produce more neutrons, some of which are very energetic. The neutron from the second branch has an energy of only 2.45 MeV (0.393 pJ), whereas the neutron from the D-T reaction has an energy of 14.1 MeV (2.26 pJ), resulting in a wider range of isotope production and material damage. When the tritons are removed quickly while allowing the ^{3}He to react, the fuel cycle is called "tritium suppressed fusion" The removed tritium decays to ^{3}He with a 12.5 year half life. By recycling the ^{3}He produced from the decay of tritium back into the fusion reactor, the fusion reactor does not require materials resistant to fast 14.1 MeV (2.26 pJ) neutrons.

Assuming complete tritium burn-up, the reduction in the fraction of fusion energy carried by neutrons would be only about 18%, so that the primary advantage of the D-D fuel cycle is that tritium breeding would not be required. Other advantages are independence from scarce lithium resources and a somewhat softer neutron spectrum. The disadvantage of D-D compared to D-T is that the energy confinement time (at a given pressure) must be 30 times longer and the power produced (at a given pressure and volume) would be 68 times less.

Assuming complete removal of tritium and recycling of ^{3}He, only 6% of the fusion energy is carried by neutrons. The tritium-suppressed D-D fusion requires an energy confinement that is 10 times longer compared to D-T and a plasma temperature that is twice as high.

Deuterium, Helium 3

A second-generation approach to controlled fusion power involves combining helium-3 (^{3}He) and deuterium (^{2}H):

$$D + {}^3He \rightarrow {}^4He + {}^1H$$

This reaction produces a helium-4 nucleus (^{4}He) and a high-energy proton. As with the p-^{11}B aneutronic fusion fuel cycle, most of the reaction energy is released as charged particles, reducing activation of the reactor housing and potentially allowing more efficient energy harvesting (via any of several speculative technologies). In practice, D-D side reactions produce a significant number of neutrons, resulting in p-^{11}B being the preferred cycle for aneutronic fusion.

Proton, Boron 11

If aneutronic fusion is the goal, then the most promising candidate may be the Hydrogen-1 (proton)/boron reaction, which releases alpha (helium) particles, but does not rely on neutron scattering for energy transfer.

$$^1H + {}^{11}B \rightarrow 3\ {}^4He$$

Under reasonable assumptions, side reactions will result in about 0.1% of the fusion power being carried by neutrons. At 123 keV, the optimum temperature for this reaction is nearly ten times higher than that for the pure hydrogen reactions, the energy confinement must be 500 times better than that required for the D-T reaction, and the power density will be 2500 times lower than for D-T.

Because the confinement properties of conventional approaches to fusion such as the tokamak and laser pellet fusion are marginal, most proposals for aneutronic fusion are based on radically different confinement concepts, such as the Polywell and the Dense Plasma Focus. Results have been extremely promising:

> "In the October 2013 edition of Nature Communications, a research team led by Christine Labaune at École Polytechnique in Palaiseau, France, reported a new record fusion rate: an estimated 80 million fusion reactions during the 1.5 nanoseconds that the laser fired, which is at least 100 times more than any previous proton-boron experiment. "

Material Selection

Considerations

Any power station using hot plasma, is going to have plasma facing walls. In even the simplest plasma approaches, the material will get blasted with matter and energy. This leads to a minimum list of considerations, including dealing with:

- A heating and cooling cycle, up to a 10 MW/m² thermal load.

- Neutron radiation, which over time leads to neutron activation and embrittlement.

- High energy ions leaving at tens to hundreds of electronvolts.

- Alpha particles leaving at millions of electronvolts.

- Electrons leaving at high energy.

- Light radiation (IR, visible, UV, X-ray).

Depending on the approach, these effects may be higher or lower than typical fission reactors like the pressurized water reactor (PWR). One estimate put the radiation at 100 times the (PWR). Materials need to be selected or developed that can withstand these basic conditions. Depending on the approach, however, there may be other considerations such as electrical conductivity, magnetic permeability and mechanical strength. There is also a need for materials whose primary components and impurities do not result in long-lived radioactive wastes.

Durability

For long term use, each atom in the wall is expected to be hit by a neutron and displaced about a hundred times before the material is replaced. High-energy neutrons will produce hydrogen and helium by way of various nuclear reactions that tends to form bubbles at grain boundaries and result in swelling, blistering or embrittlement.

Selection

One can choose either a low-Z material, such as graphite or beryllium, or a high-Z material, usually tungsten with molybdenum as a second choice. Use of liquid metals (lithium, gallium, tin) has also been proposed, e.g., by injection of 1–5 mm thick streams flowing at 10 m/s on solid substrates.

If graphite is used, the gross erosion rates due to physical and chemical sputtering would be many meters per year, so one must rely on redeposition of the sputtered material. The location of the redeposition will not exactly coincide with the location of the sputtering, so one is still left with erosion rates that may be prohibitive. An even larger problem is the tritium co-deposited with the redeposited graphite. The tritium inventory in graphite layers and dust in a reactor could quickly build up to many kilograms, representing a waste of resources and a serious radiological hazard in case of an accident. The consensus of the fusion community seems to be that graphite, although a very attractive material for fusion experiments, cannot be the primary PFC material in a commercial reactor.

The sputtering rate of tungsten by the plasma fuel ions is orders of magnitude smaller than that of carbon, and tritium is much less incorporated into redeposited tungsten, making this a more attractive choice. On the other hand, tungsten impurities in a plasma are much more damaging than carbon impurities, and self-sputtering of tungsten can be high, so it will be necessary to ensure that the plasma in contact with the tungsten is not too hot (a few tens of eV rather than hundreds of eV). Tungsten also has disadvantages in terms of eddy currents and melting in off-normal events, as well as some radiological issues.

Safety and the Environment

Accident Potential

Nuclear fusion is unlike nuclear fission: fusion requires extremely precise and controlled temperature, pressure and magnetic field parameters for any net energy to be produced. If a reactor suffers damage or loses even a small degree of required control, fusion reactions and heat generation would rapidly cease. Additionally, fusion reactors contain relatively small amounts of fuel, enough to "burn" for minutes, or in some cases, microseconds. Unless they are actively refueled, the reactions will quickly end. Therefore, fusion reactors are considered extremely safe.

Runaway reactions cannot occur in a fusion reactor. The plasma is burnt at optimal conditions, and any significant change will quench the reactions. The reaction process is so delicate that this level of safety is inherent. Although the plasma in a fusion power station is expected to have a volume of 1,000 cubic metres (35,000 cu ft) or more, the plasma density is low and the total amount of fusion fuel in the vessel typically only a few grams. If the fuel supply is closed, the reaction stops within seconds. In comparison, a fission reactor is typically loaded with enough fuel for several months or years, and no additional fuel is necessary to continue the reaction. It is this large amount of fuel that gives rise to the possibility of a meltdown; nothing analogous exists in a fusion reactor.

In the magnetic approach, strong fields are developed in coils that are held in place mechanically by the reactor structure. Failure of this structure could release this tension and allow the magnet to "explode" outward. The severity of this event would be similar to any other industrial accident or an MRI machine quench/explosion, and could be effectively stopped with a containment building similar to those used in existing (fission) nuclear generators. The laser-driven inertial approach is generally lower-stress because of the increased size of the reaction chamber. Although failure of the reaction chamber is possible, simply stopping fuel delivery would prevent any sort of catastrophic failure.

Most reactor designs rely on liquid hydrogen as both a coolant and a method for converting stray neutrons from the reaction into tritium, which is fed back into the reactor as fuel. Hydrogen is highly flammable, and in the case of a fire it is possible that the hydrogen stored on-site could be burned up and escape. In this case, the tritium contents of the hydrogen would be released into the atmosphere, posing a radiation risk. Calculations suggest that at about 1 kg the total amount of tritium and other radioactive gases in a typical power station would be so small that they would have diluted to legally acceptable limits by the time they blew as far as the station's perimeter fence.

The likelihood of *small industrial* accidents including the local release of radioactivity and injury to staff cannot be estimated yet. These would include accidental releases of lithium or tritium or mis-handling of decommissioned radioactive components of the reactor itself.

Magnet Quench

A quench is an abnormal termination of magnet operation that occurs when part of the superconducting coil enters the normal (resistive) state. This can occur because the field inside the magnet is too large, the rate of change of field is too large (causing eddy currents and resultant heating in the copper support matrix), or a combination of the two.

More rarely a defect in the magnet can cause a quench. When this happens, that particular spot is subject to rapid Joule heating from the enormous current, which raises the temperature of the surrounding regions. This pushes those regions into the normal state as well, which leads to more heating in a chain reaction. The entire magnet rapidly becomes normal (this can take several seconds, depending on the size of the superconducting coil). This is accompanied by a loud bang as the energy in the magnetic field is converted to heat, and rapid boil-off of the cryogenic fluid. The abrupt decrease of current can result in kilovolt inductive voltage spikes and arcing. Permanent damage to the magnet is rare, but components can be damaged by localized heating, high voltages, or large mechanical forces.

In practice, magnets usually have safety devices to stop or limit the current when the beginning of a quench is detected. If a large magnet undergoes a quench, the inert vapor formed by the evaporating cryogenic fluid can present a significant asphyxiation hazard to operators by displacing breathable air.

A large section of the superconducting magnets in CERN's Large Hadron Collider unexpectedly quenched during start-up operations in 2008, necessitating the replacement of a number of magnets. In order to mitigate against potentially destructive quenches, the superconducting magnets that form the LHC are equipped with fast-ramping heaters which are activated once a quench event is detected by the complex quench protection system. As the dipole bending magnets are connected in series, each power circuit includes 154 individual magnets, and should a quench event occur, the entire combined stored energy of these magnets must be dumped at once. This energy is transferred into dumps that are massive blocks of metal which heat up to several hundreds of degrees Celsius—because of resistive heating—in a matter of seconds. Although undesirable, a magnet quench is a "fairly routine event" during the operation of a particle accelerator.

Effluents

The natural product of the fusion reaction is a small amount of helium, which is completely harmless to life. Of more concern is tritium, which, like other isotopes of hydrogen, is difficult to retain completely. During normal operation, some amount of tritium will be continually released.

Although tritium is volatile and biologically active, the health risk posed by a release is much lower than that of most radioactive contaminants, because of tritium's short half-life (12.32 years) and very low decay energy (~14.95 keV), and because it does not bioaccumulate (instead being cycled out of the body as water, with a biological half-life of 7 to 14 days). Current ITER designs are investigating total containment facilities for any tritium.

Waste Management

The large flux of high-energy neutrons in a reactor will make the structural materials radioactive. The radioactive inventory at shut-down may be comparable to that of a fission reactor, but there are important differences.

The half-life of the radioisotopes produced by fusion tends to be less than those from fission, so that the inventory decreases more rapidly. Unlike fission reactors, whose waste remains radioactive for thousands of years, most of the radioactive material in a fusion reactor would be the reactor core itself, which would be dangerous for about 50 years, and low-level waste for another 100. Although this waste will be considerably more radioactive during those 50 years than fission waste, the very short half-life makes the process very attractive, as the waste management is fairly straightforward. By 500 years the material would have the same radiotoxicity as coal ash.

Additionally, the choice of materials used in a fusion reactor is less constrained than in a fission design, where many materials are required for their specific neutron cross-sections. This allows a fusion reactor to be designed using materials that are selected specifically to be "low activation", materials that do not easily become radioactive. Vanadium, for example, would become much less radioactive than stainless steel. Carbon fiber materials are also low-activation, as well as being strong and light, and are a promising area of study for laser-inertial reactors where a magnetic field is not required.

In general terms, fusion reactors would create far less radioactive material than a fission reactor, the material it would create is less damaging biologically, and the radioactivity "burns off" within a time period that is well within existing engineering capabilities for safe long-term waste storage.

Nuclear Proliferation

Although fusion power uses nuclear technology, the overlap with nuclear weapons would be limited. A huge amount of tritium could be produced by a fusion power station; tritium is used in the trigger of hydrogen bombs and in a modern boosted fission weapon, but it can also be produced by nuclear fission. The energetic neutrons from a fusion reactor could be used to breed weapons-grade plutonium or uranium for an atomic bomb (for example by transmutation of U^{238} to Pu^{239}, or Th^{232} to U^{233}).

A study conducted 2011 assessed the risk of three scenarios:

- *Use in small-scale fusion station*: As a result of much higher power consumption, heat dissipation and a more recognizable design compared to enrichment gas centrifuges this choice would be much easier to detect and therefore implausible.

- *Modifications to produce weapon-usable material in a commercial facility:* The production potential is significant. But no fertile or fissile substances necessary for the production of weapon-usable materials needs to be present at a civil fusion system at all. If not shielded, a detection of these materials can be done by their characteristic gamma radiation. The underlying redesign could be detected by regular design information verifications. In the (technically more feasible) case of solid breeder blanket modules, it would be necessary for incoming components to be inspected for the presence of fertile material, otherwise plutonium for several weapons could be produced each year.

- *Prioritizing a fast production of weapon-grade material regardless of secrecy:* The fastest way to produce weapon usable material was seen in modifying a prior civil fusion power station. Unlike in some nuclear power stations, there is no weapon compatible material during civil use. Even without the need for covert action this modification would still take about 2 months to start the production and at least an additional week to generate a significant amount for weapon production. This was seen as enough time to detect a military use and to react with diplomatic or military means. To stop the production, a military destruction of inevitable parts of the facility leaving out the reactor itself would be sufficient. This, together with the intrinsic safety of fusion power would only bear a low risk of radioactive contamination.

Another study concludes that "[..]large fusion reactors – even if not designed for fissile material breeding – could easily produce several hundred kg Pu per year with high weapon quality and very low source material requirements." It was emphasized that the implementation of features for intrinsic proliferation resistance might only be possible at this phase of research and development. The theoretical and computational tools needed for hydrogen bomb design are closely related to those needed for inertial confinement fusion, but have very little in common with the more scientifically developed magnetic confinement fusion.

Energy Source

Large-scale reactors using neutronic fuels (e.g. ITER) and thermal power production (turbine based) are most comparable to fission power from an engineering and economics viewpoint. Both fission and fusion power stations involve a relatively compact heat source powering a conventional steam turbine-based power station, while producing enough neutron radiation to make activation of the station materials problematic. The main distinction is that fusion power produces no high-level radioactive waste (though activated station materials still need to be disposed of). There are some power station ideas that may significantly lower the cost or size of such stations; however, research in these areas is nowhere near as advanced as in tokamaks.

Fusion power commonly proposes the use of deuterium, an isotope of hydrogen, as fuel and in many current designs also use lithium. Assuming a fusion energy output equal to the 1995 global power output of about 100 EJ/yr ($= 1 \times 10^{20}$ J/yr) and that this does not increase in the future, which is unlikely, then the known current lithium reserves would last 3000 years. Lithium from sea water would last 60 million years, however, and a more complicated fusion process using only deuterium from sea water would have fuel for 150 billion years. To put this in context, 150 billion years is close to 30 times the remaining lifespan of the sun, and more than 10 times the estimated age of the universe.

Economics

While fusion power is still in early stages of development, substantial sums have been and continue to be invested in research. In the EU almost €10 billion was spent on fusion research up to the end of the 1990s, and the new ITER reactor alone is budgeted at €6.6 billion total for the timeframe between 2008 and 2020.

It is estimated that up to the point of possible implementation of electricity generation by nuclear fusion, R&D will need further promotion totalling around €60–80 billion over a period of 50 years or so (of which €20–30 billion within the EU) based on a report from 2002. Nuclear fusion research receives €750 million (excluding ITER funding) from the European Union, compared with €810 million for sustainable energy research, putting research into fusion power well ahead of that of any single rivaling technology. Indeed, the size of the investments and time frame of the expected results mean that fusion research is almost exclusively publicly funded, while research in other forms of energy can be done by the private sector. In spite of that, a number of start-up companies active in the field of fusion power have managed to attract private money.

Advantages

Fusion power would provide more energy for a given weight of fuel than any fuel-consuming energy source currently in use, and the fuel itself (primarily deuterium) exists abundantly in the Earth's ocean: about 1 in 6500 hydrogen atoms in seawater is deuterium. Although this may seem a low proportion (about 0.015%), because nuclear fusion reactions are so much more energetic than chemical combustion and seawater is easier to access and more plentiful than fossil fuels, fusion could potentially supply the world's energy needs for millions of years.

Despite being technically non-renewable, fusion power has many of the benefits of renewable energy sources (such as being a long-term energy supply and emitting no greenhouse gases) as well as some of the benefits of the resource-limited energy sources as hydrocarbons and nuclear fission (without reprocessing). Like these currently dominant energy sources, fusion could provide very high power-generation density and uninterrupted power delivery (because it is not dependent on the weather, unlike wind and solar power).

Another aspect of fusion energy is that the cost of production does not suffer from diseconomies of scale. The cost of water and wind energy, for example, goes up as the optimal locations are developed first, while further generators must be sited in less ideal conditions. With fusion energy the production cost will not increase much even if large numbers of stations are built, because the raw resource (seawater) is abundant and widespread.

Some problems that are expected to be an issue in this century, such as fresh water shortages, can alternatively be regarded as problems of energy supply. For example, in desalination stations, seawater can be purified through distillation or reverse osmosis. Nonetheless, these processes are energy intensive. Even if the first fusion stations are not competitive with alternative sources, fusion could still become competitive if large-scale desalination requires more power than the alternatives are able to provide.

A scenario has been presented of the effect of the commercialization of fusion power on the future of human civilization. ITER and later Demo are envisioned to bring online the first commercial nuclear fusion energy reactor by 2050. Using this as the starting point and the history of the uptake of nuclear fission reactors as a guide, the scenario depicts a rapid take up of nuclear fusion energy starting after the middle of this century.

Fusion power could be used in interstellar space, where solar energy is not available.

Criticism

Because commercial fusion projects are very large and complex, and ongoing funding is a political issue, such projects usually involve cost overruns and missed deadlines. For example, the construction of the National Ignition Facility cost $5 billion and took seven years longer than expected. ITER's expected cost has gone from $5 billion to $20 billion, and the date for full power operation has been put back to 2027, from the original estimate of 2016.

Pinch (Plasma Physics)

A pinch is the compression of an electrically conducting filament by magnetic forces. The conductor is usually a plasma, but could also be a solid or liquid metal. Pinches were the first device used by mankind for controlled nuclear fusion.

The phenomenon may also be referred to as a Bennett pinch (after Willard Harrison Bennett), electromagnetic pinch, magnetic pinch, pinch effect or plasma pinch.

Pinches occur naturally in electrical discharges such as lightning bolts, the aurora, current sheets, and solar flares.

Lightning bolts illustrating electromagnetically pinched plasma filaments

A 1905 study of pinches, where electric lightning was used to create a Z-pinch inside a metal tube.

Basic Mechanism

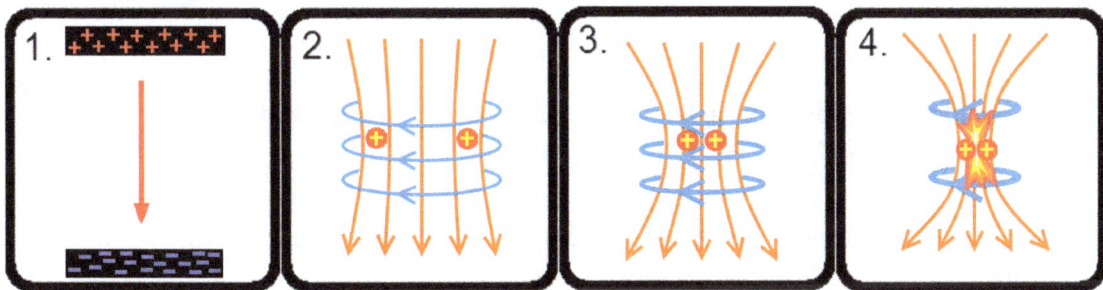

Voltage Current Magnetic Field Ion

This is a basic explanation of how a pinch works. (1) Pinches apply a huge voltage across a tube. This tube is filled with fusion fuel, typically deuterium gas. If the multiplication of the voltage & the charge is higher than the ionization energy of the gas the gas ionizes. (2) Current jumps across this gap. (3) The current makes a magnetic field which is perpendicular to the current. This magnetic field pulls the material together. (4) These atoms can get close enough to fuse.

Types

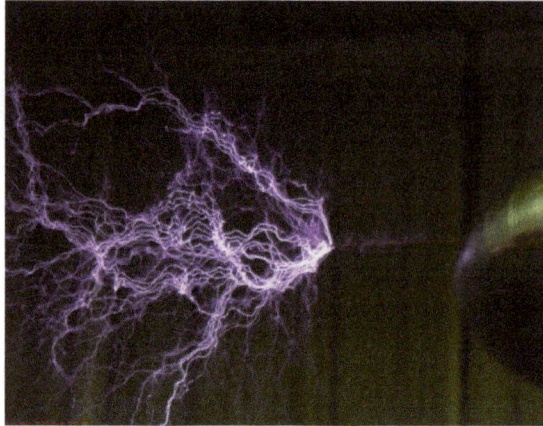

An example of a man-made pinch. Here Z-pinches constrain the plasma inside filaments of electrical discharge from a Tesla coil

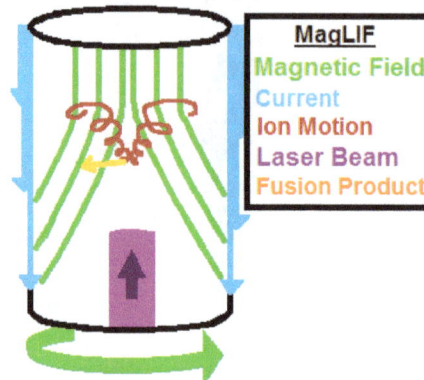

The MagLIF concept, a combination of a Z-pinch and a laser beam

Pinches exist in laboratories and in nature. Pinches differ in their geometry and operating forces. These include:

Uncontrolled

> Any time an electric current moves in large amounts (e.g., lightning, arcs, sparks, discharges) a magnetic force can pull together plasma. This can be insufficient for fusion.

Sheet pinch

> An astrophysical effect, this arises from vast sheets of charge particles.

Z-pinch

> The current runs down the axis (or walls) of the cylinder while the magnetic field is azimuthal

Theta pinch

> The magnetic field runs down the axis of the cylinder, while the electric field is in the azimuthal direction (also called a thetatron)

Screw pinch

A combination of a Z pinch and theta pinch (also called a stabilized Z-pinch, or θ-Z pinch)

Reversed field pinch

This is an attempt to do a Z-pinch inside an endless loop. The plasma has an internal magnetic field. As you move out from the center of this ring, the magnetic field reverses direction. Also called a toroidal pinch.

Inverse pinch

An early fusion concept, this device consisted of a rod surrounded by plasma. Current traveled through the plasma and returned along the center rod. This geometry was slightly different than a z-pinch in that the conductor was in the center, not the sides.

Cylindrical pinch

Orthogonal pinch effect

Ware pinch

A pinch that happens inside Tokamaks. This is when particles inside the Banana orbit condense together.

MAGLIF

This is a combination of a Z-pinch and a laser beam to stop loss of material at the end.

Common Behavior

Pinches may become unstable. They radiate energy as light across the whole electromagnetic spectrum including radio waves, x-rays, gamma rays, synchrotron radiation, and visible light. They also produce neutrons, as a product of fusion.

Applications and Devices

Pinches are used to generate X-rays and the intense magnetic fields generated are used in electromagnetic forming of metals. They also have applications in particle beams including particle beam weapons, astrophysics studies and it has been proposed to use them in space propulsion. A number of large pinch machines have been built to study fusion power; here are several:

- MAGPIE A Z-pinch at Imperial college. This dumps a massive amount of current across a wire. Under these conditions the wire becomes plasma and compresses to produce fusion.

- Z Pulsed Power Facility at Sandia National Laboratories.

- ZETA device in Culham England

- Madison Symmetric Torus at the University of Wisconsin Madison

- Reversed-Field eXperiment in Italy.

- dense plasma focus in New Jersey

- University of Nevada, Reno (USA)

- Cornell University (USA)

- University of Michigan (USA)

- University of California, San Diego (USA)

- University of Washington (USA)

- Ruhr University (Germany)

- École Polytechnique (France)

- Weizmann Institute of Science (Israel)

- Universidad Autónoma Metropolitana (Mexico).

Crushing Cans with the Pinch Effect

Many high-voltage electronics enthusiasts make their own crude electromagnetic forming devices. They use pulsed power techniques to produce a theta pinch capable of crushing an aluminium soft drink can using the Lorentz forces created when large currents are induced in the can by the strong magnetic field of the primary coil.

Pinched aluminium can, produced from a pulsed magnetic field created by rapidly discharging 2 kilojoules from a high voltage capacitor bank into a 3-turn coil of heavy gauge wire.

An electromagnetic aluminium can crusher consists of four main components: a high voltage DC power supply, which provides a source of electrical energy, a large *energy discharge* capacitor to accumulate the electrical energy, a high voltage switch or spark gap, and a robust coil (capable of surviving high magnetic pressure) through which the stored electrical energy can be quickly discharged in order to generate a correspondingly strong pinching magnetic field.

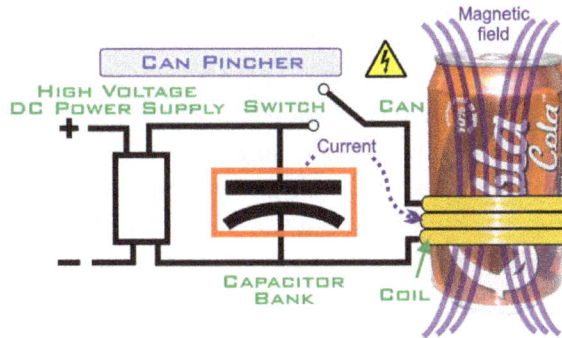

Electromagnetic pinch "can crusher": schematic diagram

In practice, such a device is somewhat more sophisticated than the schematic diagram suggests, including electrical components that control the current in order to maximize the resulting pinch, and to ensure that the device works safely.

History

The first creation of a Z-pinch in the laboratory may have occurred in 1790 in Holland when Martinus van Marum created an explosion by discharging 100 Leyden jars into a wire. The phenomenon was not understood until 1905, when Pollock and Barraclough investigated a compressed and distorted length of copper tube from a lightning rod after it had been struck by lightning. Their analysis showed that the forces due to the interaction of the large current flow with its own magnetic field could have caused the compression and distortion. A similar, and apparently independent, theoretical analysis of the pinch effect in liquid metals was published by Northrupp in 1907. The next major development was the publication in 1934 of an analysis of the radial pressure balance in a static Z-pinch by Bennett.

The Institute of Electrical and Electronics Engineers emblem shows the basic features of an azimuthal magnetic pinch.

Thereafter, the experimental and theoretical progress on pinches was driven by fusion power research. In their article on the "Wire-array Z-pinch: a powerful x-ray source for ICF", M G Haines *et al.*, wrote on the "Early history of Z-pinches".

In 1946 Thompson and Blackman submitted a patent for a fusion reactor based on a toroidal Z-pinch with an additional vertical magnetic field. But in 1954 Kruskal and Schwarzschild published their theory of MHD instabilities in a Z-pinch. In 1956 Kurchatov gave his famous Harwell lecture showing nonthermal neutrons and the presence of $m = 0$ and $m = 1$ instabilities in a deuterium pinch. In 1957 Pease and Braginskii independently predicted radiative collapse in a Z-pinch under pressure balance when in hydrogen the current exceeds 1.4 MA. (The viscous rather than resistive dissipation of magnetic energy discussed above and in would however prevent radiative collapse). In 1958, the world's first controlled fusion experiment was accomplished using a Z-pinch machine named Scylla I at the Los Alamos National Laboratory. A cylinder full of deuterium was converted into a plasma and compressed to 15 million degrees Celsius under a Z-pinch effect. Lastly, at Imperial College in 1960, led by R Latham, the Plateau-Rayleigh instability was shown, and its growth rate measured in a dynamic Z-pinch.

Equilibrium Analysis

One dimension

In plasma physics three pinch geometries are commonly studied: the θ-pinch, the Z-pinch, and the Screw Pinch. These are cylindrically shaped. The cylinder is symmetric in the axial (z) direction and the azimuthal (θ) directions. The one-dimensional pinches are named for the direction the current travels.

The θ-pinch

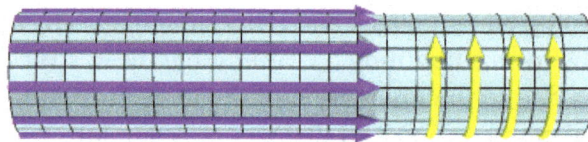

A sketch of the θ-pinch equilibrium. The ▇ z-directed magnetic field corresponds to a ▇ θ-directed plasma current.

The θ-pinch has a magnetic field directed in the z direction and a large diamagnetic current directed in the θ direction. Using Ampère's law (discarding the displacement term)

$$\vec{B} = B_z(r)\hat{z}$$

$$\mu_0 \vec{J} = \nabla \times \vec{B}$$

$$= \frac{1}{r}\frac{d}{d\theta}B_z(r)\hat{r} - \frac{d}{dr}B_z(r)\hat{\theta}$$

Since B is only a function of r we can simplify this to

$$\mu_0 \vec{J} = -\frac{d}{dr}B_z(r)\hat{\theta}$$

So J points in the θ direction.

Thus, the equilibrium condition ($\nabla p = j \times B$) for the θ-pinch reads:

$$\frac{d}{dr}\left(p + \frac{B_z^2}{2\mu_0}\right) = 0$$

θ-pinches tend to be resistant to plasma instabilities; This is due in part to Alfvén's Theorem (or, frozen in flux theorem).

The Z-pinch

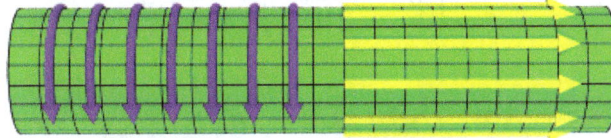

A sketch of the Z-pinch equilibrium. A ▮ θ-directed magnetic field corresponds to a ▮ z-directed plasma current.

The Z-pinch has a magnetic field in the θ direction and a current "J" flowing in the "z" direction. Again, by electrostatic Ampère's Law

$$\vec{B} = B_\theta(r)\hat{\theta}$$

$$u_0\vec{J} = \nabla \times \vec{B}$$

$$= \frac{1}{r}\frac{d}{dr}\left(rB_\theta(r)\right)\hat{z} - \frac{d}{dz}B_\theta(r)\hat{r}$$

$$= \frac{1}{r}\frac{d}{dr}\left(rB_\theta(r)\right)\hat{z}$$

Thus, the equilibrium condition, $\nabla p = j \times B$, for the Z-pinch reads:

$$\frac{d}{dr}\left(p + \frac{B_\theta^2}{2\mu_0}\right) + \frac{B_\theta^2}{\mu_0 r} = 0$$

Since particles in a plasma basically follow magnetic field lines, Z-pinches lead them around in circles. Therefore, they tend to have excellent confinement properties.

The Screw Pinch

The screw pinch is an effort to combine the stability aspects of the θ-pinch and the confinement aspects of the Z-pinch. Referring once again to Ampère's Law

$$\nabla \times \vec{B} = \mu_0\vec{J}$$

But this time, the B field has a θ component *and* a z component

$$\vec{B} = B_\theta\hat{\theta} + B_z\hat{z}$$

$$\mu_0 \vec{J} = \frac{1}{r}\frac{d}{dr}\left(rB_\theta\right)\hat{z} - \frac{d}{dr}B_z\hat{\theta}$$

So this time J has a component in the z direction and a component in the θ direction.

Finally, the equilibrium condition ($\nabla p = j \times B$) for the screw pinch reads:

$$\frac{d}{dr}\left(p + \frac{B_z^2 + B_\theta^2}{2\mu_0}\right) + \frac{B_\theta^2}{\mu_0 r} = 0$$

The screw Pinch via Colliding Optical Vortices

The *screw pinch* might be produced in laser plasma by colliding optical vortices of ultrashort duration. For this purpose optical vortices ought to be phase-conjugated. The magnetic field distribution is given here again via Ampère's law:

$$\nabla \times \vec{B} = \mu_0 \vec{J}$$

Two Dimensions

A common problem with one-dimensional pinches is the end losses. Most of the motion of particles is along the magnetic field. With the θ-pinch and the screw-pinch, this leads particles out of the end of the machine very quickly, leading to a loss of mass and energy. On top of this problem, the Z-pinch has major stability problems. Though particles can be reflected to some extent with magnetic mirrors, even these allow many particles to pass. A common method of beating these end losses, is to bend the cylinder around into a torus. Unfortunately this breaks θ symmetry, as paths on the inner portion (inboard side) of the torus are shorter than similar paths on the outer portion (outboard side). Thus, a new theory is needed. This gives rise to the famous Grad–Shafranov equation. Numerical solutions to the Grad–Shafranov equation have also yielded some equilibria, most notably that of the reversed field pinch.

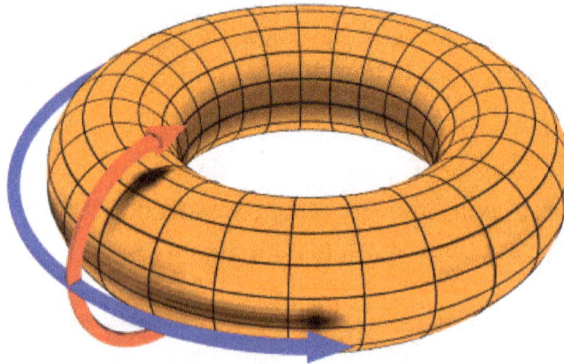

A *toroidal coordinate system* in common use in plasma physics.

The red arrow indicates the poloidal direction (θ)

The blue arrow indicates the toroidal direction (φ)

Three Dimensions

As of 2015, there is not a coherent analytical theory for three-dimensional equilibria. The general approach to finding three-dimensional equilibria is to solve the vacuum ideal MHD equations. Numerical solutions have yielded designs for stellarators. Some machines take advantage of simplification techniques such as helical symmetry (for example University of Wisconsin's Helically Symmetric eXperiment). However, for an arbitrary three-dimensional configuration an equilibrium relation, similar to that of the 1-D configurations exists:

$$\nabla_{\perp}\left(p + \frac{B^2}{2\mu_0}\right) - \frac{B^2}{\mu_0}\vec{\kappa} = 0$$

Where κ is the curvature vector defined as:

$$\vec{\kappa} = \left(\vec{b} \cdot \nabla\right)\vec{b}$$

with b the unit vector tangent to B.

Formal Treatment

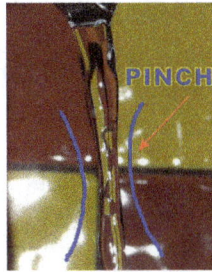

A stream of water pinching into droplets has been suggested as an analogy to the electromagnetic pinch. The gravity accelerates free-falling water which causes the water column to constrict. Then surface tension breaks the narrowing water column into droplets (not shown here) (see Plateau-Rayleigh instability), which is analogous to the magnetic field which has been suggested as the cause of pinching in bead lightning. The morphology (shape) is similar to the so-called sausage instability in plasma.

The Bennett Relation

Consider a cylindrical column of fully ionized quasineutral plasma, with an axial electric field, producing an axial current density, j, and associated azimuthal magnetic field, B. As the current flows through its own magnetic field, a pinch is generated with an inward radial force density of j x B. In a steady state with forces balancing:

$$\nabla p = \nabla(p_e + p_i) = j \times B$$

where ∇p is the magnetic pressure gradient, p_e and p_i is the electron and ion pressures. Then using Maxwell's equation $\nabla \times B = \mu_0 j$ and the ideal gas law $p = N k T$, we derive:

$$2Nk(T_e + T_i) = \frac{\mu_0}{4\pi}I^2 \text{ (the Bennett relation)}$$

where N is the number of electrons per unit length along the axis, T_e and T_i are the electron and ion temperatures, I is the total beam current, and k is the Boltzmann constant.

The Generalized Bennett Relation

The *Generalized Bennett Relation* considers a current-carrying magnetic-field-aligned cylindrical plasma pinch undergoing rotation at angular frequency ω. Along the axis of the plasma cylinder flows a current density j_z, resulting in an azimuthal magnetic field B_φ. Originally derived by Witalis, the Generalized Bennett Relation results in:

$$\frac{1}{4}\frac{\partial^2 J_0}{\partial t^2} = W_{\perp kin} + \Delta W_{E_z} + \Delta W_{B_z} + \Delta W_k - \frac{\mu_0}{8\pi}I^2(a)$$

$$-\frac{1}{2}G\bar{m}^2 N^2(a) + \frac{1}{2}\pi a^2 \epsilon_0\left(E_r^2(a) - E_\phi^2(a)\right)$$

- where a current-carrying, magnetic-field-aligned cylindrical plasma has a radius a,
- J_0 is the total moment of inertia with respect to the z axis,
- $W_{\perp kin}$ is the kinetic energy per unit length due to beam motion transverse to the beam axis
- W_{Bz} is the self-consistent B_z energy per unit length
- W_{Ez} is the self-consistent E_z energy per unit length
- W_k is thermokinetic energy per unit length
- $I(a)$ is the axial current inside the radius a (r in diagram)
- $N(a)$ is the total number of particles per unit length
- E_r is the radial electric field
- E_φ is the rotational electric field

The positive terms in the equation are expansional forces while the negative terms represent beam compressional forces.

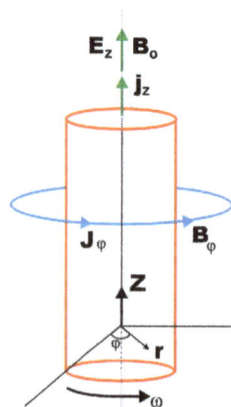

The generalized Bennett relation considers a current-carrying magnetic-field-aligned cylindrical plasma pinch undergoing rotation at angular frequency ω

The Carlqvist Relation

The Carlqvist Relation, published by Per Carlqvist in 1988, is a specialization of the Generalized Bennett Relation (above), for the case that the kinetic pressure is much smaller at the border of the pinch than in the inner parts. It takes the form

$$\frac{\mu_0}{8\pi} I^2(a) + \frac{1}{2} G\bar{m}^2 N^2(a) = \Delta W_{B_z} + \Delta W_k$$

and is applicable to many space plasmas.

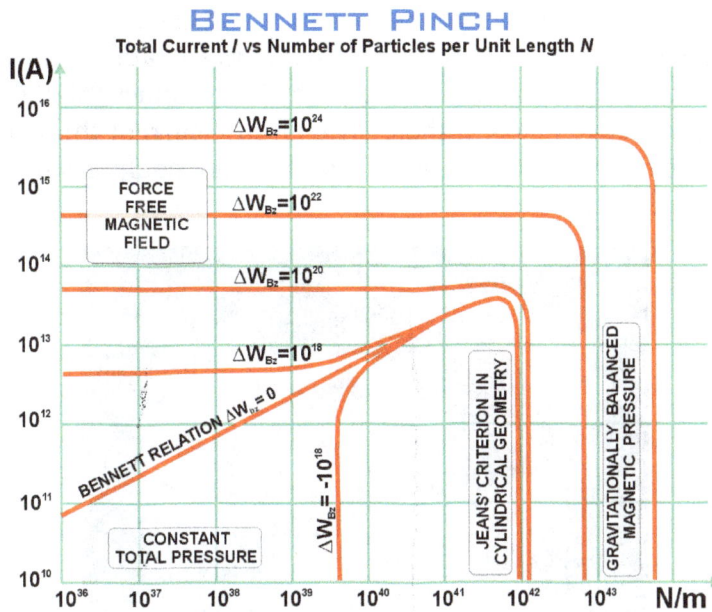

The Bennett pinch showing the total current (I) versus the number of particles per unit length (N). The chart illustrates four physically distinct regions. The plasma temperature is 20 K, the mean particle mass 3×10^{-27} kg, and ΔW_{Bz} is the excess magnetic energy per unit length due to the axial magnetic field B_z. The plasma is assumed to be non-rotational, and the kinetic pressure at the edges is much smaller than inside.

The Carlqvist Relation can be illustrated, showing the total current (*I*) versus the number of particles per unit length (*N*) in a Bennett pinch. The chart illustrates four physically distinct regions. The plasma temperature is quite cold ($T_i = T_e = T_n = 20$ K), containing mainly hydrogen with a mean particle mass 3×10^{-27} kg. The thermokinetic energy $W_k >> \pi a^2 p_k(a)$. The curves, ΔW_{Bz} show different amounts of excess magnetic energy per unit length due to the axial magnetic field B_z. The plasma is assumed to be non-rotational, and the kinetic pressure at the edges is much smaller than inside.

Chart regions: (a) In the top-left region, the pinching force dominates. (b) Towards the bottom, outward kinetic pressures balance inwards magnetic pressure, and the total pressure is constant. (c) To the right of the vertical line $\Delta W_{Bz} = 0$, the magnetic pressures balances the gravitational pressure, and the pinching force is negligible. (d) To the left of the sloping curve $\Delta W_{Bz} = 0$, the gravitational force is negligible. Note that the chart shows a special case of the Carlqvist relation, and if it is replaced by the more general Bennett relation, then the designated regions of the chart are not valid.

Carlqvist further notes that by using the relations above, and a derivative, it is possible to describe the Bennett pinch, the Jeans criterion (for gravitational instability, in one and two dimensions), force-free magnetic fields, gravitationally balanced magnetic pressures, and continuous transitions between these states.

Types of Pinch

Current Sheet

A current sheet is an electric current that is confined to a surface, rather than being spread through a volume of space. Current sheets feature in magnetohydrodynamics (MHD), the study of the behavior of electrically conductive fluids: if there is an electric current through part of the volume of such a fluid, magnetic forces tend to expel it from the fluid, compressing the current into thin layers that pass through the volume.

The heliospheric current sheet results from the influence of the Sun's rotating magnetic field on the plasma in the interplanetary medium

The largest occurring current sheet in the Solar System is the so-called Heliospheric current sheet, which is about 10,000 km thick, and extends from the Sun and out beyond the orbit of Pluto.

In astrophysical plasmas such as the solar corona, current sheets may have an aspect ratio (breadth divided by thickness) as high as 100,000:1. By contrast, the pages of most books have an aspect ratio close to 2000:1. Because current sheets are so thin in comparison to their size, they are often treated as if they have zero thickness; this is a result of the simplifying assumptions of ideal MHD. In reality, no current sheet may be infinitely thin because that would require infinitely fast motion of the charge carriers whose motion causes the current.

Current sheets in plasmas store energy by increasing the energy density of the magnetic field. Many plasma instabilities arise near strong current sheets, which are prone to collapse, causing magnetic reconnection and rapidly releasing the stored energy. This process is the cause of solar flares and is one reason for the difficulty of magnetic confinement fusion, which requires strong electric currents in a hot plasma.

Magnetic Field of An Infinite Current Sheet

An infinite current sheet can be modelled as an infinite number of parallel wires all carrying the same current. Assuming each wire carries current I, and there are N wires per unit length, the magnetic field can be derived using Ampère's law:

$$\oint_R \mathbf{B} \cdot \mathbf{ds} = \mu_0 I_{enc}$$

$$\oint_R Bds \cos \theta = \mu_0 I_{enc}$$

R is a rectangular loop surrounding the current sheet, perpendicular to the plane and perpendicular to the wires. In the two sides perpendicular to the sheet, $\mathbf{B} \cdot \mathbf{ds} = 0$ since $\cos(0) = 1$. In the other two sides, $\cos(0) = 1$, so if S is one parallel side of the rectangular loop of dimensions L x W, the integral simplifies to:

$$2\int_S Bds = \mu_0 I_{enc}$$

Since B is constant due to the chosen path, it can be pulled out of the integral:

$$2B\int_S ds = \mu_0 I_{enc}$$

The integral is evaluated:

$$2BL = \mu_0 I_{enc}$$

Solving for B, plugging in for I_{enc} (total current enclosed in path R) as $I*N*L$, and simplifying:

$$B = \frac{\mu_0 I_{enc}}{2L}$$

$$B = \frac{\mu_0 INL}{2L}$$

$$B = \frac{\mu_0 I_{enc}}{2L}$$

Notably, the magnetic field strength of an infinite current sheet does not depend on the distance from it.

The direction of B can be found via the right-hand rule.

Harris Current Sheet

A well-known one-dimensional current sheet equilibrium is Harris current sheet, which is a stationary solution to the Maxwell-Vlasov system. The magnetic field profile is given by

$$\mathbf{B} = B_0 \tanh(x/L)\mathbf{e}_z$$

Z-pinch

In fusion power research, the Z-pinch, also known as zeta pinch, is a type of plasma confinement system that uses an electrical current in the plasma to generate a magnetic field that compresses it. These systems were originally referred to simply as pinch or Bennett pinch (after Willard Harrison Bennett), but the introduction of the theta-pinch concept led to the need for increased clarity.

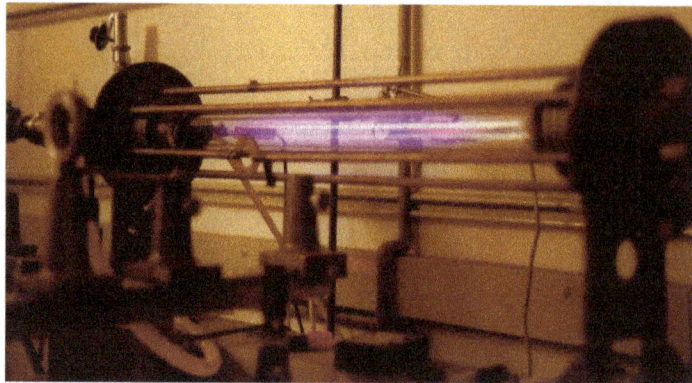

Laboratory scale Z-pinch showing glow from an expanded hydrogen plasma. Pinch and ionisation current flows through the gas and returns via the bars surrounding the plasma vessel.

The name refers to the direction of the current in the devices, the Z-axis on a normal three-dimensional graph. Any machine that causes a pinch effect due to current running in that direction is correctly referred to as a Z-pinch system, and this encompasses a wide variety of devices used for an equally wide variety of purposes. Early uses focused on fusion research in donut-shaped tubes with the Z-axis running down the inside the tube, modern devices are generally cylindrical and used to generate high-intensity x-ray sources for the study of nuclear weapons and other roles.

Physics

The Z-pinch is an application of the Lorentz force, in which a current-carrying conductor in a magnetic field experiences a force. One example of the Lorentz force is that, if two parallel wires are carrying current in the same direction, the wires will be pulled toward each other. In a Z-pinch machine the wires are replaced by a plasma, which can be thought of as many current-carrying wires. When a current is run through the plasma, the particles in plasma are pulled toward each other by the Lorentz force, thus the plasma contracts. The contraction is counteracted by the increasing gas pressure of the plasma.

As the plasma is electrically conductive, a magnetic field nearby will induce a current in it. This provides a way to run a current into the plasma without physical contact, which is important as a plasma can rapidly erode mechanical electrodes. In practical devices this was normally arranged by placing the plasma vessel inside the core of a transformer, arranged so the plasma itself would be the secondary. When current was sent into the primary side of the transformer, the magnetic field induced a current into the plasma. As induction requires a *changing* magnetic field, and the induced current is supposed to run in a single direction in most reactor designs, the current in the transformer has to be increased over time to produce the varying magnetic field. This places a limit on the product of confinement time and magnetic field, for any given source of power.

In Z-pinch machines the current is generally provided from a large bank of capacitors and triggered by a spark gap, known as a Marx Bank or Marx generator. As the conductivity of plasma is fairly good, about that of copper, the energy stored in the power source is quickly depleted by running through the plasma. Z-pinch devices are inherently pulsed in nature.

Usage

Early Machines

Pinch devices were among the earliest efforts in fusion power. The concept traces itself to work in the UK in the immediate post-war era, but a lack of interest led to little development until early 1950. Then, the announcement of the Huemul Project in early 1951 led to fusion efforts around the world, notably pinch devices in the UK and pinch and stellarators in the US. A number of small experiments were built at labs as various practical issues were addressed, but all of these machines demonstrated unexpected instabilities of the plasma that would cause it to hit the walls of the container vessel. The problem became known as the "kink instability".

Stabilized Pinch, the Race to Fusion

A number of solutions were proposed, and by 1953 the "stabilized pinch" seemed to solve the problems encountered on earlier devices. Stabilized pinch machines added a set of external magnets that created a toroidal magnetic field inside the chamber. When the device was fired, this field added to the one created by the current in the plasma. The result was that the formerly straight magnetic field was twisted into a helix, which the particles followed as they travelled around the tube under the influence of the current. A particle near the outside of the tube that would want to kink outward would travel along these lines until it found itself on the inside of the tube, where its outward-directed motion would bring it back into the centre of the plasma.

Researchers in the UK planned a major assault on the stabilized pinch field, and started construction of ZETA in 1954. ZETA was by far the largest fusion device of its era, and equipped with all of the latest equipment. At the time, almost all fusion research was classified, so progress on ZETA was generally unknown outside the labs working on it. However, in 1956 the walls started to come down, and when they visited ZETA at Harwell, US researchers became aware that they were about to be trumped. A race broke out as teams on both sides of the Atlantic rushed to be the first to complete their stabilized pinch machines.

ZETA won the race, and by the summer of 1957 it was producing bursts of neutrons on every run. Although the scientists working on the device, and similar ones in the US and UK, were careful to point out that it was not proven, the results were nevertheless released with great fanfare as the first successful step on the path to commercial fusion energy. However, further study soon demonstrated that the measurements were misleading, and none of the machines were near fusion levels. Interest in pinch devices faded, although ZETA and its cousin Sceptre would serve for many years as experimental devices.

Fusion-based Propulsion

A concept of Z-pinch fusion propulsion system has been developed through collaboration between NASA and several private companies. A magnetic field, generated by a pulse discharge, compresses a plasma to fusion temperatures using the Z-pinch effect. The energy released accelerates the lithium propellant to a high speed, resulting in a specific impulse value of 19400 s and thrust of 38 kN. A magnetic nozzle is required to convert the released energy into a useful impulse. This propulsion method can significantly reduce interplanetary travel times. For example, a mission to Mars would take about 35 days one-way with a total burn time of 20 days and a burned propellant mass of 350 tonnes.

Tokamak

Although it remained relatively unknown for years, Soviet scientists used the pinch concept to develop the tokamak device. Unlike the stabilized pinch devices in the US and UK, the tokamak used considerably more energy in the stabilizing magnets, and much less in the plasma current. This reduced the instabilities due to the large currents in the plasma, and led to great improvements in stability. The results of their experiments were so good that other researchers were skeptical of them when they were first announced in full force in 1968. Members of the still-operational ZETA team were called in to verify the results. The tokamak has since gone on to become the most studied approach to controlled fusion.

Various Z-pinch Machines

They can be found in various institutions such as University of Nevada, Reno (USA), Cornell University (USA), University of Michigan (USA), Sandia National Laboratories (USA), University of California, San Diego (USA), University of Washington (USA), Ruhr University (Germany), Imperial College (United Kingdom), École Polytechnique (France), Weizmann Institute of Science (Israel), Universidad Autónoma Metropolitana (Mexico), NSTRI (IRAN)

A Z-pinch machine at UAM, Mexico City.

Reversed Field Pinch

A reversed-field pinch (RFP) is a device used to produce and contain near-thermonuclear plasmas. It is a toroidal pinch which uses a unique magnetic field configuration as a scheme to magnetically confine a plasma, primarily to study magnetic fusion energy. Its magnetic geometry is somewhat different from that of the more common tokamak. As one moves out radially, the portion of the magnetic field pointing toroidally reverses its direction, giving rise to the term "reversed field".

This configuration can be sustained with comparatively lower fields than that of a tokamak of similar power density. One of the disadvantages of this configuration is that it tends to be more susceptible to non-linear effects and turbulence. This makes it a perfect laboratory for non-ideal (resistive) magnetohydrodynamics. RFPs are also used in the study of astrophysical plasmas as they share many features.

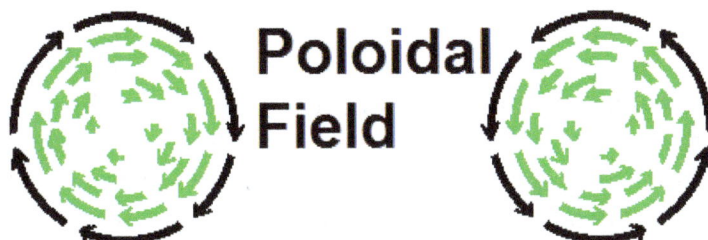

The poloidal field in a reverse field pinch

The largest Reversed Field Pinch device presently in operation is the RFX (R/a = 2/0.46) in Padua, Italy. Others include the MST (R/a = 1.5/0.5) in the United States, EXTRAP T2R (R/a = 1.24/0.18) in Sweden, TPE-RX (R = 0.51/0.25) in Japan, and KTX (R/a = 1.4/0.4) in China.

Characteristics

Unlike the Tokamak, which has a much larger magnetic field in the toroidal direction than the poloidal direction, an RFP has a comparable field strength in both directions (though the sign of the toroidal field reverses). Moreover, a typical RFP has a field strength approximately one half to one tenth that of a comparable Tokamak. The RFP also relies on driving current in the plasma to reinforce the field from the magnets through the dynamo effect.

Magnetic Topology

The reversed-field pinch works towards a state of minimum energy.

The magnetic field lines coil loosely around a center torus. They coil outwards. Near the plasma edge, the toroidal magnetic field reverses and the field lines coil in the reverse direction.

Internal fields are bigger than the fields at the magnets.

RFP in Fusion Research: Comparison with Other Confinement Configurations

The RFP has many features which make it a promising configuration for a potential fusion reactor.

Advantages

Due to the lower overall fields, an RFP reactor might not need superconducting magnets. This is a large advantage over tokamaks since superconducting magnets are delicate and expensive and so must be shielded from the Neutron rich fusion environment. RFPs are susceptible to surface instabilities and so require a close fitting shell. Some experiments (such as the Madison Symmetric Torus) use their close fitting shell as a magnetic coil by driving current through the shell itself. This is attractive from a reactor standpoint since a solid copper shell (for example) would be fairly robust against high energy neutrons, compared with superconducting magnets. There is also no established beta limit for RFPs. There exists a possibility that a reversed field pinch could achieve ignition solely with ohmic power (by driving current through the plasma and generating heat from electrical resistance, rather than through electron cyclotron resonance heating), which would be much simpler than tokamak designs, though it could not be operated in steady state.

Disadvantages

Typically RFPs require a large amount of current to be driven, and although promising experiments are underway, there is no established method of replacing ohmically driven current, which is fundamentally limited by the machine parameters. RFPs are also prone to tearing modes which lead to overlapping magnetic islands and therefore rapid transport from the core of the plasma to the edge. These problems are areas of active research in the RFP community.

The plasma confinement in the best RFP's is only about 1% as good as in the best tokamaks. One reason for this is that all existing RFP's are relatively small. MST was larger than any previous RFP device, and thus it tested this important size issue. The RFP is believed to require a shell with high electrical conductivity very close to the boundary of the plasma. This requirement is an unfortunate complication in a reactor. The Madison Symmetric Torus was designed to test this assumption and to learn how good the conductor must be and how close to the plasma it must be placed. In RFX, the thick shell was replaced with an active system of 192 coils, which cover the entire torus with their saddle shape, and response to the magnetic push of the plasma. Active control of plasma modes is also possible with this system.

Plasma Physics Research

The Reversed Field Pinch is also interesting from a physics standpoint. RFP dynamics are highly turbulent. RFPs also exhibit a strong plasma dynamo, similar to many astrophysical bodies. Basic plasma science is another important aspect of Reversed Field Pinch research.

Inertial Confinement Fusion

Inertial confinement fusion (ICF) is a type of fusion energy research that attempts to initiate nuclear fusion reactions by heating and compressing a fuel target, typically in the form of a pellet that most often contains a mixture of deuterium and tritium.

To compress and heat the fuel, energy is delivered to the outer layer of the target using high-energy beams of laser light, electrons or ions, although for a variety of reasons, almost all ICF devices as of 2015 have used lasers. The heated outer layer explodes outward, producing a reaction force against the remainder of the target, accelerating it inwards, compressing the target. This process is designed to create shock waves that travel inward through the target. A sufficiently powerful set of shock waves can compress and heat the fuel at the center so much that fusion reactions occur.

The energy released by these reactions will then heat the surrounding fuel, and if the heating is strong enough this could also begin to undergo fusion. The aim of ICF is to produce a condition known as *ignition*, where this heating process causes a chain reaction that burns a significant portion of the fuel. Typical fuel pellets are about the size of a pinhead and contain around 10 milligrams of fuel: in practice, only a small proportion of this fuel will undergo fusion, but if all this fuel were consumed it would release the energy equivalent to burning a barrel of oil.

Inertial confinement fusion using lasers rapidly progressed in the late 1970s and early 1980s from being able to deliver only a few joules of laser energy (per pulse) to being able to deliver tens of kilojoules to a target. At this point, incredibly large scientific devices were needed for experimentation. Here, a view of the 10 beam LLNL Nova laser, shown shortly after the laser's completion in 1984. Around the time of the construction of its predecessor, the Shiva laser, laser fusion had entered the realm of "big science".

ICF is one of two major branches of fusion energy research, the other being magnetic confinement fusion. When it was first proposed in the early 1970s, ICF appeared to be a practical approach to fusion power production and the field flourished. Experiments during the 1970s and '80s demonstrated that the efficiency of these devices was much lower than expected, and reaching ignition would not be easy. Throughout the 1980s and '90s, many experiments were conducted in order to understand the complex interaction of high-intensity laser light and plasma. These led to the design of newer machines, much larger, that would finally reach ignition energies.

The largest operational ICF experiment is the National Ignition Facility (NIF) in the US, designed using all of the decades-long experience of earlier experiments. Like those earlier experiments, however, NIF has failed to reach ignition and is, as of 2015, generating about $\frac{1}{3}$ of the required energy levels. As of October 7, 2013, this facility is understood to have achieved an important milestone towards commercialization of fusion, namely, for the first time a fuel capsule gave off more energy than was applied to it. This is a major step forward. A similar large-scale device in France, Laser Mégajoule, was officially inaugurated in October 2014. Experiments have started since then, albeit with low laser energies involved.

Description

Basic Fusion

Fusion reactions combine lighter atoms, such as hydrogen, together to form larger ones. Generally the reactions take place at such high temperatures that the atoms have been ionized, their electrons stripped off by the heat; thus, fusion is typically described in terms of "nuclei" instead of "atoms".

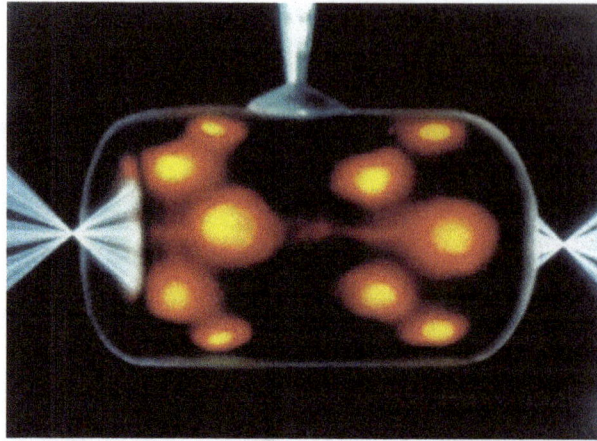

Indirect drive laser ICF uses a *hohlraum* which is irradiated with laser beam cones from either side on its inner surface to bathe a fusion microcapsule inside with smooth high intensity X-rays. The highest energy X-rays can be seen leaking through the hohlraum, represented here in orange/red.

Nuclei are positively charged, and thus repel each other due to the electrostatic force. Overcoming this repulsion costs a considerable amount of energy, which is known as the *Coulomb barrier* or *fusion barrier energy*. Generally, less energy will be needed to cause lighter nuclei to fuse, as they have less charge and thus a lower barrier energy, and when they do fuse, more energy will be released. As the mass of the nuclei increase, there is a point where the reaction no longer gives off net energy—the energy needed to overcome the energy barrier is greater than the energy released in the resulting fusion reaction.

The best fuel from an energy perspective is a one-to-one mix of deuterium and tritium; both are heavy isotopes of hydrogen. The D-T (deuterium & tritium) mix has a low barrier because of its high ratio of neutrons to protons. The presence of neutral neutrons in the nuclei helps pull them together via the nuclear force, while the presence of positively charged protons pushes the nuclei apart via electrostatic force. Tritium has one of the highest ratios of neutrons to protons of any stable or moderately unstable nuclide—two neutrons and one proton. Adding protons or removing neutrons increases the energy barrier.

A mix of D-T at standard conditions does not undergo fusion; the nuclei must be forced together before the nuclear force can pull them together into stable collections. Even in the hot, dense center of the sun, the average proton will exist for billions of years before it fuses. For practical fusion power systems, the rate must be dramatically increased; heated to tens of millions of degrees, and/or compressed to immense pressures. The temperature and pressure required for any particular fuel to fuse is known as the Lawson criterion. These conditions have been known since the 1950s when the first H-bombs were built. To meet the Lawson Criterion is extremely difficult on Earth, which explains why fusion research has taken many years to reach the current high state of technical prowess.

ICF Mechanism of Action

In a hydrogen bomb, the fusion fuel is compressed and heated with a separate fission bomb. A variety of mechanisms transfers the energy of the fission "trigger"'s explosion into the fusion fuel. The requirement of a fission bomb makes the method impractical for power generation. Not only would the triggers be prohibitively expensive to produce, but there is a minimum size that such a bomb can be built, defined roughly by the critical mass of the plutonium fuel used. Generally it seems difficult to build nuclear devices smaller than about 1 kiloton in yield, which would make it a difficult engineering problem to extract power from the resulting explosions.

As the explosion size is scaled down, so too is the amount of energy needed to start the reaction off. Studies from the late 1950s and early 1960s suggested that scaling down into the megajoule energy range would require energy levels that could be delivered by any number of means. This led to the idea of using a device that would "beam" the energy at the fusion fuel, ensuring mechanical separation. By the mid-1960s, it appeared that the laser would develop to the point where the required energy levels would be available.

Generally ICF systems use a single laser, the *driver*, whose beam is split up into a number of beams which are subsequently individually amplified by a trillion times or more. These are sent into the reaction chamber (called a target chamber) by a number of mirrors, positioned in order to illuminate the target evenly over its whole surface. The heat applied by the driver causes the outer layer of the target to explode, just as the outer layers of an H-bomb's fuel cylinder do when illuminated by the X-rays of the fission device.

The material exploding off the surface causes the remaining material on the inside to be driven inwards with great force, eventually collapsing into a tiny near-spherical ball. In modern ICF devices the density of the resulting fuel mixture is as much as one-hundred times the density of lead, around 1000 g/cm³. This density is not high enough to create any useful rate of fusion on its own. However, during the collapse of the fuel, shock waves also form and travel into the center of the fuel at high speed. When they meet their counterparts moving in from the other sides of the fuel in the center, the density of that spot is raised much further.

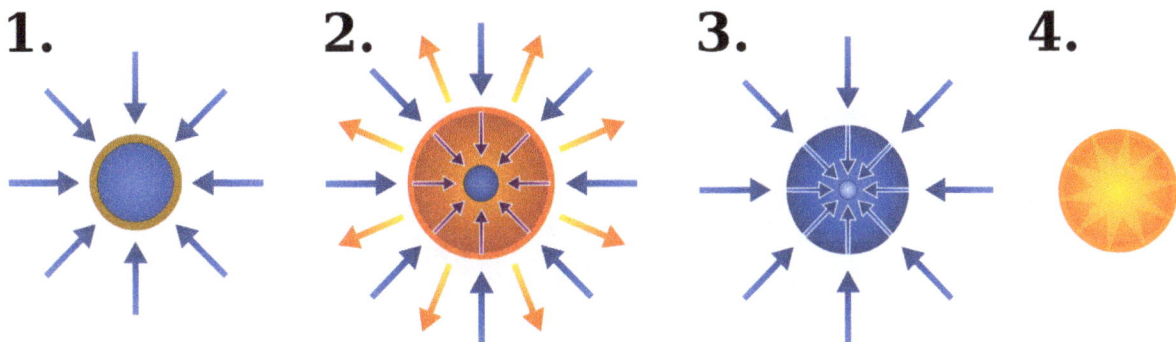

Schematic of the stages of inertial confinement fusion using lasers. The blue arrows represent radiation; orange is blowoff; purple is inwardly transported thermal energy.
1. Laser beams or laser-produced X-rays rapidly heat the surface of the fusion target, forming a surrounding plasma envelope.
2. Fuel is compressed by the rocket-like blowoff of the hot surface material.
3. During the final part of the capsule implosion, the fuel core reaches 20 times the density of lead and ignites at 100,000,000 °C.
4. Thermonuclear burn spreads rapidly through the compressed fuel, yielding many times the input energy.

Given the correct conditions, the fusion rate in the region highly compressed by the shock wave can give off significant amounts of highly energetic alpha particles. Due to the high density of the surrounding fuel, they move only a short distance before being "thermalised", losing their energy to the fuel as heat. This additional energy will cause additional fusion reactions in the heated fuel, giving off more high-energy particles. This process spreads outward from the centre, leading to a kind of self-sustaining burn known as *ignition*.

Issues with Successful Achievement

The primary problems with increasing ICF performance since the early experiments in the 1970s have been of energy delivery to the target, controlling symmetry of the imploding fuel, preventing premature heating of the fuel (before maximum density is achieved), preventing premature mixing of hot and cool fuel by hydrodynamic instabilities and the formation of a 'tight' shockwave convergence at the compressed fuel center.

In order to focus the shock wave on the center of the target, the target must be made with extremely high precision and sphericity with aberrations of no more than a few micrometres over its surface (inner and outer). Likewise the aiming of the laser beams must be extremely precise and the beams must arrive at the same time at all points on the target. Beam timing is a relatively simple issue though and is solved by using delay lines in the beams' optical path to achieve picosecond levels of timing accuracy. The other major problem plaguing the achievement of high symmetry and high temperatures/densities of the imploding target are so called "beam-beam" imbalance and beam anisotropy. These problems are, respectively, where the energy delivered by one beam may be higher or lower than other beams impinging on the target and of "hot spots" within a beam diameter hitting a target which induces uneven compression on the target surface, thereby forming Rayleigh–Taylor instabilities in the fuel, prematurely mixing it and reducing heating efficacy at the time of maximum compression. The Richtmyer-Meshkov instability is also formed during the process due to shock waves being formed.

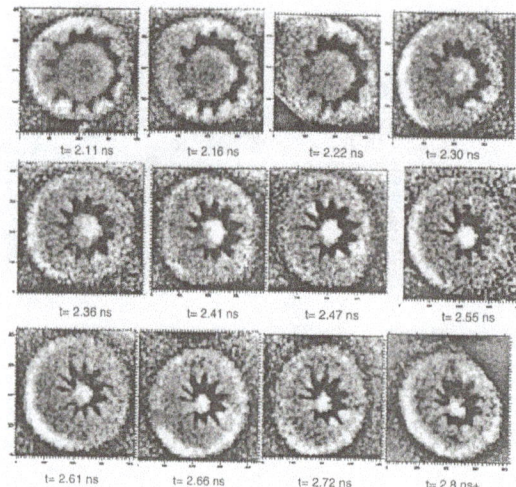

An Inertial confinement fusion target, which was a cylindrical hohlraum target of D-T, being compressed by the Nova Laser. This shot was done in 1995. The image shows the compression of the target, as well as the growth of the Rayleigh-Taylor instabilities.

All of these problems have been substantially mitigated to varying degrees in the past two decades of research by using various beam smoothing techniques and beam energy diagnostics to balance beam to beam energy; however, RT instability remains a major issue. Target design has also improved tremendously over the years. Modern cryogenic hydrogen ice targets tend to freeze a thin layer of deuterium just on the inside of a plastic sphere while irradiating it with a low power IR laser to smooth its inner surface while monitoring it with a microscope equipped camera, thereby allowing the layer to be closely monitored ensuring its "smoothness". Cryogenic targets filled with a deuterium tritium (D-T) mixture are "self-smoothing" due to the small amount of heat created by the decay of the radioactive tritium isotope. This is often referred to as "beta-layering".

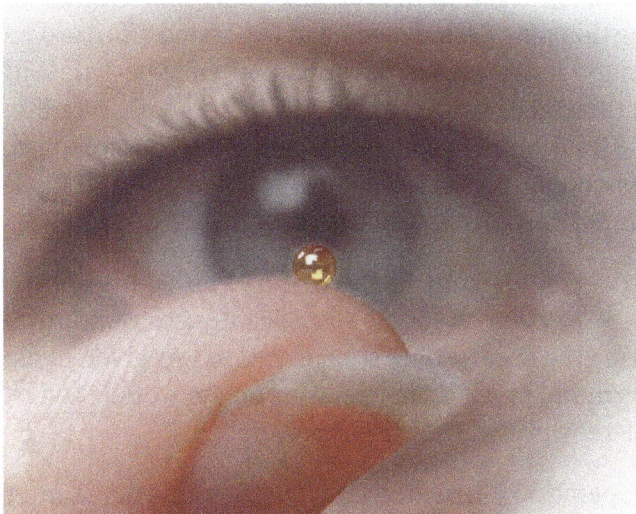

An inertial confinement fusion fuel microcapsule (sometimes called a "microballoon") of the size to be used on the NIF which can be filled with either deuterium and tritium gas or DT ice. The capsule can be either inserted in a hohlraum (as above) and imploded in the **indirect drive** mode or irradiated directly with laser energy in the **direct drive** configuration. Microcapsules used on previous laser systems were significantly smaller owing to the less powerful irradiation earlier lasers were capable of delivering to the target.

Certain targets are surrounded by a small metal cylinder which is irradiated by the laser beams instead of the target itself, an approach known as "*indirect drive*". In this approach the lasers are focused on the inner side of the cylinder, heating it to a superhot plasma which radiates mostly in X-rays. The X-rays from this plasma are then absorbed by the target surface, imploding it in the same way as if it had been hit with the lasers directly. The absorption of thermal x-rays by the target is more efficient than the direct absorption of laser light, however these *hohlraums* or "burning chambers" also take up considerable energy to heat on their own thus significantly reducing the overall efficiency of laser-to-target energy transfer. They are thus a debated feature even today; the equally numerous "*direct-drive*" design does not use them. Most often, indirect drive hohlraum targets are used to simulate thermonuclear weapons tests due to the fact that the fusion fuel in them is also imploded mainly by X-ray radiation.

A variety of ICF drivers are being explored. Lasers have improved dramatically since the 1970s, scaling up in energy and power from a few joules and kilowatts to megajoules and hundreds of terawatts, using mostly frequency doubled or tripled light from neodymium glass amplifiers.

Heavy ion beams are particularly interesting for commercial generation, as they are easy to create, control, and focus. On the downside, it is very difficult to achieve the very high energy densities required to implode a target efficiently, and most ion-beam systems require the use of a hohlraum surrounding the target to smooth out the irradiation, reducing the overall efficiency of the coupling of the ion beam's energy to that of the imploding target further.

History of ICF

First Conception

In the US

Inertial confinement feasibility began in the mid-1950s by the inventor of modern television, Dr. Philo Farnsworth. During this time period, he began concerted efforts to recreate the generation of high energy plasma he found in his earlier "Multipactor" tube design. With scant funding, he was successful in developing three generations of his "Fusor" tube. During the last several tests of this device, notable levels of excess output were produced. Patents were approved, and the science of this type of "inertial confinement" fusion is well documented. While it has been relatively dormant, it remains the first true experimentation into the concept.

A second stream of inertial confinement fusion history can be traced back to a seminal meeting called by Edward Teller in 1957 on the topic of peaceful uses of atomic explosions. Among the many topics covered during the event, some consideration was given to using a hydrogen bomb to heat a water-filled underground cavern. The resulting steam would then be used to power conventional generators, and thereby provide electrical power.

This meeting led to the Operation Plowshare efforts, given this name in 1961. Three primary concepts were studied as part of Plowshare; energy generation under Project PACER, the use of large nuclear explosions for excavation, and as a sort of nuclear fracking for the natural gas industry. PACER was directly tested in December 1961 when the 3 kt Project Gnome device was emplaced in bedded salt in New Mexico. In spite of all theorizing and attempts to stop it, radioactive steam was released from the drill shaft, some distance from the test site. Further studies as part of Project PACER led to a number of engineered cavities replacing natural ones, but through this period the entire Plowshare efforts turned from bad to worse, especially after the failure of 1962's Sedan which released huge quantities of fallout. PACER nevertheless continued to receive some funding until 1975, when a 3rd party study demonstrated that the cost of electricity from PACER would be the equivalent to conventional nuclear plants with fuel costs over ten times as great as they were.

Another outcome of the Teller meeting was to prompt John Nuckolls to start considering what happens when the fusion side of the bomb, the "secondary," was scaled down to very small size. His earliest work concerned the study of how small a fusion bomb could be made while still having a large "gain" to provide net energy output. This work suggested that at very small sizes, on the order of milligrams, very little energy would be needed to ignite it, much less than a fission "primary". He proposed building, in effect, tiny all-fusion explosives using a tiny drop of D-T fuel suspended in the center of a metal shell, today known as a hohlraum. The shell provided the same effect as the bomb casing in an H-bomb, trapping x-rays inside so they irradiated the fuel. The main difference is that the x-rays would not be supplied by a primary within the shell, but some sort of external device that heated the shell from the outside until it was glowing in the x-ray region. The power would be delivered by a then-unidentified pulsed power source he referred to using bomb terminology, the "primary".

The main advantage to this scheme is the efficiency of the fusion process at high densities. According to the Lawson criterion, the amount of energy needed to heat the D-T fuel to break-even conditions at ambient pressure is perhaps 100 times greater than the energy needed to compress it to a pressure that would deliver the same rate of fusion. So, in theory, the ICF approach would be dramatically more efficient in terms of gain. This can be understood by considering the energy losses in a conventional scenario where the fuel is slowly heated, as in the case of magnetic fusion energy; the rate of energy loss to the environment is based on the temperature difference between the fuel and its surroundings, which continues to increase as the fuel is heated. In the ICF case, the entire hohlraum is filled with high-temperature radiation, limiting losses.

In Germany

Around the same time (in 1956) a meeting was organized at the Max Planck Institute in Germany by the fusion pioneer Carl Friedrich von Weizsäcker. At this meeting Friedwardt Winterberg proposed the non-fission ignition of a thermonuclear micro-explosion by a convergent shock wave driven with high explosives. Further reference to Winterberg's work in Germany on nuclear micro explosions (mininukes) is contained in a declassified report of the former East German Stasi (Staatsicherheitsdienst).

In 1964 Winterberg proposed that ignition could be achieved by an intense beam of microparticles accelerated to a velocity of 1000 km/s. And in 1968, he proposed to use intense electron and ion beams, generated by Marx generators, for the same purpose. The advantage of this proposal is that the generation of charged particle beams is not only less expensive than the generation of laser beams but also can entrap the charged fusion reaction products due to the strong self-magnetic beam field, drastically reducing the compression requirements for beam ignited cylindrical targets.

In USSR

In 1967 research fellow Gurgen Askaryan published article with proposition to use focused laser beam in fusion lithium deuteride or deuterium.

Early Research

Through the late 1950s, Nuckolls and collaborators at the Lawrence Livermore National Laboratory (LLNL) ran a number of computer simulations of the ICF concept. In early 1960 this produced a full simulation of the implosion of 1 mg of D-T fuel inside a dense shell. The simulation suggested that a 5 MJ power input to the hohlraum would produce 50 MJ of fusion output, a gain of 10. At the time the laser had not yet been invented, and a wide variety of possible drivers were considered, including pulsed power machines, charged particle accelerators, plasma guns, and hypervelocity pellet guns.

Through the year two key theoretical advances were made. New simulations considered the timing of the energy delivered in the pulse, known as "pulse shaping", leading to better implosion. Additionally, the shell was made much larger and thinner, forming a thin shell as opposed to an almost solid ball. These two changes dramatically increased the efficiency of the implosion, and thereby greatly lowered the energy required to compress it. Using these improvements, it was calculated that a driver of about 1 MJ would be needed, a five-fold improvement. Over the next two years several other theoretical advancements were proposed, notably Ray Kidder's development of an implosion system without a hohlraum, the so-called "direct drive" approach, and Stirling Colgate and Ron Zabawski's work on very small systems with as little as 1 μg of D-T fuel.

The introduction of the laser in 1960 at Hughes Research Laboratories in California appeared to present a perfect driver mechanism. Starting in 1962, Livermore's director John S. Foster, Jr. and Edward Teller began a small-scale laser study effort directed toward the ICF approach. Even at this early stage the suitability of the ICF system for weapons research was well understood, and the primary reason for its ability to gain funding. Over the next decade, LLNL made several small experimental devices for basic laser-plasma interaction studies.

Development Begins

In 1967 Kip Siegel started KMS Industries using the proceeds of the sale of his share of an earlier company, Conductron, a pioneer in holography. In the early 1970s he formed KMS Fusion to begin development of a laser-based ICF system. This development led to considerable opposition from the weapons labs, including LLNL, who put forth a variety of reasons that KMS should not be allowed to develop ICF in public. This opposition was funnelled through the Atomic Energy Commission, who demanded funding for their own efforts. Adding to the background noise were rumours of an aggressive Soviet ICF program, new higher-powered CO_2 and glass lasers, the electron beam driver concept, and the 1970s energy crisis which added impetus to many energy projects.

In 1972 Nuckolls wrote an influential public paper in *Nature* introducing ICF and suggesting that testbed systems could be made to generate fusion with drivers in the kJ range, and high-gain systems with MJ drivers.

In spite of limited resources and numerous business problems, KMS Fusion successfully demonstrated fusion from the ICF process on 1 May 1974. However, this success was followed not long after by Siegel's death, and the end of KMS fusion about a year later, having run the company on Siegel's life insurance policy. By this point several weapons labs and universities had started their own programs, notably the solid-state lasers (Nd:glass lasers) at LLNL and the University of Rochester, and krypton fluoride excimer lasers systems at Los Alamos and the Naval Research Laboratory.

Although KMS's success led to a major development effort, the advances that followed were, and still are, hampered by the seemingly intractable problems that characterize fusion research in general.

High-Energy ICF

High energy ICF experiments (multi-hundred joules per shot and greater experiments) began in earnest in the early-1970s, when lasers of the required energy and power were first designed. This was some time after the successful design of magnetic confinement fusion systems, and around the time of the particularly successful tokamak design that was introduced in the early '70s. Nevertheless, high funding for fusion research stimulated by the multiple energy crises during the mid to late 1970s produced rapid gains in performance, and inertial designs were soon reaching the same sort of "below break-even" conditions of the best magnetic systems.

LLNL was, in particular, very well funded and started a major laser fusion development program. Their Janus laser started operation in 1974, and validated the approach of using Nd:glass lasers to generate very high power devices. Focusing problems were explored in the Long path laser and Cyclops laser, which led to the larger Argus laser. None of these were intended to be practical ICF devices, but each one advanced the state of the art to the point where there was some confidence the basic approach was valid. At the time it was believed that making a much larger device of the Cyclops type could both compress and heat the ICF targets, leading to ignition in the "short term". This was a misconception based on extrapolation of the fusion yields seen from experiments utilizing the so-called "exploding pusher" type of fuel capsules. During the period spanning the years of the late '70s and early '80s the estimates for laser energy on target needed to achieve ignition doubled almost yearly as the various plasma instabilities and laser-plasma energy coupling loss modes were gradually understood. The realization that the simple exploding pusher target designs and mere few kilojoule (kJ) laser irradiation intensities would never scale to high gain fusion yields led to the effort to increase laser energies to the 100 kJ level in the UV and to the production of advanced ablator and cryogenic DT ice target designs.

Shiva and Nova

One of the earliest serious and large scale attempts at an ICF driver design was the Shiva laser, a 20-beam neodymium doped glass laser system built at the Lawrence Livermore National Laboratory (LLNL) that started operation in 1978. Shiva was a "proof of concept" design intended to demonstrate compression of fusion fuel capsules to many times the liquid density of hydrogen. In this, Shiva succeeded and compressed its pellets to 100 times the liquid density of deuterium. However, due to the laser's strong coupling with hot electrons, premature heating of the dense plasma (ions) was problematic and fusion yields were low. This failure by Shiva to efficiently heat the compressed plasma pointed to the use of optical frequency multipliers as a solution which would frequency triple the infrared light from the laser into the ultraviolet at 351 nm. Newly discovered schemes to efficiently frequency triple high intensity laser light discovered at the Laboratory for Laser Energetics in 1980 enabled this method of target irradiation to be experimented with in the 24 beam OMEGA laser and the NOVETTE laser, which was followed by the Nova laser design with 10 times the energy of Shiva, the first design with the specific goal of reaching ignition conditions.

Nova also failed in its goal of achieving ignition, this time due to severe variation in laser intensity in its beams (and differences in intensity between beams) caused by filamentation which resulted in large non-uniformity in irradiation smoothness at the target and asymmetric implosion. The techniques pioneered earlier could not address these new issues. But again this failure led to a much greater understanding of the process of implosion, and the way forward again seemed clear, namely the increase in uniformity of irradiation, the reduction of hot-spots in the laser beams through beam smoothing techniques to reduce Rayleigh–Taylor instability imprinting on the target and increased laser energy on target by at least an order of magnitude. Funding for fusion research was severely constrained in the 80's, but Nova nevertheless successfully gathered enough information for a next generation machine.

National Ignition Facility

The resulting design, now known as the National Ignition Facility, started construction at LLNL in 1997. NIF's main objective will be to operate as the flagship experimental device of the so-called nuclear stewardship program, supporting LLNLs traditional bomb-making role. Completed in March 2009, NIF has now conducted experiments using all 192 beams, including experiments that set new records for power delivery by a laser. The first credible attempts at ignition were initially scheduled for 2010, but ignition was not achieved as of September 30, 2012. As of October 7, 2013, the facility is understood to have achieved an important milestone towards commercialization of fusion, namely, for the first time a fuel capsule gave off more energy than was applied to it. This is still a long way from satisfying the Lawson criterion, but is a major step forward.

Fast Ignition

A more recent development is the concept of "fast ignition," which may offer a way to directly heat the high density fuel after compression, thus decoupling the heating and compression phases of the implosion. In this approach the target is first compressed "normally" using a driver laser system, and then when the implosion reaches maximum density (at the stagnation point or "bang time"), a second ultra-short pulse ultra-high power petawatt (PW) laser delivers a single pulse focused on one side of the core, dramatically heating it and hopefully starting fusion ignition. The two types of fast ignition are the "plasma bore-through" method and the "cone-in-shell" method. In the first method the petawatt laser is simply expected to bore straight through the outer plasma of an imploding capsule and to impinge on and heat the dense core, whereas in the cone-in-shell method, the capsule is mounted on the end of a small high-z (high atomic number) cone such that the tip of the cone projects into the core of the capsule. In this second method, when the capsule is imploded, the petawatt has a clear view straight to the high density core and does not have to waste energy boring through a 'corona' plasma; however, the presence of the cone affects the implosion process in significant ways that are not fully understood. Several projects are currently underway to explore the fast ignition approach, including upgrades to the OMEGA laser at the University of Rochester, the GEKKO XII device in Japan, and an entirely new £500 million facility, known as HiPER, proposed for construction in the European Union. If successful, the fast ignition approach could dramatically lower the total amount of energy needed to be delivered to the target; whereas NIF uses UV beams of 2 MJ, HiPER's driver is 200 kJ and heater 70 kJ, yet the predicted fusion gains are nevertheless even higher than on NIF.

Other Projects

Laser Mégajoule, the French project, has seen its first experimental line achieved in 2002, and was finally completed in 2014.

Using a different approach entirely is the z-pinch device. Z-pinch uses massive amounts of electric current which is switched into a cylinder comprising many of extremely fine wires. The wires vaporize to form an electrically conductive plasma that carries a very high current; the resulting circumferential magnetic field squeezes the plasma cylinder, imploding it and thereby generating a high-power x-ray pulse that can be used to drive the implosion of a fuel capsule. Challenges to this approach include relatively low drive temperatures, resulting in slow implosion velocities and potentially large instability growth, and preheat caused by high-energy x-rays.

Most recently, Winterberg has proposed the ignition of a deuterium microexplosion, with a gigavolt super-Marx generator, which is a Marx generator driven by up to 100 ordinary Marx generators.

As an Energy source

Practical power plants built using ICF have been studied since the late 1970s when ICF experiments were beginning to ramp up to higher powers; they are known as inertial fusion energy, or IFE plants. These devices would deliver a successive stream of targets to the reaction chamber, several a second typically, and capture the resulting heat and neutron radiation from their implosion and fusion to drive a conventional steam turbine.

Technical Challenges

IFE faces continued technical challenges in reaching the conditions needed for ignition. But even if these were all to be solved, there are a significant number of practical problems that seem just as difficult to overcome. Laser-driven systems were initially believed to be able to generate commercially useful amounts of energy. However, as estimates of the energy required to reach ignition grew dramatically during the 1970s and '80s, these hopes were abandoned. Given the low efficiency of the laser amplification process (about 1 to 1.5%), and the losses in generation (steam-driven turbine systems are typically about 35% efficient), fusion gains would have to be on the order of 350 just to energetically break even. These sorts of gains appeared to be impossible to generate, and ICF work turned primarily to weapons research.

With the recent introduction of fast ignition and similar approaches, things have changed dramatically. In this approach gains of 100 are predicted in the first experimental device, HiPER. Given a gain of about 100 and a laser efficiency of about 1%, HiPER produces about the same amount of *fusion* energy as electrical energy was needed to create it. It also appears that an order of magnitude improvement in laser efficiency may be possible through the use of newer designs that replace the flash lamps with laser diodes that are tuned to produce most of their energy in a frequency range that is strongly absorbed. Initial experimental devices offer efficiencies of about 10%, and it is suggested that 20% is a real possibility with some additional development.

With "classical" devices like NIF about 330 MJ of electrical power are used to produce the driver beams, producing an expected yield of about 20 MJ, with the maximum credible yield of 45 MJ. Using the same sorts of numbers in a reactor combining fast ignition with newer lasers would offer dramatically improved performance. HiPER requires about 270 kJ of laser energy, so assuming a first-generation diode laser driver at 10% the reactor would require about 3 MJ of electrical power. This is expected to produce about 30 MJ of fusion power. Even a very poor conversion to electrical energy appears to offer real-world power output, and incremental improvements in yield and laser efficiency appear to be able to offer a commercially useful output.

Practical Problems

ICF systems face some of the same secondary power extraction problems as magnetic systems in generating useful power from their reactions. One of the primary concerns is how to successfully remove heat from the reaction chamber without interfering with the targets and driver beams. Another serious concern is that the huge number of neutrons released in the fusion reactions react with the plant, causing them to become intensely radioactive themselves, as well as mechanically weakening metals. Fusion plants built of conventional metals like steel would have a fairly short lifetime and the core containment vessels will have to be replaced frequently.

One current concept in dealing with both of these problems, as shown in the HYLIFE-II baseline design, is to use a "waterfall" of FLiBe, a molten mix of fluoride salts of lithium and beryllium, which both protect the chamber from neutrons and carry away heat. The FLiBe is then passed into a heat exchanger where it heats water for use in the turbines. Another, Sombrero, uses a reaction chamber built of Carbon-fiber-reinforced polymer which has a very low neutron cross section. Cooling is provided by a molten ceramic, chosen because of its ability to stop the neutrons from traveling any further, while at the same time being an efficient heat transfer agent.

Economic Viability

Even if these technical advances solve the considerable problems in IFE, another factor working against IFE is the cost of the fuel. Even as Nuckolls was developing his earliest detailed calculations on the idea, co-workers pointed this out: if an IFE machine produces 50 MJ of fusion energy, one might expect that a shot could produce perhaps 10 MJ of power for export. Converted to better known units, this is the equivalent of 2.8 kWh of electrical power. Wholesale rates for electrical power on the grid were about 0.3 cents/kWh at the time, which meant the monetary value of the shot was perhaps one cent. In the intervening 50 years the price of power has remained about even with the rate of inflation, and the rate in 2012 in Ontario, Canada was about 2.8 cents/kWh

Thus, in order for an IFE plant to be economically viable, fuel shots would have to cost considerably less than ten cents in year 2012 dollars. At the time this objection was first noted, Nuckolls suggested using liquid droplets sprayed into the hohlraum from an eye-dropper-like apparatus. Given the ever-increasing demands for higher uniformity of the targets, this approach does not appear practical, as even the inner ablator and fuel itself currently costs several orders of magnitude more than this. Moreover, Nuckolls' solution had the fuel dropped into a fixed hohlraum that would be re-used in a continual cycle, but at current energy levels the hohlraum is destroyed with every shot.

Direct-drive systems avoid the use of a hohlraum and thereby may be less expensive in fuel terms. However, these systems still require an ablator, and the accuracy and geometrical considerations are even more important. They are also far less developed than the indirect-drive systems, and face considerably more technical problems in terms of implosion physics. Currently there is no strong consensus whether a direct-drive system would actually be less expensive to operate.

Projected Development

The various phases of such a project are the following, the sequence of inertial confinement fusion development follows much the same outline:

- burning demonstration: reproducible achievement of some fusion energy release (not necessarily a Q factor of >1).

- high gain demonstration: experimental demonstration of the feasibility of a reactor with a sufficient energy gain.

- industrial demonstration: validation of the various technical options, and of the whole data needed to define a commercial reactor.

- • commercial demonstration: demonstration of the reactor ability to work over a long peri- od, while respecting all the requirements for safety, liability and cost.

At the moment, according to the available data, inertial confinement fusion experiments have not gone beyond the first phase, although Nova and others have repeatedly demonstrated operation within this realm.

In the short term a number of new systems are expected to reach the second stage.

For a true industrial demonstration, further work is required. In particular, the laser systems need to be able to run at high operating frequencies, perhaps one to ten times a second. Most of the laser systems mentioned in this article have trouble operating even as much as once a day. Parts of the HiPER budget are dedicated to research in this direction as well. Because they convert electricity into laser light with much higher efficiency, diode lasers also run cooler, which in turn allows them to be operated at much higher frequencies. HiPER is currently studying devices that operate at 1 MJ at 1 Hz, or alternately 100 kJ at 10 Hz.

Nuclear Weapons Program

The very hot and dense conditions encountered during an Inertial Confinement Fusion experi- ment are similar to those created in a thermonuclear weapon, and have applications to the nuclear weapons program. ICF experiments might be used, for example, to help determine how warhead performance will degrade as it ages, or as part of a program of designing new weapons. Retain- ing knowledge and corporate expertise in the nuclear weapons program is another motivation for pursuing ICF. Funding for the NIF in the United States is sourced from the 'Nuclear Weapons Stockpile Stewardship' program, and the goals of the program are oriented accordingly. It has been argued that some aspects of ICF research may violate the Comprehensive Test Ban Treaty or the Nuclear Non-Proliferation Treaty. In the long term, despite the formidable technical hurdles, ICF research might potentially lead to the creation of a "pure fusion weapon".

Neutron Source

Inertial confinement fusion has the potential to produce orders of magnitude more neutrons than spallation. Neutrons are capable of locating hydrogen atoms in molecules, resolving atomic ther- mal motion and studying collective excitations of photons more effectively than X-rays. Neutron scattering studies of molecular structures could resolve problems associated with protein folding, diffusion through membranes, proton transfer mechanisms, dynamics of molecular motors, etc. by modulating thermal neutrons into beams of slow neutrons. In combination with fissionable materials, neutrons produced by ICF can potentially be used in Hybrid Nuclear Fusion designs to produce electric power.

Direct Energy Conversion

Direct energy conversion (DEC) or simply direct conversion converts a charged particle's kinetic energy into a voltage. It is a scheme for power extraction from nuclear fusion.

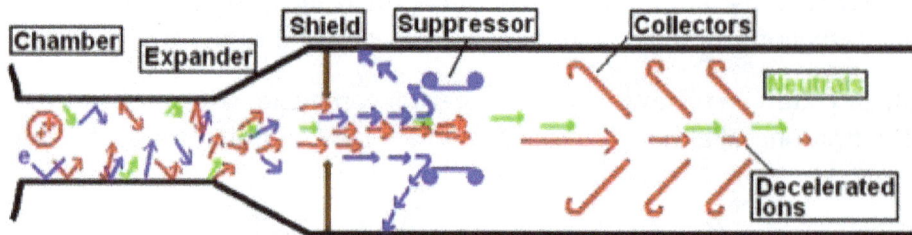

A basic direct converter

History and Theoretical Underpinnings

Electrostatic Direct Collectors

In the middle of the 1960s direct energy conversion was proposed as a method for capturing the energy from the exhaust gas in a fusion reactor. This would generate a direct current of electricity. Richard F. Post at the Lawrence Livermore National Laboratory was an early proponent of the idea. Post reasoned that capturing the energy would require five steps: (1) Ordering the charged particles into linear beam. (2) Separation of positives and negatives. (3) Separating the ions into groups, by their energy. (4) Gathering these ions as they touch collectors. (5) Using these collectors as the positive side in a circuit. Post argued that the efficiency was theoretically determined by the number of collectors.

The Venetian Blind

Designs in the early 1970s by William Barr and Ralph Moir used metal ribbons at an angle to collect these ions. This was called the Venetian Blind design, because the ribbons look like window blinds. Those metal ribbon-like surfaces are more transparent to ions going forward than to ions going backward. Ions pass through surfaces of successively increasing potential until they turn and start back, along a parabolic trajectory. They then see opaque surfaces and are caught. Thus ions are sorted by energy with high-energy ions being caught on high-potential electrodes.

William Barr and Ralph Moir then ran a group which did a series of direct energy conversion experiments through the late 1970s and early 1980s. The first experiments used beams of positives and negatives as fuel, and demonstrated energy capture at a peak efficiency of 65 percent and a minimum efficiency of 50 percent. The following experiments involved a true plasma direct converter that was tested on the Tandem Mirror Experiment (TMX), an operating magnetic mirror fusion reactor. In the experiment, the plasma moved along diverging field lines, spreading it out and converting it into a forward moving beam with a Debye length of a few centimeters. Suppressor grids then reflect the electrons, and collector anodes recovered the ion energy by slowing them down and collecting them at high-potential plates. This machine demonstrated an energy capture efficiency of 48 percent. However, Marshall Rosenbluth argued that keeping the plasma's neutral charge over the very short Debye length distance would be very challenging in practice, though he said that this problem would not occur in every version of this technology.

The Venetian Blind converter can operate with 100 to 150 keV D-T plasma, with an efficiency of about 60% under conditions compatible with economics, and an upper technical conversion efficiency up to 70% ignoring economic limitations.

Periodic Electrostatic Focusing

A second type of electrostatic converter initially proposed by Post, then developed by Barr and Moir, is the Periodic Electrostatic Focusing concept. Like the Venetian Blind concept, it is also a direct collector, but the collector plates are disposed in many stages along the longitudinal axis of an electrostatic focusing channel. As each ion is decelerated along the channel toward zero energy, the particle becomes "over-focused" and is deflected sideways from the beam, then collected. The Periodic Electrostatic Focusing converter typically operates with a 600 keV D-T plasma (as low as 400 keV and up to 800 keV) with efficiency of about 60% under conditions compatible with economics, and an upper technical conversion efficiency up to 90% ignoring economic limitations.

Magnetic Compression-expansion Converter

In addition to electrostatic converters, magnetic converters have also been proposed by Lev Artsimovich in 1963, then Alan Frederic Haught and his team from United Aircraft Research Laboratories in 1970, and Ralph Moir in 1977.

The magnetic direct energy converter is analogous to the internal combustion engine. As the hot plasma expands against a magnetic field, in a manner similar to hot gases expanding against a piston, part of the energy of the internal plasma is inductively converted to an electromagnetic coil, as an EMF (voltage) in the conductor.

This scheme is best used with pulsed devices, because the converter then works like a "magnetic four-stroke engine":

1. Compression: A column of plasma is compressed by a magnetic field that acts like a piston.

2. Thermonuclear burn: The compression heats the plasma to the thermonuclear ignition temperature.

3. Expansion/Power: The expansion of fusion reaction products (charged particles) increases the plasma pressure and pushes the magnetic field outward. A voltage is induced and collected in the electromagnetic coil.

4. Exhaust/Refuel: After expansion, the partially burned fuel is flushed out, and new fuel in the form of gas is introduced and ionized; and the cycle starts again.

In 1973, a team from Los Alamos and Argonne laboratories stated that the thermodynamic efficiency of the magnetic direct conversion cycle from alpha-particle energy to work is 62%.

Traveling-wave Direct Energy Converter

In 1992, a Japan–U.S. joint-team proposed a novel direct energy conversion system for 14.7 MeV protons produced by D-^{3}He fusion reactions, whose energy is too high for electrostatic converters.

The conversion is based on a Traveling-Wave Direct Energy Converter (TWDEC). A gyrotron converter first guides fusion product ions as a beam into a 10-meter long microwave cavity filled with a 10-tesla magnetic field, where 155 MHz microwaves are generated and converted to a high voltage DC output through rectennas.

The Field-Reversed Configuration reactor ARTEMIS in this study was designed with an efficiency of 75%. The traveling-wave direct converter has a maximum projected efficiency of 90%.

Inverse Cyclotron Converter (ICC)

Original direct converters were designed to extract the energy carried by 100 to 800 keV ions produced by D-T fusion reactions. Those electrostatic converters are not suitable for higher energy product ions above 1 MeV generated by other fusion fuels like the D-^{3}He or the p-^{11}B aneutronic fusion reactions.

A much shorter device than the Traveling-Wave Direct Energy Converter has been proposed in 1997 and patented by Tri Alpha Energy, Inc. as an Inverse Cyclotron Converter (ICC).

The ICC is able to decelerate the incoming ions based on experiments made in 1950 by Felix Bloch and Carson D. Jeffries, in order to extract their kinetic energy. The converter operates at 5 MHz and requires a magnetic field of only 0.6 tesla. The linear motion of fusion product ions is converted to circular motion by a magnetic cusp. Energy is collected from the charged particles as they spiral past quadrupole electrodes. More classical electrostatic collectors would also be used for particles with energy less than 1 MeV. The Inverse Cyclotron Converter has a maximum projected efficiency of 90%.

X-ray Photoelectric Converter

A significant amount of the energy released by fusion reactions is composed of electromagnetic radiations, essentially X-rays due to Bremsstrahlung. Those X-rays can not be converted into electric power with the various electrostatic and magnetic direct energy converters listed above, and their energy is lost.

Whereas more classical thermal conversion has been considered with the use of a radiation/boiler/energy exchanger where the X-ray energy is absorbed by a working fluid at temperatures of several thousand degrees, more recent research done by companies developing nuclear aneutronic fusion reactors, like Lawrenceville Plasma Physics (LPP) with the Dense Plasma Focus, and Tri Alpha Energy, Inc. with the Colliding Beam Fusion Reactor (CBFR), plan to harness the photoelectric and Auger effects to recover energy carried by X-rays and other high-energy photons. Those photoelectric converters are composed of X-ray absorber and electron collector sheets nested concentrically in an onion-like array. Indeed, since X-rays can go through far greater thickness of material than electrons can, many layers are needed to absorb most of the X-rays. LPP announces an overall efficiency of 81% for the photoelectric conversion scheme.

Direct Energy Conversion from Fission Products

In the early 2000s, research was undertaken by Sandia National Laboratories, Los Alamos National Laboratory, The University of Florida, Texas A&M University and General Atomics to use direct conversion to extract energy from fission reactions. Essentially, attempting to extract energy from the linear motion of charged particles coming off a fission reaction.

References

- Freidberg, Jeffrey P. (8 February 2007). Plasma Physics and Fusion Energy. Cambridge University Press. ISBN 978-0-521-85107-7.

- Atzeni, Stefano (3 June 2004). The Physics of Inertial Fusion: BeamPlasma Interaction, Hydrodynamics, Hot Dense Matter. OUP Oxford. pp. 12–13. ISBN 978-0-19-152405-9.

- McCracken, Garry; Stott, Peter (8 June 2012). Fusion: The Energy of the Universe. Academic Press. pp. 198–199. ISBN 978-0-12-384656-3. Retrieved 18 August 2012.

- Angelo, Joseph A. (30 November 2004). Nuclear Technology. Greenwood Publishing Group. p. 474. ISBN 978-1-57356-336-9. Retrieved 18 August 2012.

- Biskamp, Dieter (1997) Nonlinear Magnetohydrodynamics Cambridge University Press, Cambridge, England, page 130, ISBN 0-521-59918-0

- Moir, Ralph W. (April 1977). "Chapter 5: Direct Energy Conversion in Fusion Reactors". In Considine, Douglas M. Energy Technology Handbook (PDF). NY: McGraw-Hill. pp. 150–154. ISBN 978-0070124301.

- T. Anklam; A. J. Simon; S. Powers; W. R. Meier (December 2, 2010). "LIFE: The Case for Early Commercialization of Fusion Energy" (PDF). Livermore, LLNL-JRNL-463536. Retrieved 30 October 2014.

- Dr. Matthew McKinzie; Christopher E. Paine (2000). "When peer review fails : The Roots of the National Ignition Facility (NIF) Debacle". National Resources Defense Council. Retrieved 30 October 2014.

- "Los Alamos National Labs Aurora Laser Fusion Project | Hextek Corp". Hextek.com. 2014-06-20. Archived from the original on May 17, 2014. Retrieved 2014-08-24.

- "'Cold fusion' rebirth? New evidence for existence of controversial energy source" (Press release). American Chemical Society. Retrieved 30 October 2014.

- Schechner, Sam (2008-08-18). "Nuclear Ambitions: Amateur Scientists Get a Reaction From Fusion - WSJ". Online.wsj.com. Retrieved 2014-08-24.

- May, Kate Torgovnick (February 27, 2013). "Good energy comes in small packages: Taylor Wilson at TED2013". TED blog — Science. TED (conference). Retrieved 2014-02-10.

- "Fusion, anyone? Not quite yet, but researchers show just how close we've come". Phys.org. September 24, 2013. Retrieved 2014-08-24.

- Norris, Guy (2014-10-14). "High Hopes – Can Compact Fusion Unlock New Power For Space And Air Transport?". Aviation Week. Retrieved 2014-10-30.

- Sing Lee; Sor Heoh Saw. "Nuclear Fusion Energy-Mankind's Giant Step Forward" (PDF). HPlasmafocus.net. Retrieved 30 October 2014.

Applications of Plasma Physics

The applications of plasma physics are inductively coupled plasma mass spectrometry, reactive-ion etching, plasma propulsion engine, plasma display, plasma actuator etc. Plasma displays are flat panel display television sets whereas plasma actuators are actuators that are developed for aerodynamic flow control. The diverse applications of plasma physics in the current scenario have been thoroughly discussed in this section.

Inductively Coupled Plasma Mass Spectrometry

Inductively coupled plasma mass spectrometry (ICP-MS) is a type of mass spectrometry which is capable of detecting metals and several non-metals at concentrations as low as one part in 10^{15} (part per quadrillion, ppq) on non-interfered low-background isotopes. This is achieved by ionizing the sample with inductively coupled plasma and then using a mass spectrometer to separate and quantify those ions.

Compared to atomic absorption spectroscopy, ICP-MS has greater speed, precision, and sensitivity. However, compared with other types of mass spectrometry, such as thermal ionization mass spectrometry (TIMS) and glow discharge mass spectrometry (GD-MS), ICP-MS introduces many interfering species: argon from the plasma, component gases of air that leak through the cone orifices, and contamination from glassware and the cones.

The variety of applications exceeds that of inductively coupled plasma atomic emission spectroscopy and includes isotopic speciation. Due to possible applications in nuclear technologies, ICP-MS hardware is a subject for special exporting regulations.

Components

Inductively Coupled Plasma

An inductively coupled plasma is a plasma that is energized (ionized) by inductively heating the gas with an electromagnetic coil, and contains a sufficient concentration of ions and electrons to make the gas electrically conductive. Even a partially ionized gas in which as little as 1% of the particles are ionized can have the characteristics of a plasma (i.e., response to magnetic fields and high electrical conductivity). The plasmas used in spectrochemical analysis are essentially electrically neutral, with each positive charge on an ion balanced by a free electron. In these plasmas the positive ions are almost all singly charged and there are few negative ions, so there are nearly equal amounts of ions and electrons in each unit volume of plasma.

An inductively coupled plasma (ICP) for spectrometry is sustained in a torch that consists of three concentric tubes, usually made of quartz, although the inner tube (injector) can be sapphire if hydrofluoric acid is being used. The end of this torch is placed inside an induction coil supplied with a radio-frequency electric current. A flow of argon gas (usually 13 to 18 liters per minute) is introduced between the two outermost tubes of the torch and an electric spark is applied for a short time to introduce free electrons into the gas stream. These electrons interact with the radio-frequency magnetic field of the induction coil and are accelerated first in one direction, then the other, as the field changes at high frequency (usually 27.12 million cycles per second). The accelerated electrons collide with argon atoms, and sometimes a collision causes an argon atom to part with one of its electrons. The released electron is in turn accelerated by the rapidly changing magnetic field. The process continues until the rate of release of new electrons in collisions is balanced by the rate of recombination of electrons with argon ions (atoms that have lost an electron). This produces a 'fireball' that consists mostly of argon atoms with a rather small fraction of free electrons and argon ions. The temperature of the plasma is very high, of the order of 10,000 K. The plasma also produces ultraviolet light, so for safety should not be viewed directly.

The ICP can be retained in the quartz torch because the flow of gas between the two outermost tubes keeps the plasma away from the walls of the torch. A second flow of argon (around 1 liter per minute) is usually introduced between the central tube and the intermediate tube to keep the plasma away from the end of the central tube. A third flow (again usually around 1 liter per minute) of gas is introduced into the central tube of the torch. This gas flow passes through the centre of the plasma, where it forms a channel that is cooler than the surrounding plasma but still much hotter than a chemical flame. Samples to be analyzed are introduced into this central channel, usually as a mist of liquid formed by passing the liquid sample into a nebulizer.

To maximise plasma temperature (and hence ionisation efficiency) and stability, the sample should be introduced through the central tube with as little liquid (solvent load) as possible, and with consistent droplet sizes. A nebuliser can be used for liquid samples, followed by a spray chamber to remove larger droplets, or a desolvating nebuliser can be used to evaporate most of the solvent before it reaches the torch. Solid samples can also be introduced using laser ablation. The sample enters the central channel of the ICP, evaporates, molecules break apart, and then the constituent atoms ionise. At the temperatures prevailing in the plasma a significant proportion of the atoms of many chemical elements are ionized, each atom losing its most loosely bound electron to form a singly charged ion. The plasma temperature is selected to maximise ionisation efficiency for elements with a high first ionisation energy, while minimising second ionisation (double charging) for elements that have a low second ionisation energy.

Mass Spectrometry

For coupling to mass spectrometry, the ions from the plasma are extracted through a series of cones into a mass spectrometer, usually a quadrupole. The ions are separated on the basis of their mass-to-charge ratio and a detector receives an ion signal proportional to the concentration.

The concentration of a sample can be determined through calibration with certified reference material such as single or multi-element reference standards. ICP-MS also lends itself to quantitative determinations through isotope dilution, a single point method based on an isotopically enriched standard.

Other mass analyzers coupled to ICP systems include double focusing magnetic-electrostatic sector systems with both single and multiple collector, as well as time of flight systems (both axial and orthogonal accelerators have been used).

Applications

One of the largest volume uses for ICP-MS is in the medical and forensic field, specifically, toxicology. A physician may order a metal assay for a number of reasons, such as suspicion of heavy metal poisoning, metabolic concerns, and even hepatological issues. Depending on the specific parameters unique to each patient's diagnostic plan, samples collected for analysis can range from whole blood, urine, plasma, serum, to even packed red blood cells. Another primary use for this instrument lies in the environmental field. Such applications include water testing for municipalities or private individuals all the way to soil, water and other material analysis for industrial purposes. In the forensic field, glass ICP-MS is popular for glass analysis. Trace elements on glass can be detected using the LA-ICP-MS. The trace elements from the glass can be used to match a sample found at the crime scene to a suspect.

In recent years, industrial and biological monitoring has presented another major need for metal analysis via ICP-MS. Individuals working in plants where exposure to metals is likely and unavoidable, such as a battery factory, are required by their employer to have their blood or urine analyzed for metal toxicity on a regular basis. This monitoring has become a mandatory practice implemented by OSHA, in an effort to protect workers from their work environment and ensure proper rotation of work duties (i.e. rotating employees from a high exposure position to a low exposure position).

Regardless of the sample type, blood, water, etc., it is important that it be free of clots or other particulate matter, as even the smallest clot can disrupt sample flow and block or clog the sample tips within the spray chamber. Very high concentrations of salts, e.g. sodium chloride in sea water, can eventually lead to blockages as some of the ions reunite after leaving the torch and build up around the orifice of the skimmer cone. This can be avoided by diluting samples whenever high salt concentrations are suspected, though at a cost to detection limits.

ICP-MS is also used widely in the geochemistry field for radiometric dating, in which it is used to analyze relative abundance of different isotopes, in particular uranium and lead. ICP-MS is more suitable for this application than the previously used thermal ionization mass spectrometry, as species with high ionization energy such as osmium and tungsten can be easily ionized. For high precision ratio work, multiple collector instruments are normally used to reduce the effect noise on the calculated ratios.

In the field of flow cytometry, a new technique uses ICP-MS to replace the traditional fluorochromes. Briefly, instead of labelling antibodies (or other biological probes) with fluorochromes, each antibody is labelled with a distinct combinations of lanthanides. When the sample of interest is analysed by ICP-MS in a specialised flow cytometer, each antibody can be identified and quantitated by virtue of a distinct ICP "footprint". In theory, hundreds of different biological probes can thus be analysed in an individual cell, at a rate of ca. 1,000 cells per second. Because elements are easily distinguished in ICP-MS, the problem of compensation in multiplex flow cytometry is effectively eliminated.

In the pharmaceutical industry, ICP-MS is used for detecting inorganic impurities in pharmaceuticals and their ingredients. New and reduced maximum permitted exposure levels of heavy metals form dietary supplements, introduced in USP (United States Pharmacopeia) <232>Elemental Impurities—Limits and USP <233> Elemental Impurities—Procedures, will increase the need for ICP-MS technology, where, previously, other analytic methods have been sufficient.

Metal Speciation

A growing trend in the world of elemental analysis has revolved around the speciation, or determination of oxidation state of certain metals such as chromium and arsenic. One of the primary techniques to achieve this is to separate the chemical species with high-performance liquid chromatography (HPLC) or field flow fractionation (FFF) and then measure the concentrations with ICP-MS.

Quantification of Proteins and Biomolecules

There is an increasing trend of using ICP-MS as a tool in speciation analysis, which normally involves a front end chromatograph separation and an elemental selective detector, such as AAS and ICP-MS. For example, ICP-MS may be combined with size exclusion chromatography and quantitative preparative native continuous polyacrylamide gel electrophoresis (QPNC-PAGE) for identifying and quantifying native metal cofactor containing proteins in biofluids. Also the phosphorylation status of proteins can be analyzed.

In 2007, a new type of protein tagging reagents called metal-coded affinity tags (MeCAT) were introduced to label proteins quantitatively with metals, especially lanthanides. The MeCAT labelling allows relative and absolute quantification of all kind of proteins or other biomolecules like peptides. MeCAT comprises a site-specific biomolecule tagging group with at least a strong chelate group which binds metals. The MeCAT labelled proteins can be accurately quantified by ICP-MS down to low attomol amount of analyte which is at least 2–3 orders of magnitude more sensitive than other mass spectrometry based quantification methods. By introducing several MeCAT labels to a biomolecule and further optimization of LC-ICP-MS detection limits in the zeptomol range are within the realm of possibility. By using different lanthanides MeCAT multiplexing can be used for pharmacokinetics of proteins and peptides or the analysis of the differential expression of proteins (proteomics) e.g. in biological fluids. Breakable PAGE SDS-PAGE (DPAGE, dissolvable PAGE), two-dimensional gel electrophoresis or chromatography is used for separation of MeCAT labelled proteins. Flow-injection ICP-MS analysis of protein bands or spots from DPAGE SDS-PAGE gels can be easily performed by dissolving the DPAGE gel after electrophoresis and staining of the gel. MeCAT labelled proteins are identified and relatively quantified on peptide level by MALDI-MS or ESI-MS.

Elemental Analysis

The ICP-MS allows determination of elements with atomic mass ranges 7 to 250 (Li to U), and sometimes higher. Some masses are prohibited such as 40 due to the abundance of argon in the sample. Other blocked regions may include mass 80 (due to the argon dimer), and mass 56 (due to ArO), the latter of which greatly hinders Fe analysis unless the instrumentation is fitted with a reaction chamber. Such interferences can be reduced by using a high resolution ICP-MS (HR-ICP-MS) which uses two or more slits to constrict the beam and distinguish between nearby peaks. This comes at the cost of sensitivity. For example, distinguishing iron from argon requires a resolving power of about 10,000, which may reduce the iron sensitivity by around 99%.

A single collector ICP-MS may use a multiplier in pulse counting mode to amplify very low signals, an attenuation grid or a multiplier in analogue mode to detect medium signals, and a Faraday cup/ bucket to detect larger signals. A multi-collector ICP-MS may have more than one of any of these, normally Faraday buckets which are much less expensive. With this combination, a dynamic range of 12 orders of magnitude, from 1 ppq to 100 ppm is possible.

ICP-MS is a method of choice for the determination of cadmium in biological samples.

Unlike atomic absorption spectroscopy, which can only measure a single element at a time, ICP-MS has the capability to scan for all elements simultaneously. This allows rapid sample processing. A simultaneous ICP-MS that can record the entire analytical spectrum from lithium to uranium in every analysis won the Silver Award at the 2010 Pittcon Editors' Awards. An ICP-MS may use multiple scan modes, each one striking a different balance between speed and precision. Using the magnet alone to scan is slow, due to hysteresis, but is precise. Electrostatic plates can be used in addition to the magnet to increase the speed, and this, combined with multiple collectors, can allow a scan of every element from Lithium 6 to Uranium Oxide 256 in less than a quarter of a second. For low detection limits, interfering species and high precision, the counting time can increase substantially. The rapid scanning, large dynamic range and large mass range is ideally suited to measuring multiple unknown concentrations and isotope ratios in samples that have had minimal preparation (an advantage over TIMS), for example seawater, urine, and digested whole rock samples. It also lends well to laser ablated rock samples, where the scanning rate is so quick that a real time plot of any number of isotopes is possible.This also allows easy spatial mapping of mineral grains.

Hardware

In terms of input and output, ICP-MS instrument consumes prepared sample material and translates it into mass-spectral data. Actual analytical procedure takes some time; after that time the instrument can be switched to work on the next sample. Series of such sample measurements requires the instrument to have plasma ignited, meanwhile a number of technical parameters has to be stable in order for the results obtained to have feasibly accurate and precise interpretation. Maintaining the plasma requires a constant supply of carrier gas (usually, pure argon) and increased power consumption of the instrument. When these additional running costs are not considered justified, plasma and most of auxiliary systems can be turned off. In such standby mode only pumps are working to keep proper vacuum in mass-spectrometer.

The constituents of ICP-MS instrument are designed to allow for reproducible and/or stable operation.

Sample Introduction

The first step in analysis is the introduction of the sample. This has been achieved in ICP-MS through a variety of means.

The most common method is the use of *analytical nebulizers*. Nebulizer converts liquids into an aerosol, and that aerosol can then be swept into the plasma to create the ions. Nebulizers work best with simple liquid samples (i.e. solutions). However, there have been instances of their use with more complex materials like a slurry. Many varieties of nebulizers have been coupled to ICP-MS, including pneumatic, cross-flow, Babington, ultrasonic, and desolvating types. The aerosol generated is often treated to limit it to only smallest droplets, commonly by means of a Peltier cooled double pass or cyclonic spray chamber. Use of autosamplers makes this easier and faster, especially for routine work and large numbers of samples. A Desolvating Nebuliser (DSN) may also be used; this uses a long heated capillary, coated with a fluoropolymer membrane, to remove most of the solvent and reduce the load on the plasma. Matrix removal introduction systems are sometimes used for samples, such as seawater, where the species of interest are at trace levels, and are surrounded by much more abundant contaminants.

Laser ablation is another method. While being less common in the past, is rapidly becoming popular has been used as a means of sample introduction, thanks to increased ICP-MS scanning speeds. In this method, a pulsed UV laser is focused on the sample and creates a plume of ablated material which can be swept into the plasma. This allows geochemists to spacially map the isotope composition in cross-sections of rock samples, a tool which is lost if the rock is digested and introduced as a liquid sample. Lasers for this task are built to have highly controllable power outputs and uniform radial power distributions, to produce craters which are flat bottomed and of a chosen diameter and depth.

For both Laser Ablation and Desolvating Nebulisers, a small flow of Nitrogen may also be introduced into the Argon flow. Nitrogen exists as a dimer, so has more vibrational modes and is more efficient at receiving energy from the RF coil around the torch.

Other methods of sample introduction are also utilized. Electrothermal vaporization (ETV) and in torch vaporization (ITV) use hot surfaces (graphite or metal, generally) to vaporize samples for introduction. These can use very small amounts of liquids, solids, or slurries. Other methods like vapor generation are also known.

Plasma Torch

The plasma used in an ICP-MS is made by partially ionizing argon gas ($Ar \rightarrow Ar^+ + e^-$). The energy required for this reaction is obtained by pulsing an alternating electric current in wires that surround the argon gas.

After the sample is injected, the plasma's extreme temperature causes the sample to separate into individual atoms (atomization). Next, the plasma ionizes these atoms ($M \rightarrow M^+ + e^-$) so that they can be detected by the mass spectrometer.

An inductively coupled plasma (ICP) for spectrometry is sustained in a torch that consists of three concentric tubes, usually made of quartz. The two major designs are the Fassel and Greenfield torches. The end of this torch is placed inside an induction coil supplied with a radio-frequency electric current. A flow of argon gas (usually 14 to 18 liters per minute) is introduced between the two outermost tubes of the torch and an electrical spark is applied for a short time to introduce free electrons into the gas stream. These electrons interact with the radio-frequency magnetic field of the induction coil and are accelerated first in one direction, then the other, as the field changes at high frequency (usually 27.12 MHz). The accelerated electrons collide with argon atoms, and sometimes a collision causes an argon atom to part with one of its electrons. The released electron is in turn accelerated by the rapidly changing magnetic field. The process continues until the rate of release of new electrons in collisions is balanced by the rate of recombination of electrons with argon ions (atoms that have lost an electron). This produces a 'fireball' that consists mostly of argon atoms with a rather small fraction of free electrons and argon ions.

Advantage of Argon

Making the plasma from argon, instead of other gases, has several advantages. First, argon is abundant (in the atmosphere, as a result of the radioactive decay of potassium) and therefore cheaper than other noble gases. Argon also has a higher first ionization potential than all other elements except He, F, and Ne. Because of this high ionization energy, the reaction ($Ar^+ + e^- \rightarrow Ar$) is more energetically favorable than the reaction ($M^+ + e^- \rightarrow M$). This ensures that the sample remains ionized (as M^+) so that the mass spectrometer can detect it.

Argon can be purchased for use with the ICP-MS in either a refrigerated liquid or a gas form. However it is important to note that whichever form of argon purchased, it should have a guaranteed purity of 99.9% Argon at a minimum. It is important to determine which type of argon will be best suited for the specific situation. Liquid argon is typically cheaper and can be stored in a greater quantity as opposed to the gas form, which is more expensive and takes up more tank space. If the instrument is in an environment where it gets infrequent use, then buying argon in the gas state will be most appropriate as it will be more than enough to suit smaller run times and gas in the cylinder will remain stable for longer periods of time, whereas liquid argon will suffer loss to the environment due to venting of the tank when stored over extended time frames. However, if the ICP-MS is to be used routinely and is on and running for eight or more hours each day for several days a week, then going with liquid argon will be the most suitable. If there are to be multiple ICP-MS instruments running for long periods of time, then it will most likely be beneficial for the laboratory to install a bulk or micro bulk argon tank which will be maintained by a gas supply company, thus eliminating the need to change out tanks frequently as well as minimizing loss of argon that is left over in each used tank as well as down time for tank changeover.

There are rare ICP-MS solutions that utilize helium for plasma generation.

Transfer of Ions into Vacuum

The carrier gas is sent through the central channel and into the very hot plasma. The sample is then exposed to radio frequency which converts the gas into a plasma. The high temperature of the plasma is sufficient to cause a very large portion of the sample to form ions. This fraction of ionization can approach 100% for some elements (e.g. sodium), but this is dependent on the ionization potential. A fraction of the formed ions passes through a ~1 mm hole (sampler cone) and then a ~0.4 mm hole (skimmer cone). The purpose of which is to allow a vacuum that is required by the mass spectrometer.

The vacuum is created and maintained by a series of pumps. The first stage is usually based on a roughing pump, most commonly a standard rotary vane pump. This removes most of the gas and typically reaches a pressure of around 133 Pa. Later stages have their vacuum generated by more powerful vacuum systems, most often turbomolecular pumps. Older instruments may have used oil diffusion pumps for high vacuum regions.

Ion Optics

Before mass separation, a beam of positive ions has to be extracted from the plasma and focused into the mass-analyzer. It is important to separate the ions from UV photons, energetic neutrals and from any solid particles that may have been carried into the instrument from the ICP. Traditionally, ICP-MS instruments have used transmitting ion lens arrangements for this purpose. Examples include the Einzel lens, the Barrel lens, Agilent's Omega Lens and Perkin-Elmer's Shadow Stop. Another approach is to use ion guides (quadrupoles, hexapoles, or octopoles) to guide the ions into mass analyzer along a path away from the trajectory of photons or neutral particles. Yet another approach is Varian patented used by Analytik Jena ICP-MS 90 degrees reflecting parabolic "Ion Mirror" optics, which are claimed to provide more efficient ion transport into the mass-analyzer, resulting in better sensitivity and reduced background. Baffled flight tubes and off-axis detectors are also used. Analytik Jena ICP-MS is the most sensitive instrument on the market.

A sector ICP-MS will commonly have four sections: an extraction acceleration region, steering lenses, an electrostatic sector and a magnetic sector. The first region takes ions from the plasma and accelerates them using a high voltage. The second uses may use a combination of parallel plates, rings, quadropoles, hexapoles and octopoles to steer, shape and focus the beam so that the resulting peaks are symmetrical, flat topped and have high transmission. The electrostatic sector may be before or after the magnetic sector depending on the particular instrument, and reduces the spread in kinetic energy caused by the plasma. This spread is particularly large for ICP-MS, being larger than Glow Discharge and much larger than TIMS. The geometry of the instrument is chosen so that the instrument the combined focal point of the electrostatic and magnetic sectors is at the collector, known as Double Focussing (or Double Foccussing).

If the mass of interest has a low sensitivity and is just below a much larger peak, the low mass tail from this larger peak can intrude onto the mass of interest. A Retardation Filter might be used to reduce this tail. This sits near the collector, and applies a voltage equal but opposite to the accelerating voltage; any ions that have lost energy while flying around the instrument will be decelerated to rest by the filter.

Collision Reaction Cell and CRI

The collision/reaction cell is used to remove interfering ions through ion/neutral reactions. Collision/reaction cells are known under several names. The dynamic reaction cell is located before the quadrupole in the ICP-MS device. The chamber has a quadrupole and can be filled with reaction (or collision) gases (ammonia, methane, oxygen or hydrogen), with one gas type at a time or a mixture of two of them, which reacts with the introduced sample, eliminating some of the interference.

The collisional reaction interface (CRI) is a mini-collision cell installed in front of the parabolic ion mirror optics that removes interfering ions by injecting a collisional gas (He), or a reactive gas (H_2), or a mixture of the two, directly into the plasma as it flows through the skimmer cone and/or the sampler cone. The CRI removed interfering ions using a collisional kinetic energy discrimination (KED) phenomenon and chemical reactions with interfering ions similarly to traditionally used larger collision cells.

Routine Maintenance

As with any piece of instrumentation or equipment, there are many aspects of maintenance that need to be encompassed by daily, weekly and annual procedures. The frequency of maintenance is typically determined by the sample volume and cumulative run time that the instrument is subjected to.

One of the first things that should be carried out before the calibration of the ICP-MS is a sensitivity check and optimization. This ensures that the operator is aware of any possible issues with the instrument and if so, may address them before beginning a calibration. Typical indicators of sensitivity are Rhodium levels, Cerium/Oxide ratios and DI water blanks.

One of the most frequent forms of routine maintenance is replacing sample and waste tubing on the peristaltic pump, as these tubes can get worn fairly quickly resulting in holes and clogs in the sample line, resulting in skewed results. Other parts that will need regular cleaning and/or replacing are sample tips, nebulizer tips, sample cones, skimmer cones, injector tubes, torches and lenses. It may also be necessary to change the oil in the interface roughing pump as well as the vacuum backing pump, depending on the workload put on the instrument.

Sample Preparation

For most clinical methods using ICP-MS, there is a relatively simple and quick sample prep process. The main component to the sample is an internal standard, which also serves as the diluent. This internal standard consists primarily of deionized water, with nitric or hydrochloric acid, and Indium and/or Gallium. Depending on the sample type, usually 5 ml of the internal standard is added to a test tube along with 10–500 microliters of sample. This mixture is then vortexed for several seconds or until mixed well and then loaded onto the autosampler tray. For other applications that may involve very viscous samples or samples that have particulate matter, a process known as sample digestion may have to be carried out, before it can be pipetted and analyzed. This adds an extra first step to the above process, and therefore makes the sample prep more lengthy.

Reactive-ion Etching

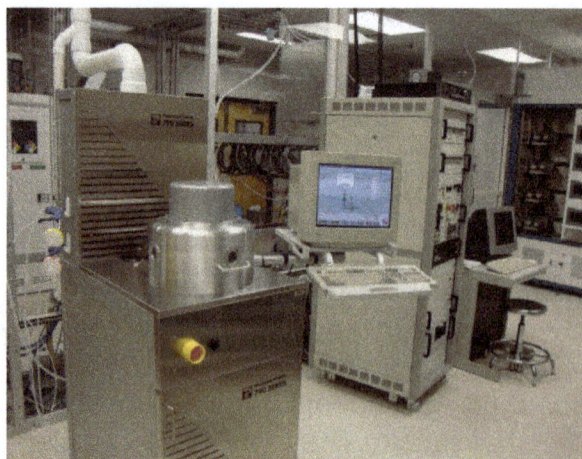

A commercial reactive-ion etching setup in a cleanroom

Reactive-ion etching (RIE) is an etching technology used in microfabrication. RIE is a type of dry etching which has different characteristics than wet etching. RIE uses chemically reactive plasma to remove material deposited on wafers. The plasma is generated under low pressure (vacuum) by an electromagnetic field. High-energy ions from the plasma attack the wafer surface and react with it.

Equipment

A typical (parallel plate) RIE system consists of a cylindrical vacuum chamber, with a wafer platter situated in the bottom portion of the chamber. The wafer platter is electrically isolated from the rest of the chamber. Gas enters through small inlets in the top of the chamber, and exits to the vacuum pump system through the bottom. The types and amount of gas used vary depending upon the etch process; for instance, sulfur hexafluoride is commonly used for etching silicon. Gas pressure is typically maintained in a range between a few millitorr and a few hundred millitorr by adjusting gas flow rates and/or adjusting an exhaust orifice.

Other types of RIE systems exist, including inductively coupled plasma (ICP) RIE. In this type of system, the plasma is generated with an RF powered magnetic field. Very high plasma densities can be achieved, though etch profiles tend to be more isotropic.

A combination of parallel plate and inductively coupled plasma RIE is possible. In this system, the ICP is employed as a high density source of ions which increases the etch rate, whereas a separate RF bias is applied to the substrate (silicon wafer) to create directional electric fields near the substrate to achieve more anisotropic etch profiles.

Method of Operation

Plasma is initiated in the system by applying a strong RF (radio frequency) electromagnetic field to the wafer platter. The field is typically set to a frequency of 13.56 Megahertz, applied at a few hundred watts. The oscillating electric field ionizes the gas molecules by stripping them of electrons, creating a plasma.

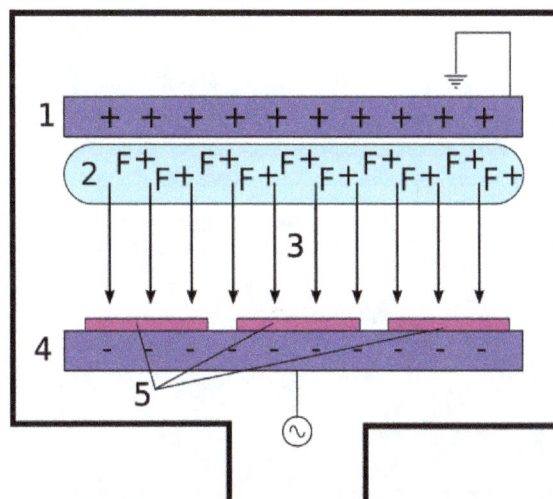

A diagram of a common RIE setup. An RIE consists of two electrodes (1 and 4) that create an electric field (3) meant to accelerate ions (2) toward the surface of the samples (5).

In each cycle of the field, the electrons are electrically accelerated up and down in the chamber, sometimes striking both the upper wall of the chamber and the wafer platter. At the same time, the much more massive ions move relatively little in response to the RF electric field. When electrons are absorbed into the chamber walls they are simply fed out to ground and do not alter the electronic state of the system. However, electrons deposited on the wafer platter cause the platter to build up charge due to its DC isolation. This charge build up develops a large negative voltage on the platter, typically around a few hundred volts. The plasma itself develops a slightly positive charge due to the higher concentration of positive ions compared to free electrons.

Because of the large voltage difference, the positive ions tend to drift toward the wafer platter, where they collide with the samples to be etched. The ions react chemically with the materials on the surface of the samples, but can also knock off (sputter) some material by transferring some of their kinetic energy. Due to the mostly vertical delivery of reactive ions, reactive-ion etching can produce very anisotropic etch profiles, which contrast with the typically isotropic profiles of wet chemical etching.

Etch conditions in an RIE system depend strongly on the many process parameters, such as pressure, gas flows, and RF power. A modified version of RIE is deep reactive-ion etching, used to excavate deep features.

Plasma Propulsion Engine

An early plasma propulsion engine from the Lewis Research Center in Cleveland, Ohio in 1961

A plasma thruster during test firing

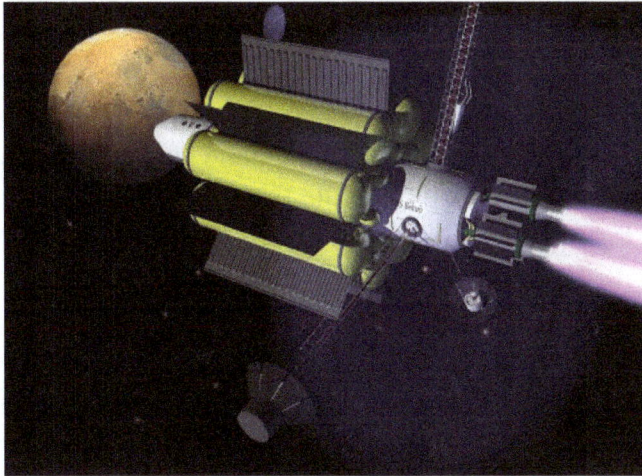

Artist rendition of VASIMR plasma engine

A plasma propulsion engine is a type of electric propulsion that generates thrust from a quasi-neutral plasma. This is in contrast to ion thruster engines, which generates thrust through extracting an ion current from plasma source, which is then accelerated to high velocities using grids/anodes. These exist in many forms. Plasma thrusters do not typically use high voltage grids or anodes/cathodes to accelerate the charged particles in the plasma, but rather uses currents and potentials which are generated internally in the plasma to accelerate the plasma ions. While this results in a lower exhaust velocities by virtue of the lack of high accelerating voltages, this type of thruster has a number of interesting advantages. The lack of high voltage grids of anodes removes a possible limiting element as a result of grid ion erosion. The plasma exhaust is 'quasi- neutral', which means that ion and electrons exist in equal number, which allows simply ion- electron recombination in the exhaust to neutralise the exhaust plume, removing the need for an electron gun (hollow cathode). This type of thruster often generates the source plasma using radio frequency of microwave energy, using an external antenna. This fact, combined with the absence of hollow cathodes (which are very sensitive to all but the few noble gases) allows the intriguing possibility of being able to use this type of thruster on a huge range of propellants, from argon, to carbon dioxide, air mixtures to astronaut urine.

Plasma engines are better suited for long-distance interplanetary space travel missions.

In recent years, many agencies have developed several forms of plasma-fueled engines, including the European Space Agency, Iranian Space Agency and Australian National University, which have co-developed a more advanced type described as a double layer thruster. However, this form of plasma engine is only one of many types.

Engine Types

Helicon Double Layer Thrusters

Helicon double-layer thrusters use radio waves to create a plasma and a magnetic nozzle to focus and accelerate the plasma away from the rocket engine. A Mini-Helicon Plasma Thruster, ideal for space maneuvers, runs off of nitrogen, and the fuel has an exhaust velocity (specific impulse) 10 times that of chemical rockets.

Magnetoplasmadynamic Thrusters

Magnetoplasmadynamic thrusters (MPD) use the Lorentz force (a force resulting from the inter-action between a magnetic field and an electric current) to generate thrust - The electric charge flowing through the plasma in the presence of a magnetic field causing the plasma to accelerate due to the generated magnetic force. The Lorentz force is also crucial to the operation of most pulsed plasma thruster

Pulsed Inductive Thrusters

Pulsed inductive thrusters (PIT) also use the Lorentz force to generate thrust, but unlike the mag-netoplasmadynamic thruster, they do not use any electrode, preventing their erosion. Ionization and electric currents in the plasma are induced by a rapidly varying magnetic field.

Electrodeless Plasma Thrusters

Electrodeless plasma thrusters use the ponderomotive force which acts on any plasma or charged particle when under the influence of a strong electromagnetic energy density gradient to acceler-ate both electrons and ions of the plasma in the same direction, thereby able to operate without neutralizer.

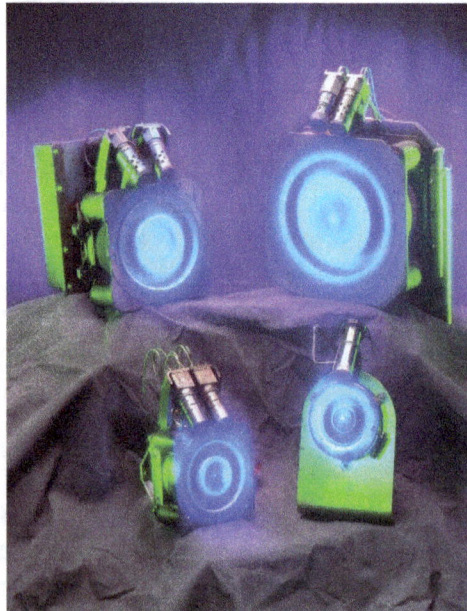

SPT

Hall Effect Thrusters

Hall effect thrusters (also called stationary plasma thrusters SPT) combine a strong localized static magnetic field perpendicular to the electric field created between an upstream anode and a down-stream cathode called neutralizer, to create a "virtual cathode" (area of high electron density) at the exit of the device. This virtual cathode then attracts the ions formed inside the thruster closer to the anode. Finally the accelerated ion beam is neutralized by some of the electrons emitted by the neutralizer. Serial production of Hall effect thruster started in Soviet Union in the 1970s. One of the early variants, SPT-100 is now produced under license by European Snecma Moteurs under the name PPS-1350. Similarly BPT-4000 and PPS-5000 are closely related to SPT-140. SPT-290 has a thrust of 1.5N, 5-30 kW power and specific impulse 30 km/s, efficiency 65% and weight 23 kg.

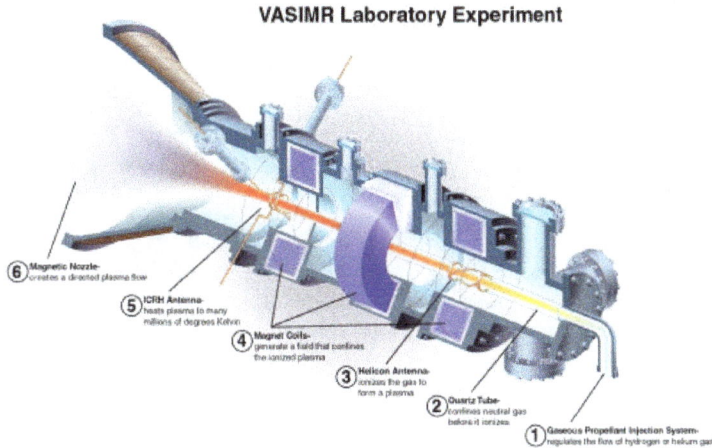

VASIMR Laboratory Experiment

6 Magnetic Nozzle- creates a directed plasma flow

5 ICRH Antenna- heats plasma to many millions of degrees Kelvin

4 Magnet Coils- generate a field that confines the ionized plasma

3 Helicon Antenna- ionizes the gas to form a plasma

2 Quartz Tube- confines neutral gas before it ionizes

1 Gaseous Propellant Injection System- regulates the flow of hydrogen or helium gas.

VASIMR

VASIMR

VASIMR, short for Variable Specific Impulse Magnetoplasma Rocket, uses radio waves to ionize a propellant into a plasma. Then, a magnetic field accelerates the plasma from the rocket engine, generating thrust. The VASIMR is being developed by Ad Astra Rocket Company, headquartered in Houston, TX. A Nova Scotia, Canada-based company Nautel, is producing the 200 kW RF generators required to ionize the propellant. Some component tests and "Plasma Shoot" experiments are performed in a Liberia, Costa Rica laboratory. This project is led by former NASA astronaut Dr. Franklin Chang-Díaz (CRC-USA).

The Costa Rican Aerospace Alliance has announced development of an exterior support for the VASIMR to be fitted outside the International Space Station. This phase of the plan to test the VASIMR in space is expected to be conducted in 2016. A projected 200 megawatt VASIMR engine could reduce the time to travel from Earth to Jupiter or Saturn from six years to fourteen months, and from Earth to Mars from 6 months to 39 days.

Plasma Display

A plasma display panel (PDP) is a type of flat panel display common to large TV displays 30 inches (76 cm) or larger. They are called "plasma" displays because they use small cells containing electrically charged ionized gases, which are plasmas.

Plasma displays have lost nearly all market share, mostly due to competition from low-cost LCD and more expensive but high-contrast OLED flat-panel displays; manufacturing for the United States retail market ended in 2014, and manufacturing for the Chinese market is expected to end in 2016.

A 103" plasma display panel by Panasonic, on display in August 2009.

General Characteristics

Plasma displays are bright (1,000 lux or higher for the module), have a wide color gamut, and can be produced in fairly large sizes—up to 3.8 metres (150 in) diagonally. They had a very low-luminance "dark-room" black level compared with the lighter grey of the unilluminated parts of an LCD screen at least in the early history of the competing technologies (in the early history of plasma panels the blacks were blacker on plasmas and greyer on LCDs). LED-backlit LCD televisions have been developed to reduce this distinction. The display panel itself is about 6 cm (2.4 in) thick, generally allowing the device's total thickness (including electronics) to be less than 10 cm (3.9 in). Power consumption varies greatly with picture content, with bright scenes drawing significantly more power than darker ones – this is also true for CRTs as well as modern LCDs where LED backlight brightness is adjusted dynamically. The plasma that illuminates the screen can reach a temperature of at least 1200 °C (2200 °F). Typical power consumption is 400 watts for a 127 cm (50 in) screen. 200 to 310 watts for a 127 cm (50 in) display when set to cinema mode. Most screens are set to "shop" mode by default, which draws at least twice the power (around 500–700 watts) of a "home" setting of less extreme brightness. Panasonic has greatly reduced power consumption ("1/3 of 2007 models"). Panasonic states that PDPs will consume only half the power of their previous series of plasma sets to achieve the same overall brightness for a given display size. The lifetime of the latest generation of plasma displays is estimated at 100,000 hours of actual display time, or 27 years at 10 hours per day. This is the estimated time over which maximum picture brightness degrades to half the original value.

Plasma screens are made out of glass.This may causes glare from reflected objects in the viewing area. Companies such as Panasonic coat their newer plasma screens with an anti-glare filter material. Currently, plasma panels cannot be economically manufactured in screen sizes smaller than 82 centimetres (32 in). Although a few companies have been able to make plasma enhanced-definition televisions (EDTV) this small, even fewer have made 32 inch plasma HDTVs. With the trend toward large-screen television technology, the 32 inch screen size is rapidly disappearing. Though considered bulky and thick compared with their LCD counterparts, some sets such as Panasonic's Z1 and Samsung's B860 series are as slim as 2.5 cm (1 in) thick making them comparable to LCDs in this respect.

Competing display technologies include cathode ray tube (CRT), organic light-emitting diode (OLED), AMLCD, Digital Light Processing DLP, SED-tv, LED display, field emission display (FED), and quantum dot display (QLED).

Plasma Display Advantages and Disadvantages

Advantages

- Capable of producing deeper blacks allowing for superior contrast ratio

- Wider viewing angles than those of LCD; images do not suffer from degradation at less than straight ahead angles like LCDs. LCDs using IPS technology have the widest angles, but they do not equal the range of plasma primarily due to "IPS glow", a generally whitish haze that appears due to the nature of the IPS pixel design.

- Less visible motion blur, thanks in large part to very high refresh rates and a faster response time, contributing to superior performance when displaying content with significant amounts of rapid motion.

- Superior uniformity. LCD panel backlights nearly always produce uneven brightness levels, although this is not always noticeable. High-end computer monitors have technologies to try to compensate for the uniformity problem.

- Unaffected by clouding from the polishing process. Some LCD panel types, like IPS, require a polishing process that can introduce a haze usually referred to as "clouding".

- Less expensive for the buyer per square inch than LCD, particularly when equivalent performance is considered.

Disadvantages

- Earlier generation displays were more susceptible to screen burn-in and image retention. Recent models have a pixel orbiter that moves the entire picture slower than is noticeable to the human eye, which reduces the effect of burn-in but does not prevent it.

- Due to the bistable nature of the colour and intensity generating method, some people will notice that plasma displays have a shimmering or flickering effect with a number of hues, intensities and dither patterns.

- Earlier generation displays (circa 2006 and prior) had phosphors that lost luminosity over time, resulting in gradual decline of absolute image brightness. Newer models have advertised lifespans exceeding 100 000 hours, far longer than older CRTs

- Uses more electrical power, on average, than an LCD TV using an LED backlight. Older CCFL backlights for LCD panels used quite a bit more power, and older plasma TVs used quite a bit more power than recent models.

- Does not work as well at high altitudes above 6,500 feet (2,000 metres) due to pressure differential between the gases inside the screen and the air pressure at altitude. It may cause a buzzing noise. Manufacturers rate their screens to indicate the altitude parameters.

- For those who wish to listen to AM radio, or are amateur radio operators (hams) or short-wave listeners (SWL), the radio frequency interference (RFI) from these devices can be irritating or disabling.

- Plasma displays are generally heavier than LCD, and may require more careful handling such as being kept upright.

Native Plasma Television Resolutions

Fixed-pixel displays such as plasma TVs scale the video image of each incoming signal to the native resolution of the display panel. The most common native resolutions for plasma display panels are 853×480 (EDTV), 1,366×768 or 1,920×1,080 (HDTV). As a result, picture quality varies depending on the performance of the video scaling processor and the upscaling and downscaling algorithms used by each display manufacturer.

Enhanced-definition Plasma Television

Early plasma televisions were enhanced-definition (ED) with a native resolution of 840×480 (discontinued) or 853×480, and down-scaled their incoming High-definition video signals to match their native display resolution.

ED Resolutions

The following ED resolutions were common prior to the introduction of HD displays, but have long been phased out in favor of HD displays, as well as because the overall pixel count in ED displays is lower than the pixel count on SD PAL displays (853x480 vs 720x576, respectively).

- 840×480p

- 853×480p

High-definition Plasma Television

Early high-definition (HD) plasma displays had a resolution of 1024x1024 and were alternate lighting of surfaces (ALiS) panels made by Fujitsu/Hitachi. These were interlaced displays, with non-square pixels.

Modern HDTV plasma televisions usually have a resolution of 1,024×768 found on many 42 inch plasma screens, 1,280×768, 1,366×768 found on 50 in, 60 in, and 65 in plasma screens, or 1,920×1,080 found in plasma screen sizes from 42 inch to 103 inch. These displays are usually progressive displays, with square pixels, and will up-scale their incoming standard-definition signals to match their native display resolution. 1024x768 resolution requires that 720p content be downscaled.

HD Resolutions

- 1024×1024

- 1024×768

- 1280×768

- 1366×768

- 1280×1080

- 1920×1080

Design

A panel of a plasma display typically comprises millions of tiny compartments in between two panels of glass. These compartments, or "bulbs" or "cells", hold a mixture of noble gases and a minuscule amount of another gas (e.g., mercury vapor). Just as in the fluorescent lamps over an office desk, when a high voltage is applied across the cell, the gas in the cells forms a plasma. With flow of electricity (electrons), some of the electrons strike mercury particles as the electrons move through the plasma, momentarily increasing the energy level of the atom until the excess energy is shed. Mercury sheds the energy as ultraviolet (UV) photons. The UV photons then strike phosphor that is painted on the inside of the cell. When the UV photon strikes a phosphor molecule, it momentarily raises the energy level of an outer orbit electron in the phosphor molecule, moving the electron from a stable to an unstable state; the electron then sheds the excess energy as a photon at a lower energy level than UV light; the lower energy photons are mostly in the infrared range but about 40% are in the visible light range. Thus the input energy is converted to mostly infrared but also as visible light. The screen heats up to between 30 and 41 °C (86 and 106 °F) during operation. Depending on the phosphors used, different colors of visible light can be achieved. Each pixel in a plasma display is made up of three cells comprising the primary colors of visible light. Varying the voltage of the signals to the cells thus allows different perceived colors.

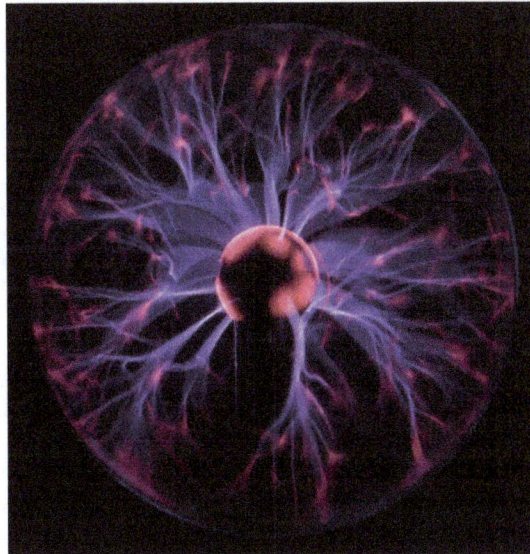

Ionized gases such as the ones shown here are confined to millions of tiny individual compartments across the face of a plasma display, to collectively form a visual image.

The long electrodes are stripes of electrically conducting material that also lies between the glass plates in front of and behind the cells. The "address electrodes" sit behind the cells, along the rear glass plate, and can be opaque. The transparent display electrodes are mounted in front of the cell, along the front glass plate. As can be seen in the illustration, the electrodes are covered by an insulating protective layer.

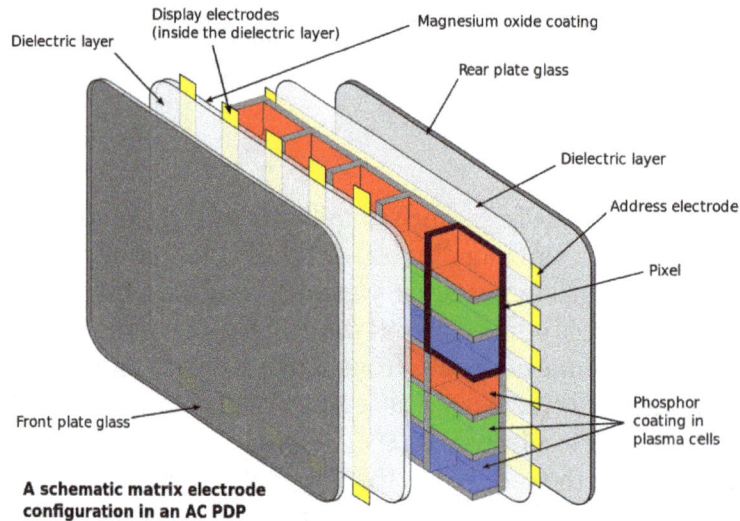

Composition of plasma display panel

Control circuitry charges the electrodes that cross paths at a cell, creating a voltage difference between front and back. Some of the atoms in the gas of a cell then lose electrons and become ionized, which creates an electrically conducting plasma of atoms, free electrons, and ions. The collisions of the flowing electrons in the plasma with the inert gas atoms leads to light emission; such light-emitting plasmas are known as glow discharges.

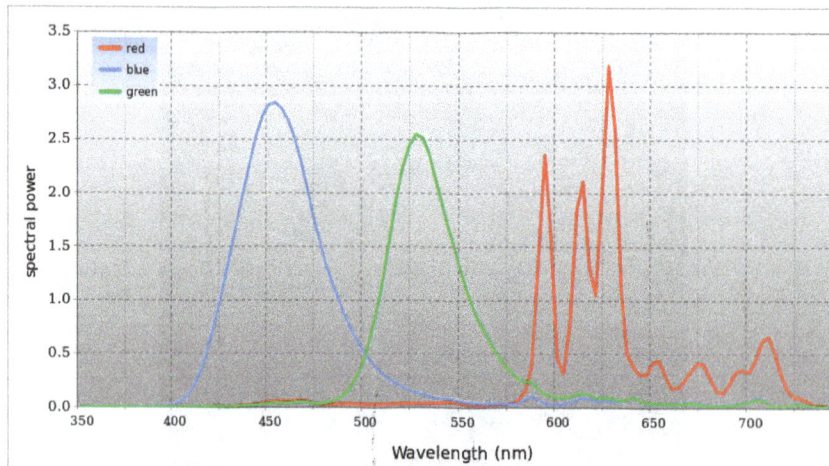

Relative spectral power of Red, Green and Blue phosphors of a common plasma display. The units of spectral power are simply raw sensor values (with a linear response at specific wavelengths).

In a monochrome plasma panel, the gas is mostly neon, and the color is the characteristic orange of a neon-filled lamp (or sign). Once a glow discharge has been initiated in a cell, it can be maintained by applying a low-level voltage between all the horizontal and vertical electrodes–even after the ionizing voltage is removed. To erase a cell all voltage is removed from a pair of electrodes. This type of panel has inherent memory. A small amount of nitrogen is added to the neon to increase hysteresis. In color panels, the back of each cell is coated with a phosphor. The ultraviolet photons emitted by the plasma excite these phosphors, which give off visible light with colors determined by the phosphor materials. This aspect is comparable to fluorescent lamps and to the neon signs that use colored phosphors.

Every pixel is made up of three separate subpixel cells, each with different colored phosphors. One subpixel has a red light phosphor, one subpixel has a green light phosphor and one subpixel has a blue light phosphor. These colors blend together to create the overall color of the pixel, the same as a triad of a shadow mask CRT or color LCD. Plasma panels use pulse-width modulation (PWM) to control brightness: by varying the pulses of current flowing through the different cells thousands of times per second, the control system can increase or decrease the intensity of each subpixel color to create billions of different combinations of red, green and blue. In this way, the control system can produce most of the visible colors. Plasma displays use the same phosphors as CRTs, which accounts for the extremely accurate color reproduction when viewing television or computer video images (which use an RGB color system designed for CRT displays).

Plasma displays are different from liquid crystal displays (LCDs), another lightweight flat-screen display using very different technology. LCDs may use one or two large fluorescent lamps as a backlight source, but the different colors are controlled by LCD units, which in effect behave as gates that allow or block light through red, green, or blue filters on the front of the LCD panel.

Contrast Ratio

Contrast ratio is the difference between the brightest and darkest parts of an image, measured in discrete steps, at any given moment. Generally, the higher the contrast ratio, the more realistic the image is (though the "realism" of an image depends on many factors including color accuracy, luminance linearity, and spatial linearity.) Contrast ratios for plasma displays are often advertised as high as 5,000,000:1. On the surface, this is a significant advantage of plasma over most other current display technologies, a notable exception being organic light-emitting diode. Although there are no industry-wide guidelines for reporting contrast ratio, most manufacturers follow either the ANSI standard or perform a full-on-full-off test. The ANSI standard uses a checkered test pattern whereby the darkest blacks and the lightest whites are simultaneously measured, yielding the most accurate "real-world" ratings. In contrast, a full-on-full-off test measures the ratio using a pure black screen and a pure white screen, which gives higher values but does not represent a typical viewing scenario. Some displays, using many different technologies, have some "leakage" of light, through either optical or electronic means, from lit pixels to adjacent pixels so that dark pixels that are near bright ones appear less dark than they do during a full-off display. Manufacturers can further artificially improve the reported contrast ratio by increasing the contrast and brightness settings to achieve the highest test values. However, a contrast ratio generated by this method is misleading, as content would be essentially unwatchable at such settings.

Each cell on a plasma display must be precharged before it is lit, otherwise the cell would not respond quickly enough. This precharging means the cells cannot achieve a true black, whereas an LED backlit LCD panel can actually turn off parts of the backlight, in "spots" or "patches" (this technique, however, does not prevent the large accumulated passive light of adjacent lamps, and the reflection media, from returning values from within the panel). Some manufacturers have reduced the precharge and the associated background glow, to the point where black levels on modern plasmas are starting to become close to some high-end CRTs Sony and Mitsubishi produced before ten years before the comparable plasma displays. It is important to note that plasma displays were developed for ten more years than CRTs; it is almost certain that if CRTs had been developed for as long as plasma displays were, the contrast on CRTs would have been far better than contrast on the plasma displays. With an LCD, black pixels are generated by a light polarization method; many panels are unable to completely block the underlying backlight. More recent LCD panels using LED illumination can automatically reduce the backlighting on darker scenes, though this method cannot be used in high-contrast scenes, leaving some light showing from black parts of an image with bright parts, such as (at the extreme) a solid black screen with one fine intense bright line. This is called a "halo" effect which has been minimized on newer LED-backlit LCDs with local dimming. Edgelit models cannot compete with this as the light is reflected via a light guide to distribute the light behind the panel.

Screen Burn-in

Image burn-in occurs on CRTs and plasma panels when the same picture is displayed for long periods. This causes the phosphors to overheat, losing some of their luminosity and producing a "shadow" image that is visible with the power off. Burn-in is especially a problem on plasma panels because they run hotter than CRTs. Early plasma televisions were plagued by burn-in, making it impossible to use video games or anything else that displayed static images.

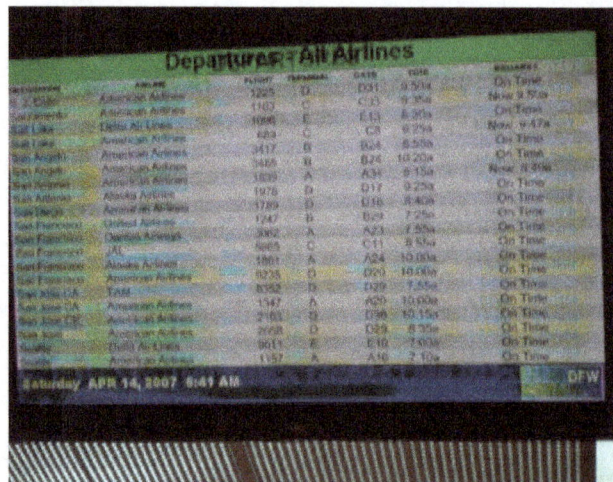

An example of a plasma display that has suffered severe burn-in from static text

Plasma displays also exhibit another image retention issue which is sometimes confused with screen burn-in damage. In this mode, when a group of pixels are run at high brightness (when displaying white, for example) for an extended period, a charge build-up in the pixel structure occurs and a ghost image can be seen. However, unlike burn-in, this charge build-up is transient and self-corrects after the image condition that caused the effect has been removed and a long enough period has passed (with the display either off or on).

Plasma manufacturers have tried various ways of reducing burn-in such as using gray pillarboxes, pixel orbiters and image washing routines, but none to date have eliminated the problem and all plasma manufacturers continue to exclude burn-in from their warranties.

Environmental Impact

Plasma screens have been lagging behind CRT and LCD screens in terms of energy consumption efficiency. To reduce the energy consumption, new technologies are also being found. Although it can be expected that plasma screens will continue to become more energy efficient in the future, a growing problem is that people tend to keep their old TVs running and an increasing trend to escalating screen sizes.

History

In 1936, Kálmán Tihanyi, a Hungarian engineer, described the principle of "plasma television" and conceived the first flat-panel display system.

Plasma displays were first used in PLATO computer terminals. This PLATO V model illustrates the display's monochromatic orange glow seen in 1981.

The monochrome plasma video display was co-invented in 1964 at the University of Illinois at Urbana-Champaign by Donald Bitzer, H. Gene Slottow, and graduate student Robert Willson for the PLATO Computer System. The original neon orange monochrome Digivue display panels built by glass producer Owens-Illinois were very popular in the early 1970s because they were rugged and needed neither memory nor circuitry to refresh the images. A long period of sales decline occurred in the late 1970s because semiconductor memory made CRT displays cheaper than the 2500 USD 512 x 512 PLATO plasma displays. Nonetheless, the plasma displays' relatively large screen size and 1 inch thickness made them suitable for high-profile placement in lobbies and stock exchanges.

Burroughs Corporation, a maker of adding machines and computers, developed the Panaplex display in the early 1970s. The Panaplex display, generically referred to as a gas-discharge or gas-plasma display, uses the same technology as later plasma video displays, but began life as seven-segment display for use in adding machines. They became popular for their bright orange luminous look and found nearly ubiquitous use in cash registers, calculators, pinball machines, aircraft avionics such as radios, navigational instruments, and stormscopes; test equipment such as frequency counters and multimeters; and generally anything that previously used nixie tube or numitron displays with a high digit-count throughout the late 1970s and into the 1990s. These displays remained popular until LEDs gained popularity because of their low-current draw and module-flexibility, but are still found in some applications where their high-brightness is desired, such as pinball machines and avionics. Pinball displays started with six- and seven-digit seven-segment displays and later evolved into 16-segment alphanumeric displays, and later into 128x32 dot-matrix displays in 1990, which are still used today.

1983

In 1983, IBM introduced a 19 inches (48 cm) orange-on-black monochrome display (model 3290 'information panel') which was able to show up to four simultaneous IBM 3270 terminal sessions. Due to heavy competition from monochrome LCDs, in 1987 IBM planned to shut down its factory in upstate New York, the largest plasma plant in the world, in favor of manufacturing mainframe computers. Consequently, Larry Weber co-founded a startup company Plasmaco with Stephen Globus, as well as James Kehoe, who was the IBM plant manager, and bought the plant from IBM. Weber stayed in Urbana as CTO until 1990, then moved to upstate New York to work at Plasmaco.

1990s

1992

In 1992, Fujitsu introduced the world's first 21-inch (53 cm) full-color display. It was a hybrid, the plasma display created at the University of Illinois at Urbana-Champaign and NHK Science & Technology Research Laboratories.

1994

In 1994, Weber demonstrated a color plasma display at an industry convention in San Jose. Panasonic Corporation began a joint development project with Plasmaco, which led in 1996 to the purchase of Plasmaco, its color AC technology, and its American factory.

1995

In 1995, Fujitsu introduced the first 42-inch (107 cm) plasma display; it had 852x480 resolution and was progressively scanned. Also in 1997, Philips introduced a 42-inch (107 cm) display, with 852x480 resolution. It was the only plasma to be displayed to the retail public in four Sears locations in the US. The price was US$14,999 and included in-home installation. Later in 1997, Pioneer started selling their first plasma television to the public, and others followed.

2000s

2006–2009

In late 2006, analysts noted that LCDs overtook plasmas, particularly in the 40-inch (1.0 m) and above segment where plasma had previously gained market share. Another industry trend is the consolidation of manufacturers of plasma displays, with around 50 brands available but only five manufacturers. In the first quarter of 2008 a comparison of worldwide TV sales breaks down to 22.1 million for direct-view CRT, 21.1 million for LCD, 2.8 million for Plasma, and 0.1 million for rear-projection.

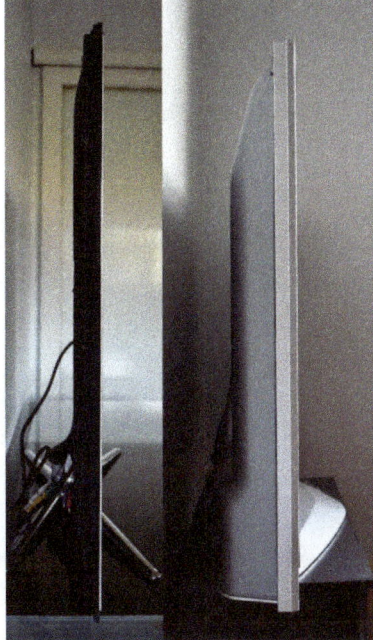

Average plasma displays have become one quarter the thickness from 2006 to 2011

Until the early 2000s, plasma displays were the most popular choice for HDTV flat panel display as they had many benefits over LCDs. Beyond plasma's deeper blacks, increased contrast, faster response time, greater color spectrum, and wider viewing angle; they were also much bigger than LCDs, and it was believed that LCDs were suited only to smaller sized televisions. However, improvements in VLSI fabrication have since narrowed the technological gap. The increased size, lower weight, falling prices, and often lower electrical power consumption of LCDs now make them competitive with plasma television sets.

Screen sizes have increased since the introduction of plasma displays. The largest plasma video display in the world at the 2008 Consumer Electronics Show in Las Vegas, Nevada, was a 150 inches (380 cm) unit manufactured by Matsushita Electric Industrial (Panasonic) standing 6 ft (180 cm) tall by 11 ft (330 cm) wide.

2010s

At the 2010 Consumer Electronics Show in Las Vegas, Panasonic introduced their 152" 2160p 3D plasma. In 2010 Panasonic shipped 19.1 million plasma TV panels.

In 2010, the shipments of plasma TVs reached 18.2 million units globally. Since that time, shipments of plasma TVs have declined substantially. This decline has been attributed to the competition from liquid crystal (LCD) televisions, whose prices have fallen more rapidly than those of the plasma TVs. In late 2013, Panasonic announced that they would stop producing plasma TVs from March 2014 onwards. In 2014, LG and Samsung discontinued plasma TV production as well, effectively killing the technology, probably because of lowering demand.

Display Manufacturers

Most have discontinued doing so, but at one time or another all of these companies have produced plasma displays:

- Panasonic
- Samsung
- LG
- Pioneer
- Toshiba
- Gradiente
- Lanix
- ProScan
- Sanyo
- Funai
- Magnavox
- JVC
- Sony
- Orion
- Philips
- Protron
- Apple
- Beko (known sometimes as grundig)
- Vestel (both under Vestel name but also under various brands)

Panasonic was the biggest plasma display manufacturer until 2013, when it decided to discontinue plasma production. In the following months, Samsung and LG also ceased production of plasma sets. Panasonic, Samsung and LG were the last plasma manufacturers for the U.S. retail market.

Plasma Actuator

Plasma actuators are a type of actuator currently being developed for aerodynamic flow control. Plasma actuators impart force in a similar way to ionocraft.

The working of these actuators is based on the formation of a low-temperature plasma between a pair of asymmetric electrodes by application of a high-voltage AC signal across the electrodes.

Consequently, air molecules from the air surrounding the electrodes are ionized, and are accelerated through the electric field.

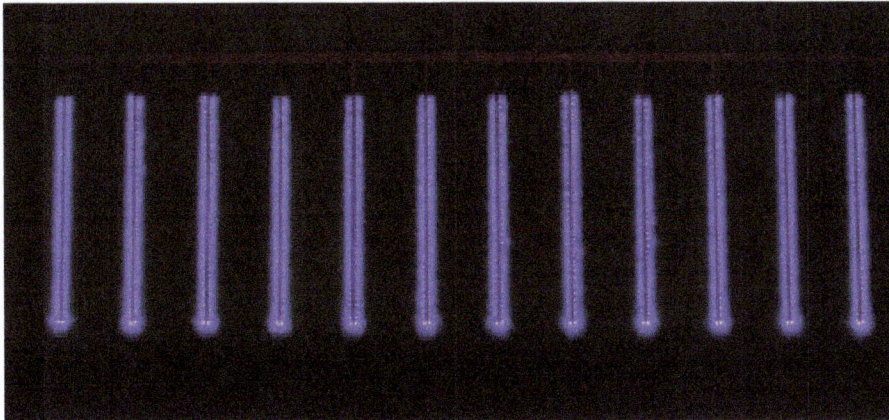

Glow of plasma actuator discharges

Introduction

Plasma actuators operating at the atmospheric conditions are promising for flow control, mainly for their physical properties, such as the induced body force by a strong electric field and the generation of heat during an electric arc, and the simplicity of their constructions and placements. In particular, the recent invention of glow discharge plasma actuators by Roth (2003) that can produce sufficient quantities of glow discharge plasma in the atmosphere pressure air helps to yield an increase in flow control performance.

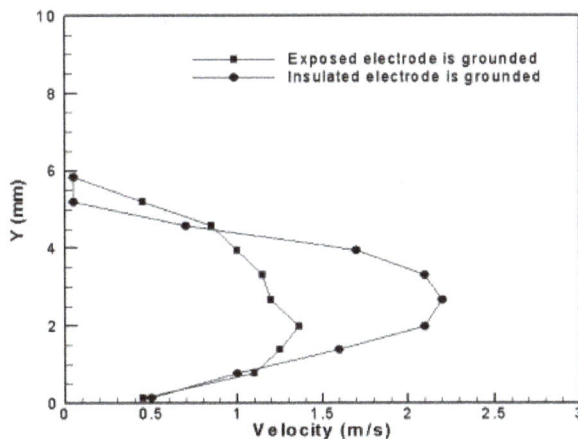

Local flow speed induced by a plasma actuator

Power Supply and Electrode Layouts

Either a direct current (DC) or an alternating current(AC) power supply or a microwave microdischarge can be used for different configurations of plasma actuators. One schematic of an AC power supply design for a dielectric barrier discharge plasma actuator is given here as an example. The performance of plasma actuators is determined by dielectric materials and power inputs, later is limited by the qualities of MOSFET or IGBT.

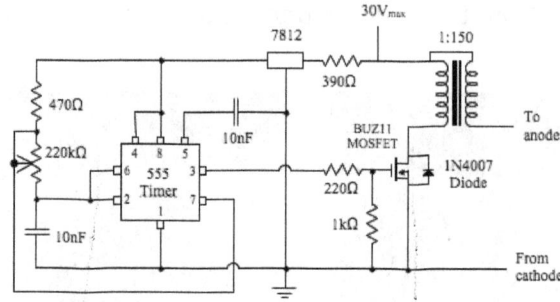

Driving circuits (E-type) of a power supply

The driving waveforms can be optimized to achieve a better actuation (induced flow speed). However, a sinusoidal waveform may be more preferable for the simplicity in power supply construction. The additional benefit is the relatively less electromagnetic interference. Pulse width modulation can be adopted to instantaneously adjust the strength of actuation.

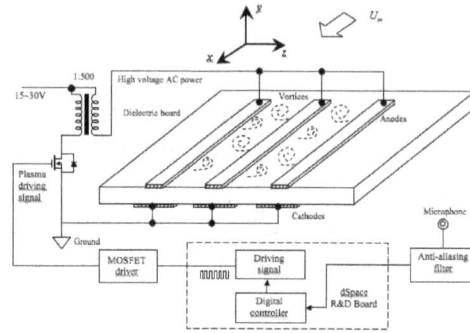

Pulse width modulation of plasma power input

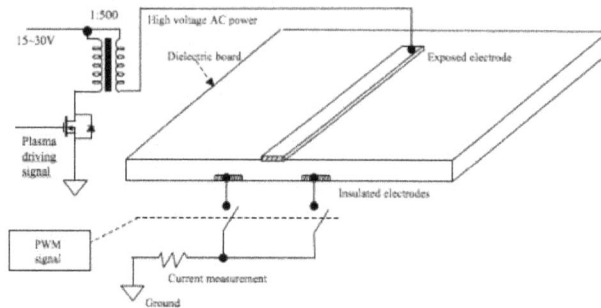

One configuration of DBD plasma actuator

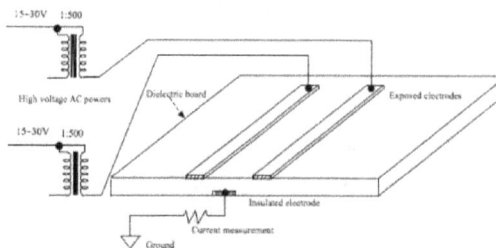

One configuration of DBD plasma actuator

Manipulation of the encapsulated electrode and distributing the encapsulated electrode throughout the dielectric layer has been shown to alter the performance of the dielectric barrier discharge (DBD) plasma actuator. Locating the initial encapsulated electrode closer to the dielectric surface results in induced velocities higher than the baseline case for a given voltage. In addition, Actuators with a shallow initial electrode are able to impart more momentum and mechanical power into the flow.

No matter how much funding has been invested and the number of various private claims of a high induced speed, the maximum, average speed induced by plasma actuators on an atmospheric pressure conviction, without any assistant of mechanical amplifier (chamber, cavity etc.), is still less than 10 m/s.

Influence of Temperature

When dealing with real life aircraft equipped with plasma actuators, it is important to consider the effect of temperature. The temperature variations encountered during a flight envelope may have adverse effects in actuator performance. It is found that for a constant peak-to-peak voltage the maximum velocity produced by the actuator depends directly on the dielectric surface temperature. The findings suggest that by changing the actuator temperature the performance can be maintained or even altered at different environmental conditions. Increasing dielectric surface temperature can increase the plasma actuator performance by increasing the momentum flux whilst consuming slightly higher energy.

Flow Control Applications

Some recent applications of plasma actuation include high-speed flow control using localized arc filament plasma actuators, and low-speed flow control using dielectric barrier discharges and sliding discharges. The present research of plasma actuators is mainly focused on three directions: (1) various designs of plasma actuators; (2) flow control applications; and (3) control-oriented modeling of flow applications under plasma actuation. In addition, new experimental and numerical methods are being developed to provide physical insights.

Vortex Generator

A plasma actuator induces a local flow speed perturbation, which will be developed downstream to a vortex sheet. As a result, plasma actuators can behave as vortex generators. The difference between this and traditional vortex generation is that there are no mechanical moving parts or any drilling holes on aerodynamic surfaces, demonstrating an important benefit of plasma actuators.

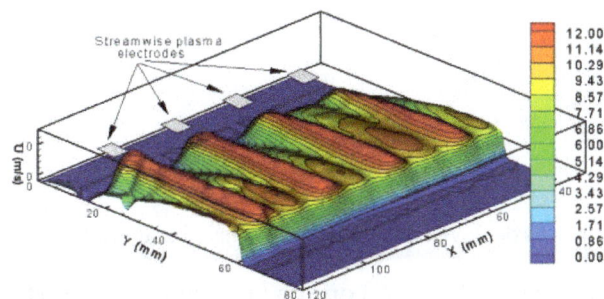

Plasma induced flow field

Active Noise Control

Active noise control normally denotes noise cancellation, that is, a noise-cancellation speaker emits a sound wave with the same amplitude but with inverted phase (also known as antiphase) to the original sound. However, active noise control with plasma adopts different strategies. The first one uses the discovery that sound pressure could be attenuated when it passes through a plasma sheet. The second one, and being more widely used, is to actively suppress the flow-field that is responsible to flow-induced noise (also known as aeroacoustics), using plasma actuators. It has been demonstrated that both tonal noise and broadband noise (difference can refer to tonal versus broadband) can be actively attenuated by a carefully designed plasma actuator.

Supersonic and Hypersonic Flow Control

Plasma has been introduced to hypersonic flow control. Firstly, plasma could be much easier generated for hypersonic vehicle at high altitude with quite low atmospheric pressure and high surface temperature. Secondly, the classical aerodynamic surface has little actuation for the case.

Interest in plasma actuators as active flow control devices is growing rapidly due to their lack of mechanical parts, light weight and high response frequency. The characteristics of a dielectric barrier discharge (DBD) plasma actuator when exposed to an unsteady flow generated by a shock tube is examined. A Study shows that not only is the shear layer outside of the shock tube affected by the plasma but the passage of the shock front and high-speed flow behind it also greatly influences the properties of the plasma

Flight Control

Plasma actuators could be mounted on the airfoil to control flight attitude and thereafter flight trajectory. The cumbersome design and maintenance efforts of mechanical and hydraulic transmission systems in a classical rudder can thus be saved. The price to pay is that one should design a suitable high voltage/power electric system satisfying EMC rule. Hence, in addition to flow control, plasma actuators hold potential in top-level flight control, in particular for UAV and extraterrestrial planet (with suitable atmospheric conditions) investigations.

On the other hand, the whole flight control strategy should be reconsidered taking account of characteristics of plasma actuators. One preliminary roll control system with DBD plasma actuators is shown in the figure.

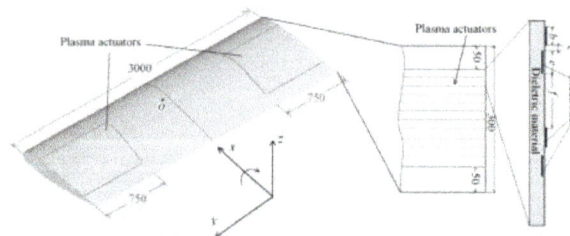

DBD Plasma actuators deployed on a NACA 0015 airfoil to do rudderless flight control

It can be seen that plasma actuators deployed on the both sides of an airfoil. The roll control can be controlled by activating plasma actuators according to the roll angle feedback. After studying various feedback control methodologies, the bang–bang control method was chosen to design the roll control system based on plasma actuators. The reason is that bang-bang control is time optimal and insensitive to plasma actuations, which quickly vary in difference atmospheric and electric conditions.

Modeling

Various numerical models have been proposed to simulate plasma actuations in flow control. They are listed below according to the computational cost, from the most expensive to the cheapest.

- Monte carlo method plus particle-in-cell;

- Electricity modeling coupled with Navier-Stokes equations;

- Lumped element model coupled with Navier-Stokes equations

- Surrogate model to simulate plasma actuation.

The most important potential of plasma actuators is its ability to bridge fluids and electricity. A modern closed-loop control system and the following information theoretical methods can be applied to the relatively classical aerodynamic sciences. A control-oriented model for plasma actuation in flow control has been proposed for a cavity flow control case.

Plasma Deep Drilling Technology

Plasma deep drilling technology is one of several different variants of recently explored new drilling technologies which would be able to substitute conventional, contact-based rotary systems. These new technologies, including plasma deep drilling, water jet, hydrothermal spallation or laser, are matter of active research. Only a very small number of companies have embraced plasma-drilling method, e.g. GA Drilling, headquartered in Bratislava, Slovakia.

Plasmatorch using water steam as plasma-creation gas

High-energetic Electrical Plasma

High-energetic electrical plasma is a technology currently being developed in deep drilling applications to address many issues related to drilling in water environment or in the production of boreholes with a wide range of diameters.

Physical Principle of Electrical Plasma

An electric arc is an electrical breakdown of a gas that produces an ongoing plasma discharge, resulting from a current flowing through normally nonconductive media such as air or gas. An arc discharge is characterized by a lower voltage than a glow discharge, and relies on thermionic emission of electrons from the electrodes supporting the arc. The electric arc is influenced by factors such as: the gas flow, inner and outer magnetic fields, and construction elements of the chamber which confine the arc. The development of highly effective plasma torches to be used as a source of the thermal plasma, demands a deep understanding of a wide spectrum of the processes taking place in the discharge chamber.

Drilling using electrical plasma

Advantages of Plasma Deep Drilling Technology

1. Higher drilling energy efficiency

2. Continuous drilling process without replacement of mechanical parts

3. Constant casing diameter

4. Effective transport of disintegrated rock

References

* Myers, Robert L. (2002). Display interfaces: fundamentals and standards. John Wiley and Sons. pp. 69–71. ISBN 978-0-471-49946-6. Plasma displays are closely related to the simple neon lamp.

* Rasool Erfani, Zare-Behtash H.; Hale, C.; Kontis, K. (2015). "Development of DBD plasma actuators: The double encapsulated electrode". Acta Astronautica. 109: 132–143. doi:10.1016/j.actaastro.2014.12.016.

* Thomas, Robert (2001). "A Beginner's Guide to ICP-MS" (PDF). Spectroscopy. Advanstar Communications. Retrieved 2014-05-09.

* G4techTV – Plasma vs LCD power consumption shootout Archived March 5, 2012, at the Wayback Machine.‹The template Wayback is being considered for merging.›

* Rasool Erfani, Zare-Behtash H.; Kontis, K. (2012). "Plasma actuator: Influence of dielectric surface temperature". Experimental Thermal and Fluid Science. 42: 258–264. doi:10.1016/j.expthermflusci.2012.04.023.

- LCD TV Market Ten Times Larger Than Plasma TVs On Units-Shipped Basis, 20 February 2011, Jonathan Sutton, hdtvtest.co.uk, retrieved at September 12, 2011

- Huang, X., Zhang, X., and Li, Y. (2010) Broadband Flow-Induced Sound Control using Plasma Actuators, Journal of Sound and Vibration, Vol 329, No 13, pp. 2477–2489.

- Li, Y.; Zhang, X.; Huang, X. (2010). "The Use of Plasma Actuators for Bluff Body Broadband Noise Control". Experiments in Fluids. 49 (2): 367–377. doi:10.1007/s00348-009-0806-3.

- Peers, Ed; Huang, Xun; Ma, Zhaokai (2010). "A numerical model of plasma effects in flow control". Physics Letters A. 374 (13-14): 1501–1504.

Allied Fields of Plasma Physics

The allied fields of plasma physics are magnetohydrodynamics, fluid dynamics, physical cosmology and electrodynamics. Plasma physics is a vast subject that branches out into significant fields that have been thoroughly discussed in this chapter.

Magnetohydrodynamics

Magnetohydrodynamics (MHD; also magneto fluid dynamics or hydromagnetics) is the study of the magnetic properties of electrically conducting fluids. Examples of such magneto-fluids include plasmas, liquid metals, and salt water or electrolytes. The word *magnetohydrodynamics (MHD)* is derived from *magneto-* meaning magnetic field, *hydro-* meaning water, and *-dynamics* meaning movement. The field of MHD was initiated by Hannes Alfvén, for which he received the Nobel Prize in Physics in 1970.

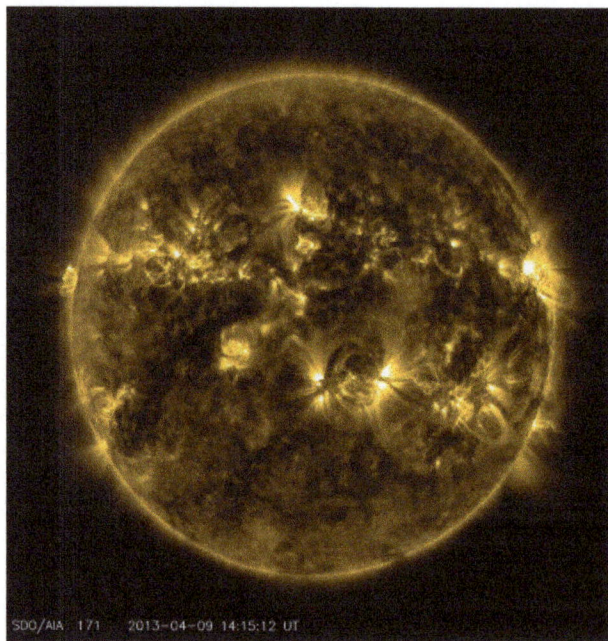

The sun is an MHD system that is not well understood.

The fundamental concept behind MHD is that magnetic fields can induce currents in a moving conductive fluid, which in turn polarizes the fluid and reciprocally changes the magnetic field itself. The set of equations that describe MHD are a combination of the Navier-Stokes equations of fluid dynamics and Maxwell's equations of electromagnetism. These differential equations must be solved simultaneously, either analytically or numerically.

History

The first recorded use of the word *magnetohydrodynamics* is by Hannes Alfvén in 1942:

> "At last some remarks are made about the transfer of momentum from the Sun to the planets, which is fundamental to the theory (§11). The importance of the magnetohydrodynamic waves in this respect are pointed out."

Michael Faraday

The ebbing salty water flowing past London's Waterloo Bridge interacts with the Earth's magnetic field to produce a potential difference between the two river-banks. Michael Faraday tried this experiment in 1832 but the current was too small to measure with the equipment at the time, and the river bed contributed to short-circuit the signal. However, by a similar process the voltage induced by the tide in the English Channel was measured in 1851.

Ideal and Resistive MHD

The simplest form of MHD, Ideal MHD, assumes that the fluid has so little resistivity that it can be treated as a perfect conductor. This is the limit of infinite magnetic Reynolds number. In ideal MHD, Lenz's law dictates that the fluid is in a sense *tied* to the magnetic field lines. To explain, in ideal MHD a small rope-like volume of fluid surrounding a field line will continue to lie along a magnetic field line, even as it is twisted and distorted by fluid flows in the system. This is sometimes referred to as the magnetic field lines being "frozen" in the fluid. The connection between magnetic field lines and fluid in ideal MHD fixes the topology of the magnetic field in the fluid—for example, if a set of magnetic field lines are tied into a knot, then they will remain so as long as the fluid/plasma has negligible resistivity. This difficulty in reconnecting magnetic field lines makes it possible to store energy by moving the fluid or the source of the magnetic field. The energy can then become available if the conditions for ideal MHD break down, allowing magnetic reconnection that releases the stored energy from the magnetic field.

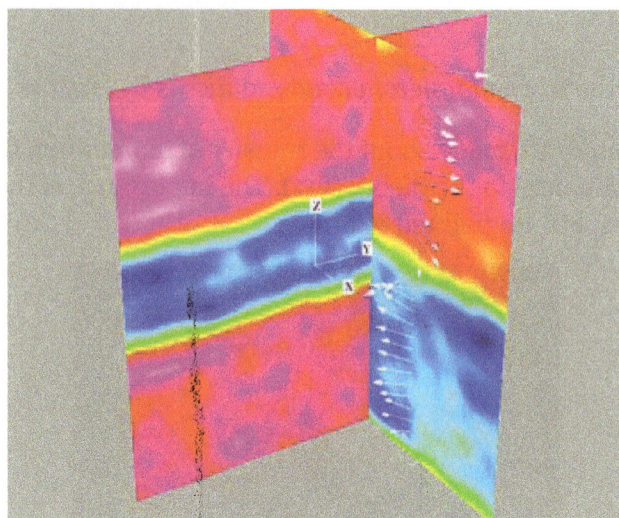

MHD Simulation of the Solar Wind

Ideal MHD Equations

The ideal MHD equations consist of the continuity equation, the Cauchy momentum equation, Ampere's Law neglecting displacement current, and a temperature evolution equation. As with any fluid description to a kinetic system, a closure approximation must be applied to highest moment of the particle distribution equation. This is often accomplished with approximations to the heat flux through a condition of adiabaticity or isothermality.

In the following, $\mathbf{B}$ is the magnetic field, $\mathbf{E}$ is the electric field, $\mathbf{v}$ is the bulk plasma velocity, $\mathbf{J}$ is the current density, ρ is the mass density, p is the plasma pressure, $\nabla\cdot$ is divergence, and t is time. The continuity equation is

$$\frac{\partial \rho}{\partial t} + \nabla \cdot (\rho \mathbf{v}) = 0.$$

The Cauchy momentum equation is

$$\rho\left(\frac{\partial}{\partial t} + \mathbf{v}\cdot\nabla\right)\mathbf{v} = \mathbf{J}\times\mathbf{B} - \nabla p.$$

The Lorentz force term $\mathbf{J}\times\mathbf{B}$ can be expanded using Ampere's law and the identity

$$\frac{1}{2}\nabla(\mathbf{B}\cdot\mathbf{B}) = (\mathbf{B}\cdot\nabla)\mathbf{B} + \mathbf{B}\times(\nabla\times\mathbf{B})$$

to give

$$\mathbf{J}\times\mathbf{B} = \frac{(\mathbf{B}\cdot\nabla)\mathbf{B}}{\mu_0} - \nabla\left(\frac{B^2}{2\mu_0}\right),$$

where the first term on the right hand side is the magnetic tension force and the second term is the magnetic pressure force. The ideal Ohm's law for a plasma is given by

$$\mathbf{E} + \mathbf{v} \times \mathbf{B} = 0.$$

Faraday's law is

$$\frac{\partial \mathbf{B}}{\partial t} = -\nabla \times \mathbf{E}.$$

The low-frequency Ampere's law neglects displacement current and is given by

$$\mu_0 \mathbf{J} = \nabla \times \mathbf{B}.$$

The magnetic divergence constraint is

$$\nabla \cdot \mathbf{B} = 0.$$

The energy equation is given by

$$\frac{\mathrm{d}}{\mathrm{d}t}\left(\frac{p}{\rho^\gamma}\right) = 0,$$

where $\gamma = 5/3$ is the ratio of specific heats for an adiabatic equation of state. This energy equation is, of course, only applicable in the absence of shocks or heat conduction as it assumes that the entropy of a fluid element does not change.

Applicability of Ideal MHD to Plasmas

Ideal MHD is only strictly applicable when:

1. The plasma is strongly collisional, so that the time scale of collisions is shorter than the other characteristic times in the system, and the particle distributions are therefore close to Maxwellian.

2. The resistivity due to these collisions is small. In particular, the typical magnetic diffusion times over any scale length present in the system must be longer than any time scale of interest.

3. Interest in length scales much longer than the ion skin depth and Larmor radius perpendicular to the field, long enough along the field to ignore Landau damping, and time scales much longer than the ion gyration time (system is smooth and slowly evolving).

Importance of Resistivity

In an imperfectly conducting fluid the magnetic field can generally move through the fluid following a diffusion law with the resistivity of the plasma serving as a diffusion constant. This means that solutions to the ideal MHD equations are only applicable for a limited time for a region of a given size before diffusion becomes too important to ignore. One can estimate the diffusion time across a solar active region (from collisional resistivity) to be hundreds to thousands of years, much longer than the actual lifetime of a sunspot—so it would seem reasonable to ignore the resistivity. By contrast, a meter-sized volume of seawater has a magnetic diffusion time measured in milliseconds.

Even in physical systems – which are large and conductive enough that simple estimates of the Lundquist number suggest that the resistivity can be ignored – resistivity may still be important: many instabilities exist that can increase the effective resistivity of the plasma by factors of more than a billion. The enhanced resistivity is usually the result of the formation of small scale structure like current sheets or fine scale magnetic turbulence, introducing small spatial scales into the system over which ideal MHD is broken and magnetic diffusion can occur quickly. When this happens, magnetic reconnection may occur in the plasma to release stored magnetic energy as waves, bulk mechanical acceleration of material, particle acceleration, and heat.

Magnetic reconnection in highly conductive systems is important because it concentrates energy in time and space, so that gentle forces applied to a plasma for long periods of time can cause violent explosions and bursts of radiation.

When the fluid cannot be considered as completely conductive, but the other conditions for ideal MHD are satisfied, it is possible to use an extended model called resistive MHD. This includes an extra term in Ohm's Law which models the collisional resistivity. Generally MHD computer simulations are at least somewhat resistive because their computational grid introduces a numerical resistivity.

Importance of Kinetic Effects

Another limitation of MHD (and fluid theories in general) is that they depend on the assumption that the plasma is strongly collisional (this is the first criterion listed above), so that the time scale of collisions is shorter than the other characteristic times in the system, and the particle distributions are Maxwellian. This is usually not the case in fusion, space and astrophysical plasmas. When this is not the case, or the interest is in smaller spatial scales, it may be necessary to use a kinetic model which properly accounts for the non-Maxwellian shape of the distribution function. However, because MHD is relatively simple and captures many of the important properties of plasma dynamics it is often qualitatively accurate and is therefore often the first model tried.

Effects which are essentially kinetic and not captured by fluid models include double layers, Landau damping, a wide range of instabilities, chemical separation in space plasmas and electron runaway. In the case of ultra-high intensity laser interactions, the incredibly short timescales of energy deposition mean that hydrodynamic codes fail to capture the essential physics.

Structures in MHD Systems

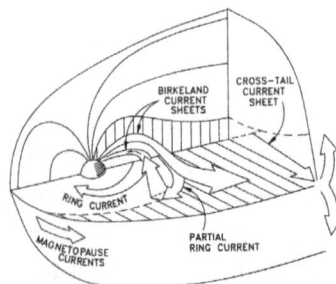

Schematic view of the different current systems which shape the Earth's magnetosphere

In many MHD systems most of the electric current is compressed into thin nearly-two-dimensional ribbons termed current sheets. These can divide the fluid into magnetic domains, inside of which the currents are relatively weak. Current sheets in the solar corona are thought to be between a few meters and a few kilometers in thickness, which is quite thin compared to the magnetic domains (which are thousands to hundreds of thousands of kilometers across). Another example is in the Earth's magnetosphere, where current sheets separate topologically distinct domains, isolating most of the Earth's ionosphere from the solar wind.

Waves

The wave modes derived using MHD plasma theory are called magnetohydrodynamic waves or MHD waves. In general there are three MHD wave modes:

- Pure (or oblique) Alfvén wave

- Slow MHD wave

- Fast MHD wave

All these waves have constant phase velocities for all frequencies, and hence there is no dispersion. At the limits when the angle between the wave propagation vector k and magnetic field B is either 0 (180) or 90 degrees, the wave modes are called:

name	type	propagation	phase velocity	association	medium	other names
Sound wave	longitudinal	$\vec{k} \backslash \vec{B}$	adiabatic sound velocity	none	compressible, nonconducting fluid	
Alfvén wave	transverse	$\vec{k} \backslash \vec{B}$	Alfvén velocity	B		shear Alfvén wave, the slow Alfvén wave, torsional Alfvén wave
Magnetosonic wave	longitudinal	$\vec{k} \perp \vec{B}$		B , E		compressional Alfvén wave, fast Alfvén wave, magnetoacoustic wave

The phase velocity depends on the angle between the wave vector k and the magnetic field B. An MHD wave propagating at an arbitrary angle θ with respect to the time independent or bulk field $\mathbf{B}_0$ will satisfy the dispersion relation

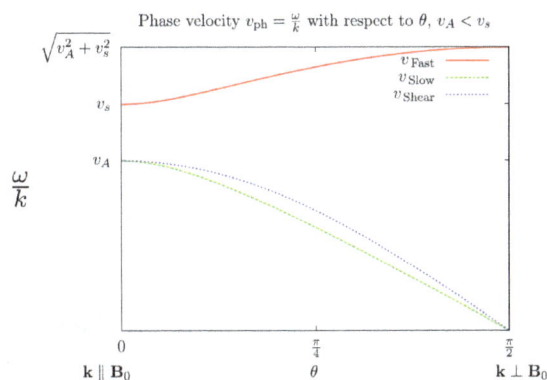

Phase velocity $v_{\mathrm{ph}} = \frac{\omega}{k}$ with respect to θ, $v_A < v_s$

Phase velocity plotted with respect to θ for $v_A > v_s$.

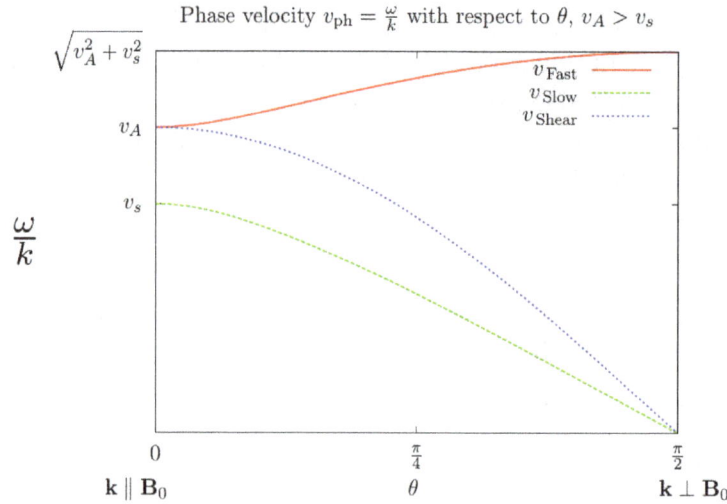

Phase velocity $v_{ph} = \frac{\omega}{k}$ with respect to θ, $v_A > v_s$

Phase velocity plotted with respect to θ for $v_A < v_s$.

$$\frac{\omega}{k} = v_A \cos\theta$$

where

$$v_A = \frac{B_0}{\sqrt{\mu_0 \rho}}$$

is the Alfvén speed. This branch corresponds to the shear Alfvén mode. Additionally the dispersion equation gives

$$\frac{\omega}{k} = \left[\frac{1}{2}(v_A^2 + v_s^2) \pm \frac{1}{2}\sqrt{(v_A^2 + v_s^2)^2 - 4v_s^2 v_A^2 \cos^2\theta} \right]^{\frac{1}{2}}$$

where

$$v_s = \sqrt{\frac{\gamma p}{\rho}}$$

is the ideal gas sound speed. The plus branch corresponds to the fast-MHD wave mode and the minus branch corresponds to the slow-MHD wave mode.

The MHD oscillations will be damped if the fluid is not perfectly conducting but has a finite conductivity, or if viscous effects are present.

MHD waves and oscillations are a popular tool for the remote diagnostics of laboratory and astrophysical plasmas, e.g. the corona of the Sun (Coronal seismology).

Extensions

Resistive

Resistive MHD describes magnetized fluids with finite electron diffusivity (). This diffusivity leads to a breaking in the magnetic topology; magnetic field lines can 'reconnect' when they collide. Usually this term is small and reconnections can be handled by thinking of them as not dissimilar to shocks; this process has been shown to be important in the Earth-Solar magnetic interactions.

Extended

Extended MHD describes a class of phenomena in plasmas that are higher order than resistive MHD, but which can adequately be treated with a single fluid description. These include the effects of Hall physics, electron pressure gradients, finite Larmor Radii in the particle gyromotion, and electron inertia.

Two-fluid

Two-fluid MHD describes plasmas that include a non-negligible Hall electric field. As a result, the electron and ion momenta must be treated separately. This description is more closely tied to Maxwell's equations as an evolution equation for the electric field exists.

Hall

In 1960, M. J. Lighthill criticized the applicability of ideal or resistive MHD theory for plasmas. It concerned the neglect of the "Hall current term", a frequent simplification made in magnetic fusion theory. Hall-magnetohydrodynamics (HMHD) takes into account this electric field description of magnetohydrodynamics. The most important difference is that in the absence of field line breaking, the magnetic field is tied to the electrons and not to the bulk fluid.

Collisionless

MHD is also often used for collisionless plasmas. In that case the MHD equations are derived from the Vlasov equation.

Reduced

By using a multiscale analysis the (resistive) MHD equations can be reduced to a set of four closed scalar equations. This allows e.g. for more efficient numerical calculations.

Applications

Geophysics

Beneath the Earth's mantle lies the core, which is made up of two parts: the solid inner core and liquid outer core. Both have significant quantities of iron. The liquid outer core moves in the presence of the magnetic field and eddies are set up into the same due to the Coriolis effect. These eddies develop a magnetic field which boosts Earth's original magnetic field—a process which is self-sustaining and is called the geomagnetic dynamo.

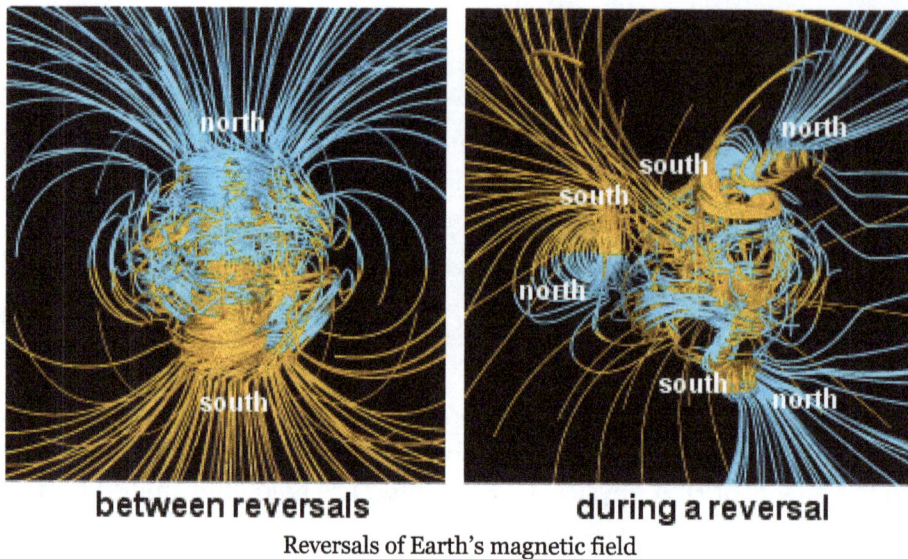

between reversals   **during a reversal**

Reversals of Earth's magnetic field

Based on the MHD equations, Glatzmaier and Paul Roberts have made a supercomputer model of the Earth's interior. After running the simulations for thousands of years in virtual time, the changes in Earth's magnetic field can be studied. The simulation results are in good agreement with the observations as the simulations have correctly predicted that the Earth's magnetic field flips every few hundred thousands of years. During the flips, the magnetic field does not vanish altogether—it just gets more complex.

Earthquakes

Some monitoring stations have reported that earthquakes are sometimes preceded by a spike in ultra low frequency (ULF) activity. A remarkable example of this occurred before the 1989 Loma Prieta earthquake in California, although a subsequent study indicates that this was little more than a sensor malfunction. On December 9, 2010, geoscientists announced that the DEMETER satellite observed a dramatic increase in ULF radio waves over Haiti in the month before the magnitude 7.0 M_w 2010 earthquake. Researchers are attempting to learn more about this correlation to find out whether this method can be used as part of an early warning system for earthquakes.

Astrophysics

MHD applies to astrophysical, including stars, the interplanetary medium (space between the planets), and possibly within the interstellar medium (space between the stars) and jets. Most astrophysical systems are not in local thermal equilibrium, and therefore require an additional kinematic treatment to describe all the phenomena within the system.

Sunspots are caused by the Sun's magnetic fields, as Joseph Larmor theorized in 1919. The solar wind is also governed by MHD. The differential solar rotation may be the long-term effect of magnetic drag at the poles of the Sun, an MHD phenomenon due to the Parker spiral shape assumed by the extended magnetic field of the Sun.

Previously, theories describing the formation of the Sun and planets could not explain how the Sun has 99.87% of the mass, yet only 0.54% of the angular momentum in the solar system. In a closed system such as the cloud of gas and dust from which the Sun was formed, mass and angular momentum are both conserved. That conservation would imply that as the mass concentrated in the center of the cloud to form the Sun, it would spin faster, much like a skater pulling their arms in. The high speed of rotation predicted by early theories would have flung the proto-Sun apart before it could have formed. However, magnetohydrodynamic effects transfer the Sun's angular momentum into the outer solar system, slowing its rotation.

Breakdown of ideal MHD (in the form of magnetic reconnection) is known to be the likely cause of solar flares. The magnetic field in a solar active region over a sunspot can store energy that is released suddenly as a burst of motion, X-rays, and radiation when the main current sheet collapses, reconnecting the field.

Sensors

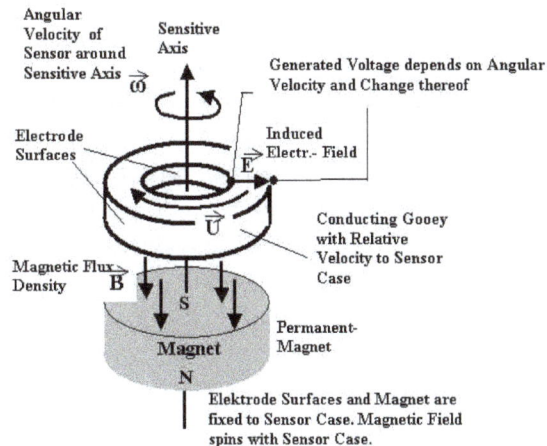

Principle of MHD sensor for angular velocity measurement

Magnetohydrodynamic sensors are used for precision measurements of angular velocities in inertial navigation systems such as in aerospace engineering. Accuracy improves with the size of the sensor. The sensor is capable of surviving in harsh environments.

Engineering

MHD is related to engineering problems such as plasma confinement, liquid-metal cooling of nuclear reactors, and electromagnetic casting (among others).

A magnetohydrodynamic drive or MHD propulsor is a method for propelling seagoing vessels using only electric and magnetic fields with no moving parts, using magnetohydrodynamics. The working principle involves electrification of the propellant (gas or water) which can then be directed by a magnetic field, pushing the vehicle in the opposite direction. Although some working prototypes exist, MHD drives remain impractical.

The first prototype of this kind of propulsion was built and tested in 1965 by Steward Way, a professor of mechanical engineering at the University of California, Santa Barbara. Way, on leave from his job at Westinghouse Electric, assigned his senior-year undergraduate students to develop a submarine with this new propulsion system. In the early 1990s, a foundation in Japan (Ship & Ocean Foundation (Minato-ku, Tokyo)) built a experimental boat, the 'Yamato-1,' which used a magnetohydrodynamic drive incorporating a superconductor cooled by liquid helium, and could travel at 15 km/h.

MHD power generation fueled by potassium-seeded coal combustion gas showed potential for more efficient energy conversion (the absence of solid moving parts allows operation at higher temperatures), but failed due to cost-prohibitive technical difficulties. One major engineering problem was the failure of the wall of the primary-coal combustion chamber due to abrasion.

In microfluidics, MHD is studied as a fluid pump for producing a continuous, nonpulsating flow in a complex microchannel design.

MHD can be implemented in the continuous casting process of metals to suppress instabilities and control the flow.

Magnetic Drug Targeting

An important task in cancer research is developing more precise methods for delivery of medicine to affected areas. One method involves the binding of medicine to biologically compatible magnetic particles (e.g. ferrofluids), which are guided to the target via careful placement of permanent magnets on the external body. Magnetohydrodynamic equations and finite element analysis are used to study the interaction between the magnetic fluid particles in the bloodstream and the external magnetic field.

Fluid Dynamics

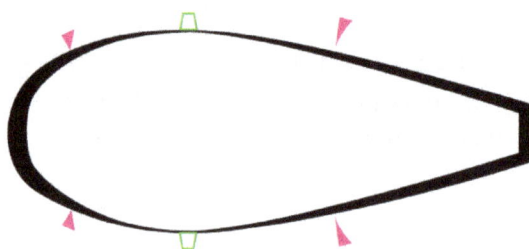

Typical aerodynamic teardrop shape, assuming a viscous medium passing from left to right, the diagram shows the pressure distribution as the thickness of the black line and shows the velocity in the boundary layer as the violet triangles. The green vortex generators prompt the transition to turbulent flow and prevent back-flow also called flow separation from the high pressure region in the back. The surface in front is as smooth as possible or even employs shark-like skin, as any turbulence here increases the energy of the airflow. The truncation on the right, known as a Kammback, also prevents backflow from the high pressure region in the back across the spoilers to the convergent part.

In physics, fluid dynamics is a subdiscipline of fluid mechanics that deals with fluid flow—the science of fluids (liquids and gases) in motion. It has several subdisciplines itself, including aerodynamics (the study of air and other gases in motion) and hydrodynamics (the study of liquids in motion). Fluid dynamics has a wide range of applications, including calculating forces and moments on aircraft, determining the mass flow rate of petroleum through pipelines, predicting weather patterns, understanding nebulae in interstellar space and modelling fission weapon detonation. Some of its principles are even used in traffic engineering, where traffic is treated as a continuous fluid, and crowd dynamics.

Fluid dynamics offers a systematic structure—which underlies these practical disciplines—that embraces empirical and semi-empirical laws derived from flow measurement and used to solve practical problems. The solution to a fluid dynamics problem typically involves calculating various properties of the fluid, such as flow velocity, pressure, density, and temperature, as functions of space and time.

Before the twentieth century, *hydrodynamics* was synonymous with fluid dynamics. This is still reflected in names of some fluid dynamics topics, like magnetohydrodynamics and hydrodynamic stability, both of which can also be applied to gases.

Equations of Fluid Dynamics

The foundational axioms of fluid dynamics are the conservation laws, specifically, conservation of mass, conservation of linear momentum (also known as Newton's Second Law of Motion), and conservation of energy (also known as First Law of Thermodynamics). These are based on classical mechanics and are modified in quantum mechanics and general relativity. They are expressed using the Reynolds Transport Theorem.

In addition to the above, fluids are assumed to obey the *continuum assumption*. Fluids are composed of molecules that collide with one another and solid objects. However, the continuum assumption assumes that fluids are continuous, rather than discrete. Consequently, it is assumed that properties such as density, pressure, temperature, and flow velocity are well-defined at infinitesimally small points and vary continuously from one point to another. The fact that the fluid is made up of discrete molecules is ignored.

For fluids that are sufficiently dense to be a continuum, do not contain ionized species, and have flow velocities small in relation to the speed of light, the momentum equations for Newtonian fluids are the Navier–Stokes equations—which is a non-linear set of differential equations that describes the flow of a fluid whose stress depends linearly on flow velocity gradients and pressure. The unsimplified equations do not have a general closed-form solution, so they are primarily of use in Computational Fluid Dynamics. The equations can be simplified in a number of ways, all of which make them easier to solve. Some of the simplifications allow appropriate fluid dynamics problems to be solved in closed form.

In addition to the mass, momentum, and energy conservation equations, a thermodynamical equation of state giving the pressure as a function of other thermodynamic variables for the fluid is required to completely specify the problem. An example of this would be the perfect gas equation of state:

$$p = \frac{\rho R_u T}{M}$$

where p is pressure, ρ is density, R_u is the gas constant, M is molar mass and T is temperature.

Conservation Laws

Three conservation laws are used to solve fluid dynamics problems, and may be written in integral or differential form. Mathematical formulations of these conservation laws may be interpreted by considering the concept of a *control volume*. A control volume is a specified volume in space through which air can flow in and out. Integral formulations of the conservation laws consider the change in mass, momentum, or energy within the control volume. Differential formulations of the conservation laws apply Stokes' theorem to yield an expression which may be interpreted as the integral form of the law applied to an infinitesimal volume at a point within the flow.

- Mass continuity (conservation of mass): The rate of change of fluid mass inside a control volume must be equal to the net rate of fluid flow into the volume. Physically, this statement requires that mass is neither created nor destroyed in the control volume, and can be translated into the integral form of the continuity equation:

$$\frac{\partial}{\partial t} \iiint_V \rho \, dV = - \oiint_S \rho \mathbf{u} \cdot d\mathbf{S}$$

Above, ρ is the fluid density, u is the flow velocity vector, and t is time. The left-hand side of the above expression contains a triple integral over the control volume, whereas the right-hand side contains a surface integral over the surface of the control volume. The differential form of the continuity equation is, by the divergence theorem:

$$\frac{\partial \rho}{\partial t} + \nabla \cdot (\rho \mathbf{u}) = 0$$

- Conservation of momentum: This equation applies Newton's second law of motion to the control volume, requiring that any change in momentum of the air within a control volume be due to the net flow of air into the volume and the action of external forces on the air within the volume. In the integral formulation of this equation, body forces here are represented by f_{body}, the body force per unit mass. Surface forces, such as viscous forces, are represented by $\mathbf{F}_{surf}$, the net force due to stresses on the control volume surface.

$$\frac{\partial}{\partial t} \iiint_V \rho \mathbf{u} \, dV = - \oiint_S (p\mathbf{u}.d\mathbf{S})\mathbf{u} - \oiint_S p \, d\mathbf{S} + \iiint_V \rho f_{body} \, dV + \mathbf{F}_{surf}$$

The differential form of the momentum conservation equation is as follows. Here, both surface and body forces are accounted for in one total force, F. For example, F may be expanded into an expression for the frictional and gravitational forces acting on an internal flow.

$$\frac{D\mathbf{u}}{Dt} = \mathbf{F} - \frac{\nabla p}{\rho}$$

In aerodynamics, air is assumed to be a Newtonian fluid, which posits a linear relationship between the shear stress (due to internal friction forces) and the rate of strain of the fluid.

The equation above is a vector equation: in a three-dimensional flow, it can be expressed as three scalar equations. The conservation of momentum equations for the compressible, viscous flow case are called the Navier–Stokes equations.

- Conservation of energy: Although energy can be converted from one form to another, the total energy in a given closed system remains constant.

$$\rho \frac{Dh}{Dt} = \frac{Dp}{Dt} + \nabla \cdot (k \nabla T) + \Phi$$

Above, h is enthalpy, k is the thermal conductivity of the fluid, T is temperature, and is the viscous dissipation function. The viscous dissipation function governs the rate at which mechanical energy of the flow is converted to heat. The second law of thermodynamics requires that the dissipation term is always positive: viscosity cannot create energy within the control volume. The expression on the left side is a material derivative.

Compressible vs Incompressible Flow

All fluids are compressible to some extent; that is, changes in pressure or temperature cause changes in density. However, in many situations the changes in pressure and temperature are sufficiently small that the changes in density are negligible. In this case the flow can be modelled as an incompressible flow. Otherwise the more general compressible flow equations must be used.

Mathematically, incompressibility is expressed by saying that the density ρ of a fluid parcel does not change as it moves in the flow field, i.e.,

$$\frac{D\rho}{Dt} = 0,$$

where D/Dt is the substantial derivative, which is the sum of local and convective derivatives. This additional constraint simplifies the governing equations, especially in the case when the fluid has a uniform density.

For flow of gases, to determine whether to use compressible or incompressible fluid dynamics, the Mach number of the flow is evaluated. As a rough guide, compressible effects can be ignored at Mach numbers below approximately 0.3. For liquids, whether the incompressible assumption is valid depends on the fluid properties (specifically the critical pressure and temperature of the fluid) and the flow conditions (how close to the critical pressure the actual flow pressure becomes). Acoustic problems always require allowing compressibility, since sound waves are compression waves involving changes in pressure and density of the medium through which they propagate.

Inviscid vs Newtonian and Non-Newtonian Fluids

All fluids are viscous, meaning that they exert some resistance to deformation: neighbouring parcels of fluid moving at different velocities exert viscous forces on each other. The velocity gradient is referred to as a strain rate; it has dimensions T^{-1}. Isaac Newton showed that for many familiar fluids such as water and air, the stress due to these viscous forces is linearly related to the strain rate. Such fluids are called Newtonian fluids. The coefficient of proportionality is called the fluid's viscosity; for Newtonian fluids, it is a fluid property independent of the strain rate.

Potential flow around an airfoil

Non-Newtonian fluids have a more complicated, non-linear stress-strain behaviour. The sub-discipline of rheology studies the stress-strain behaviours of these fluids, which include emulsions and slurries, some viscoelastic materials such as blood and some polymers, and *sticky liquids* such as latex, honey and lubricants.

The dynamic of fluid parcels is described with the help of Newton's second law. An accelerating parcel of fluid is subject to inertial effects.

The Reynolds number is a dimensionless quantity which characterises the magnitude of inertial effects compared to the magnitude of viscous effects. A low Reynolds number ($Re<<1$) indicates that viscous forces are very strong compared to inertial forces. In such cases, inertial forces are sometimes neglected; this flow regime is called Stokes or creeping flow.

On the contrary, high Reynolds numbers ($Re>>1$) indicate that the inertial effects have more effect on the velocity field than the viscous (friction) effects. In high Reynolds number flows, the flow is often modeled as an inviscid flow, an approximation in which viscosity is completely neglected. The Navier-Stokes equations then simplify into the Euler equations. Integrating these along a streamline in an inviscid flow yields Bernoulli's equation. When, in addition to being inviscid, the flow is irrotational everywhere, Bernoulli's equation can be used throughout the flow field. Such flows are called potential flows, because the velocity field may be expressed as the gradient of a potential.

This idea can work fairly well when the Reynolds number is high. However, problems such as those involving solid boundaries may require that the viscosity be included. Viscosity cannot be neglected near solid boundaries because the no-slip condition generates a thin region of large strain rate, the boundary layer, in which viscosity effects dominate and which thus generates vorticity. Therefore, to calculate net forces on bodies (such as wings), viscous flow equations must be used: inviscid flow theory fails to predict drag forces, a limitation known as the d'Alembert's paradox.

A commonly used model, especially in computational fluid dynamics, is to use two flow models: the Euler equations away from the body, and boundary layer equations in a region close to the body. The two solutions can then be matched with each other, using the method of matched asymptotic expansions.

Steady vs Unsteady Flow

When all the time derivatives of a flow field vanish, the flow is considered steady flow. Steady-state flow refers to the condition where the fluid properties at a point in the system do not change over time. Otherwise, flow is known as unsteady (also called transient). Whether a particular flow is steady or unsteady, can depend on the chosen frame of reference. For instance, laminar flow over a sphere is steady in the frame of reference that is stationary with respect to the sphere. In a frame of reference that is stationary with respect to a background flow, the flow is unsteady.

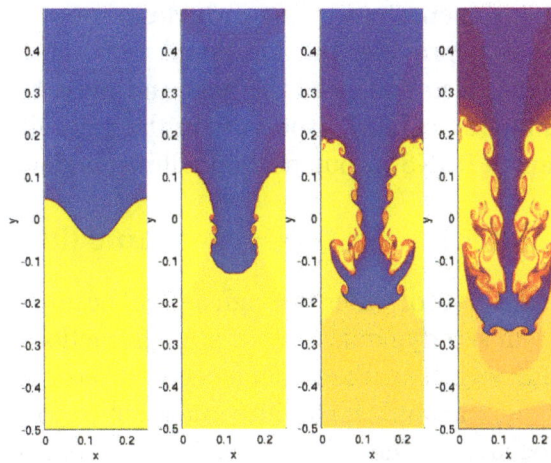

Hydrodynamics simulation of the Rayleigh–Taylor instability

Turbulent flows are unsteady by definition. A turbulent flow can, however, be statistically stationary. According to Pope:

The random field $U(x,t)$ is statistically stationary if all statistics are invariant under a shift in time.

This roughly means that all statistical properties are constant in time. Often, the mean field is the object of interest, and this is constant too in a statistically stationary flow.

Steady flows are often more tractable than otherwise similar unsteady flows. The governing equations of a steady problem have one dimension fewer (time) than the governing equations of the same problem without taking advantage of the steadiness of the flow field.

Laminar vs Turbulent Flow

Turbulence is flow characterized by recirculation, eddies, and apparent randomness. Flow in which turbulence is not exhibited is called laminar. It should be noted, however, that the presence of eddies or recirculation alone does not necessarily indicate turbulent flow—these phenomena may be present in laminar flow as well. Mathematically, turbulent flow is often represented via a Reynolds decomposition, in which the flow is broken down into the sum of an average component and a perturbation component.

It is believed that turbulent flows can be described well through the use of the Navier–Stokes equations. Direct numerical simulation (DNS), based on the Navier–Stokes equations, makes it possible to simulate turbulent flows at moderate Reynolds numbers. Restrictions depend on the power of the computer used and the efficiency of the solution algorithm. The results of DNS have been found to agree well with experimental data for some flows.

Most flows of interest have Reynolds numbers much too high for DNS to be a viable option, given the state of computational power for the next few decades. Any flight vehicle large enough to carry a human (L > 3 m), moving faster than 72 km/h (20 m/s) is well beyond the limit of DNS simulation (Re = 4 million). Transport aircraft wings (such as on an Airbus A300 or Boeing 747) have Reynolds numbers of 40 million (based on the wing chord). Solving these real-life flow problems requires turbulence models for the foreseeable future. Reynolds-averaged Navier–Stokes equations (RANS) combined with turbulence modelling provides a model of the effects of the turbulent flow. Such a modelling mainly provides the additional momentum transfer by the Reynolds stresses, although the turbulence also enhances the heat and mass transfer. Another promising methodology is large eddy simulation (LES), especially in the guise of detached eddy simulation (DES)—which is a combination of RANS turbulence modelling and large eddy simulation.

Subsonic vs Transonic, Supersonic and Hypersonic flows

While many terrestrial flows (e.g. flow of water through a pipe) occur at low mach numbers, many flows of practical interest (e.g. in aerodynamics) occur at high fractions of the Mach Number M=1 or in excess of it (supersonic flows). New phenomena occur at these Mach number regimes (e.g. shock waves for supersonic flow, transonic instability in a regime of flows with M nearly equal to 1, non-equilibrium chemical behaviour due to ionization in hypersonic flows) and it is necessary to treat each of these flow regimes separately.

Magnetohydrodynamics

Magnetohydrodynamics is the multi-disciplinary study of the flow of electrically conducting fluids in electromagnetic fields. Examples of such fluids include plasmas, liquid metals, and salt water. The fluid flow equations are solved simultaneously with Maxwell's equations of electromagnetism.

Other Approximations

There are a large number of other possible approximations to fluid dynamic problems. Some of the more commonly used are listed below.

- The *Boussinesq approximation* neglects variations in density except to calculate buoyancy forces. It is often used in free convection problems where density changes are small.

- *Lubrication theory* and *Hele–Shaw flow* exploits the large aspect ratio of the domain to show that certain terms in the equations are small and so can be neglected.

- *Slender-body theory* is a methodology used in Stokes flow problems to estimate the force on, or flow field around, a long slender object in a viscous fluid.

- The *shallow-water equations* can be used to describe a layer of relatively inviscid fluid with a free surface, in which surface gradients are small.

- The *Boussinesq equations* are applicable to surface waves on thicker layers of fluid and with steeper surface slopes.

- *Darcy's law* is used for flow in porous media, and works with variables averaged over several pore-widths.

- In rotating systems, the *quasi-geostrophic equations* assume an almost perfect balance between pressure gradients and the Coriolis force. It is useful in the study of atmospheric dynamics.

Terminology in Fluid Dynamics

The concept of pressure is central to the study of both fluid statics and fluid dynamics. A pressure can be identified for every point in a body of fluid, regardless of whether the fluid is in motion or not. Pressure can be measured using an aneroid, Bourdon tube, mercury column, or various other methods.

Some of the terminology that is necessary in the study of fluid dynamics is not found in other similar areas of study. In particular, some of the terminology used in fluid dynamics is not used in fluid statics.

Terminology in Incompressible Fluid Dynamics

The concepts of total pressure and dynamic pressure arise from Bernoulli's equation and are significant in the study of all fluid flows. (These two pressures are not pressures in the usual sense—they cannot be measured using an aneroid, Bourdon tube or mercury column.) To avoid potential ambiguity when referring to pressure in fluid dynamics, many authors use the term static pressure to distinguish it from total pressure and dynamic pressure. Static pressure is identical to pressure and can be identified for every point in a fluid flow field.

In *Aerodynamics*, L.J. Clancy writes: *To distinguish it from the total and dynamic pressures, the actual pressure of the fluid, which is associated not with its motion but with its state, is often referred to as the static pressure, but where the term pressure alone is used it refers to this static pressure.*

A point in a fluid flow where the flow has come to rest (i.e. speed is equal to zero adjacent to some solid body immersed in the fluid flow) is of special significance. It is of such importance that it is given a special name—a stagnation point. The static pressure at the stagnation point is of special significance and is given its own name—stagnation pressure. In incompressible flows, the stagnation pressure at a stagnation point is equal to the total pressure throughout the flow field.

Terminology in Compressible Fluid Dynamics

In a compressible fluid, such as air, the temperature and density are essential when determining the state of the fluid. In addition to the concept of total pressure (also known as stagnation pressure), the concepts of total (or stagnation) temperature and total (or stagnation) density are also essential in any study of compressible fluid flows. To avoid potential ambiguity when referring to temperature and density, many authors use the terms static temperature and static density. Static temperature is identical to temperature; and static density is identical to density; and both can be identified for every point in a fluid flow field.

The temperature and density at a stagnation point are called stagnation temperature and stagnation density.

A similar approach is also taken with the thermodynamic properties of compressible fluids. Many authors use the terms total (or stagnation) enthalpy and total (or stagnation) entropy. The terms *static enthalpy* and *static entropy* appear less common, but where they are used they mean enthalpy and entropy respectively, using the prefix "static" to avoid ambiguity with their 'total' or 'stagnation' counterparts. Because the 'total' flow conditions are defined by isentropically bringing the fluid to rest, the total (or stagnation) entropy is by definition always equal to the "static" entropy.

Physical Cosmology

Physical cosmology is the study of the largest-scale structures and dynamics of the Universe and is concerned with fundamental questions about its origin, structure, evolution, and ultimate fate. For most of human history, it was a branch of metaphysics and religion. Cosmology as a science originated with the Copernican principle, which implies that celestial bodies obey identical physical laws to those on Earth, and Newtonian mechanics, which first allowed us to understand those physical laws.

Physical cosmology, as it is now understood, began with the development in 1915 of Albert Einstein's general theory of relativity, followed by major observational discoveries in the 1920s: first, Edwin Hubble discovered that the universe contains a huge number of external galaxies beyond our own Milky Way; then, work by Vesto Slipher and others showed that the universe is expanding. These advances made it possible to speculate about the origin of the universe, and allowed the establishment of the Big Bang Theory, by Georges Lemaitre, as the leading cosmological model. A few researchers still advocate a handful of alternative cosmologies; however, most cosmologists agree that the Big Bang theory explains the observations better.

Dramatic advances in observational cosmology since the 1990s, including the cosmic microwave background, distant supernovae and galaxy redshift surveys, have led to the development of a standard model of cosmology. This model requires the universe to contain large amounts of dark matter and dark energy whose nature is currently not well understood, but the model gives detailed predictions that are in excellent agreement with many diverse observations.

Cosmology draws heavily on the work of many disparate areas of research in theoretical and applied physics. Areas relevant to cosmology include particle physics experiments and theory, theoretical and observational astrophysics, general relativity, quantum mechanics, and plasma physics.

Modern cosmology developed along tandem tracks of theory and observation. In 1916, Albert Einstein published his theory of general relativity, which provided a unified description of gravity as a geometric property of space and time. At the time, Einstein believed in a static universe, but found that his original formulation of the theory did not permit it. This is because masses distributed throughout the universe gravitationally attract, and move toward each other over time. However, he realized that his equations permitted the introduction of a constant term which could counter-act the attractive force of gravity on the cosmic scale. Einstein published his first paper on rela-tivistic cosmology in 1917, in which he added this *cosmological constant* to his field equations in order to force them to model a static universe. However, this so-called Einstein model is unstable to small perturbations—it will eventually start to expand or contract. The Einstein model describes a static universe; space is finite and unbounded (analogous to the surface of a sphere, which has a finite area but no edges). It was later realized that Einstein's model was just one of a larger set of possibilities, all of which were consistent with general relativity and the cosmological principle. The cosmological solutions of general relativity were found by Alexander Friedmann in the early 1920s. His equations describe the Friedmann–Lemaître–Robertson–Walker universe, which may expand or contract, and whose geometry may be open, flat, or closed.

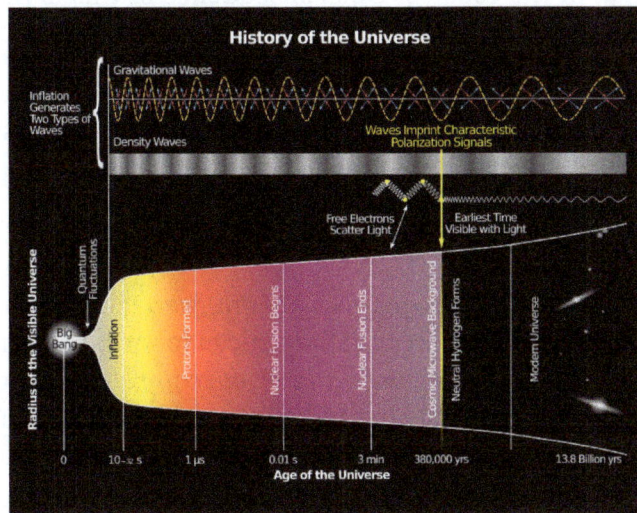

History of the Universe – gravitational waves are hypothesized to arise from cosmic inflation, a faster-than-light expansion just after the Big Bang

In the 1910s, Vesto Slipher (and later Carl Wilhelm Wirtz) interpreted the red shift of spiral neb-ulae as a Doppler shift that indicated they were receding from Earth. However, it is difficult to determine the distance to astronomical objects. One way is to compare the physical size of an object to its angular size, but a physical size must be assumed to do this. Another method is to measure the brightness of an object and assume an intrinsic luminosity, from which the distance may be determined using the inverse square law. Due to the difficulty of using these methods, they did not realize that the nebulae were actually galaxies outside our own Milky Way, nor did they speculate about the cosmological implications. In 1927, the Belgian Roman Catholic priest Georg-es Lemaître independently derived the Friedmann–Lemaître–Robertson–Walker equations and proposed, on the basis of the recession of spiral nebulae, that the universe began with the "explo-sion" of a "primeval atom"—which was later called the Big Bang. In 1929, Edwin Hubble provided an observational basis for Lemaître's theory. Hubble showed that the spiral nebulae were galaxies by determining their distances using measurements of the brightness of Cepheid variable stars. He discovered a relationship between the redshift of a galaxy and its distance. He interpreted this as evidence that the galaxies are receding from Earth in every direction at speeds proportional to their distance. This fact is now known as Hubble's law, though the numerical factor Hubble found relating recessional velocity and distance was off by a factor of ten, due to not knowing about the types of Cepheid variables.

Given the cosmological principle, Hubble's law suggested that the universe was expanding. Two primary explanations were proposed for the expansion. One was Lemaître's Big Bang theory, advocated and developed by George Gamow. The other explanation was Fred Hoyle's steady state model in which new matter is created as the galaxies move away from each other. In this model, the universe is roughly the same at any point in time.

For a number of years, support for these theories was evenly divided. However, the observational evidence began to support the idea that the universe evolved from a hot dense state. The discovery of the cosmic microwave background in 1965 lent strong support to the Big Bang model, and since the precise measurements of the cosmic microwave background by the Cosmic Background Explorer in the early 1990s, few cosmologists have seriously proposed other theories of the origin and evolution of the cosmos. One consequence of this is that in standard general relativity, the universe began with a singularity, as demonstrated by Roger Penrose and Stephen Hawking in the 1960s.

An alternative view to extend the Big Bang model, suggesting the universe had no beginning or singularity and the age of the universe is infinite, has been presented.

Energy of the Cosmos

Light chemical elements, primarily hydrogen and helium, were created in the Big Bang process. The small atomic nuclei combined into larger atomic nuclei to form heavier elements such as iron and nickel, which are more stable. This caused a *later energy release*. Such reactions of nuclear particles inside stars continue to contribute to *sudden energy releases*, such as in nova stars. Gravitational collapse of matter into black holes is also thought to power the most energetic processes, generally seen at the centers of galaxies.

Cosmologists cannot explain all cosmic phenomena exactly, such as those related to the accelerating expansion of the universe, using conventional forms of energy. Instead, cosmologists propose a new form of energy called dark energy that permeates all space. One hypothesis is that dark energy is the energy of virtual particles, which are believed to exist in a vacuum due to the uncertainty principle.

There is no clear way to define the total energy in the universe using the most widely accepted theory of gravity, general relativity. Therefore, it remains controversial whether the total energy is conserved in an expanding universe. For instance, each photon that travels through intergalactic space loses energy due to the redshift effect. This energy is not obviously transferred to any other system, so seems to be permanently lost. On the other hand, some cosmologists insist that energy is conserved in some sense; this follows the law of conservation of energy.

Thermodynamics of the universe is a field of study that explores which form of energy dominates the cosmos – relativistic particles which are referred to as radiation, or non-relativistic particles referred to as matter. Relativistic particles are particles whose rest mass is zero or negligible compared to their kinetic energy, and so move at the speed of light or very close to it; non-relativistic particles have much higher rest mass than their energy and so move much slower than the speed of light.

As the universe expands, both matter and radiation in it become diluted. However, the energy densities of radiation and matter dilute at different rates. As a particular volume expands, mass energy density is changed only by the increase in volume, but the energy density of radiation is changed both by the increase in volume and by the increase in the wavelength of the photons that make it up. Thus the energy of radiation becomes a smaller part of the universe's total energy than that of matter as it expands. The very early universe is said to have been 'radiation dominated' and radiation controlled the deceleration of expansion. Later, as the average energy per photon becomes roughly 10 eV and lower, matter dictates the rate of deceleration and the universe is said to be 'matter dominated'. The intermediate case is not treated well analytically. As the expansion of the universe continues, matter dilutes even further and the cosmological constant becomes dominant, leading to an acceleration in the universe's expansion.

History of the Universe

The history of the universe is a central issue in cosmology. The history of the universe is divided into different periods called epochs, according to the dominant forces and processes in each period. The standard cosmological model is known as the Lambda-CDM model.

Equations of Motion

The equations of motion governing the universe as a whole are derived from general relativity with a small, positive cosmological constant. The solution is an expanding universe; due to this expansion, the radiation and matter in the universe cool down and become diluted. At first, the expansion is slowed down by gravitation attracting the radiation and matter in the universe. However, as these become diluted, the cosmological constant becomes more dominant and the expansion of the universe starts to accelerate rather than decelerate. In our universe this happened billions of years ago.

Timeline of the Big Bang

Observations suggest that the universe began around 13.8 billion years ago. Since then, the evolution of the universe has passed through three phases. The very early universe, which is still poorly understood, was the split second in which the universe was so hot that particles had energies higher than those currently accessible in particle accelerators on Earth. Therefore, while the basic features of this epoch have been worked out in the Big Bang theory, the details are largely based on educated guesses. Following this, in the early universe, the evolution of the universe proceeded according to known high energy physics. This is when the first protons, electrons and neutrons formed, then nuclei and finally atoms. With the formation of neutral hydrogen, the cosmic microwave background was emitted. Finally, the epoch of structure formation began, when matter started to aggregate into the first stars and quasars, and ultimately galaxies, clusters of galaxies and superclusters formed. The future of the universe is not yet firmly known, but according to the ΛCDM model it will continue expanding forever.

Areas of Study

Below, some of the most active areas of inquiry in cosmology are described, in roughly chronological order. This does not include all of the Big Bang cosmology, which is presented in *Timeline of the Big Bang*.

Very Early Universe

The early, hot universe appears to be well explained by the Big Bang from roughly 10^{-33} seconds onwards, but there are several problems. One is that there is no compelling reason, using current particle physics, for the universe to be flat, homogeneous, and isotropic. Moreover, grand unified theories of particle physics suggest that there should be magnetic monopoles in the universe, which have not been found. These problems are resolved by a brief period of cosmic inflation, which drives the universe to flatness, smooths out anisotropies and inhomogeneities to the observed level, and exponentially dilutes the monopoles. The physical model behind cosmic inflation is extremely simple, but it has not yet been confirmed by particle physics, and there are difficult problems reconciling inflation and quantum field theory. Some cosmologists think that string theory and brane cosmology will provide an alternative to inflation.

Another major problem in cosmology is what caused the universe to contain far more matter than antimatter. Cosmologists can observationally deduce that the universe is not split into regions of matter and antimatter. If it were, there would be X-rays and gamma rays produced as a result of annihilation, but this is not observed. Therefore, some process in the early universe must have created a small excess of matter over antimatter, and this (currently not understood) process is called *baryogenesis*. Three required conditions for baryogenesis were derived by Andrei Sakharov in 1967, and requires a violation of the particle physics symmetry, called CP-symmetry, between matter and antimatter. However, particle accelerators measure too small a violation of CP-symmetry to account for the baryon asymmetry. Cosmologists and particle physicists look for additional violations of the CP-symmetry in the early universe that might account for the baryon asymmetry.

Both the problems of baryogenesis and cosmic inflation are very closely related to particle physics, and their resolution might come from high energy theory and experiment, rather than through observations of the universe.

Big Bang Theory

Big Bang nucleosynthesis is the theory of the formation of the elements in the early universe. It finished when the universe was about three minutes old and its temperature dropped below that at which nuclear fusion could occur. Big Bang nucleosynthesis had a brief period during which it could operate, so only the very lightest elements were produced. Starting from hydrogen ions (protons), it principally produced deuterium, helium-4, and lithium. Other elements were produced in only trace abundances. The basic theory of nucleosynthesis was developed in 1948 by George Gamow, Ralph Asher Alpher, and Robert Herman. It was used for many years as a probe of physics at the time of the Big Bang, as the theory of Big Bang nucleosynthesis connects the abundances of primordial light elements with the features of the early universe. Specifically, it can be used to test the equivalence principle, to probe dark matter, and test neutrino physics. Some cosmologists have proposed that Big Bang nucleosynthesis suggests there is a fourth "sterile" species of neutrino.

Standard Model of Big Bang Cosmology

The ΛCDM (Lambda cold dark matter) or Lambda-CDM model is a parametrization of the Big Bang cosmological model in which the universe contains a cosmological constant, denoted by

Lambda (Greek Λ), associated with dark energy, and cold dark matter (abbreviated CDM). It is frequently referred to as the standard model of Big Bang cosmology.

Cosmic Microwave Background

The cosmic microwave background is radiation left over from decoupling after the epoch of recombination when neutral atoms first formed. At this point, radiation produced in the Big Bang stopped Thomson scattering from charged ions. The radiation, first observed in 1965 by Arno Penzias and Robert Woodrow Wilson, has a perfect thermal black-body spectrum. It has a temperature of 2.7 kelvins today and is isotropic to one part in 10^5. Cosmological perturbation theory, which describes the evolution of slight inhomogeneities in the early universe, has allowed cosmologists to precisely calculate the angular power spectrum of the radiation, and it has been measured by the recent satellite experiments (COBE and WMAP) and many ground and balloon-based experiments (such as Degree Angular Scale Interferometer, Cosmic Background Imager, and Boomerang). One of the goals of these efforts is to measure the basic parameters of the Lambda-CDM model with increasing accuracy, as well as to test the predictions of the Big Bang model and look for new physics. The recent measurements made by WMAP, for example, have placed limits on the neutrino masses.

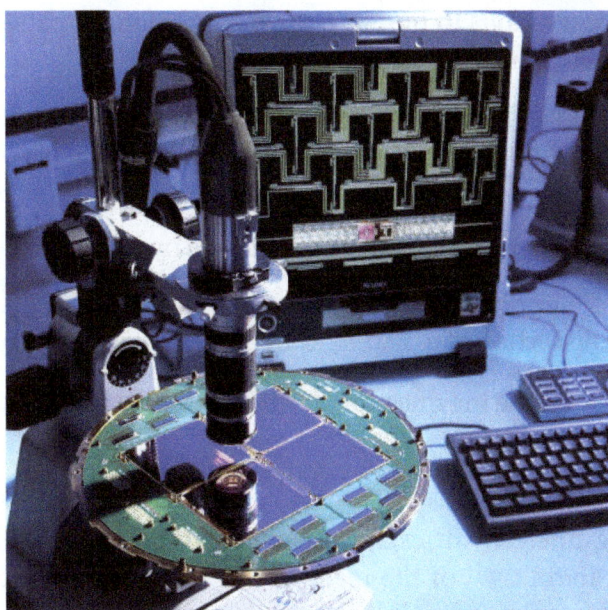

Evidence of gravitational waves in the infant universe may have been uncovered by the microscopic examination of the focal plane of the BICEP2 radio telescope.

Newer experiments, such as QUIET and the Atacama Cosmology Telescope, are trying to measure the polarization of the cosmic microwave background. These measurements are expected to provide further confirmation of the theory as well as information about cosmic inflation, and the so-called secondary anisotropies, such as the Sunyaev-Zel'dovich effect and Sachs-Wolfe effect, which are caused by interaction between galaxies and clusters with the cosmic microwave background.

On 17 March 2014, astronomers at the Harvard–Smithsonian Center for Astrophysics announced the apparent detection of gravitational waves, which, if confirmed, may provide strong evidence for inflation and the Big Bang. However, on 19 June 2014, lowered confidence in confirming the cosmic inflation findings was reported.

Formation and Evolution of Large-scale Structure

Understanding the formation and evolution of the largest and earliest structures (i.e., quasars, galaxies, clusters and superclusters) is one of the largest efforts in cosmology. Cosmologists study a model of hierarchical structure formation in which structures form from the bottom up, with smaller objects forming first, while the largest objects, such as superclusters, are still assembling. One way to study structure in the universe is to survey the visible galaxies, in order to construct a three-dimensional picture of the galaxies in the universe and measure the matter power spectrum. This is the approach of the *Sloan Digital Sky Survey* and the 2dF Galaxy Redshift Survey.

Another tool for understanding structure formation is simulations, which cosmologists use to study the gravitational aggregation of matter in the universe, as it clusters into filaments, super-clusters and voids. Most simulations contain only non-baryonic cold dark matter, which should suffice to understand the universe on the largest scales, as there is much more dark matter in the universe than visible, baryonic matter. More advanced simulations are starting to include baryons and study the formation of individual galaxies. Cosmologists study these simulations to see if they agree with the galaxy surveys, and to understand any discrepancy.

Other, complementary observations to measure the distribution of matter in the distant universe and to probe reionization include:

- The Lyman-alpha forest, which allows cosmologists to measure the distribution of neutral atomic hydrogen gas in the early universe, by measuring the absorption of light from distant quasars by the gas.

- The 21 centimeter absorption line of neutral atomic hydrogen also provides a sensitive test of cosmology

- Weak lensing, the distortion of a distant image by gravitational lensing due to dark matter.

These will help cosmologists settle the question of when and how structure formed in the universe.

Dark Matter

Evidence from Big Bang nucleosynthesis, the cosmic microwave background and structure formation suggests that about 23% of the mass of the universe consists of non-baryonic dark matter, whereas only 4% consists of visible, baryonic matter. The gravitational effects of dark matter are well understood, as it behaves like a cold, non-radiative fluid that forms haloes around galaxies. Dark matter has never been detected in the laboratory, and the particle physics nature of dark matter remains completely unknown. Without observational constraints, there are a number of candidates, such as a stable supersymmetric particle, a weakly interacting massive particle, an axion, and a massive compact halo object. Alternatives to the dark matter hypothesis include a modification of gravity at small accelerations (MOND) or an effect from brane cosmology.

Dark Energy

If the universe is flat, there must be an additional component making up 73% (in addition to the 23% dark matter and 4% baryons) of the energy density of the universe. This is called dark energy.

In order not to interfere with Big Bang nucleosynthesis and the cosmic microwave background, it must not cluster in haloes like baryons and dark matter. There is strong observational evidence for dark energy, as the total energy density of the universe is known through constraints on the flatness of the universe, but the amount of clustering matter is tightly measured, and is much less than this. The case for dark energy was strengthened in 1999, when measurements demonstrated that the expansion of the universe has begun to gradually accelerate.

Apart from its density and its clustering properties, nothing is known about dark energy. *Quantum field theory* predicts a cosmological constant (CC) much like dark energy, but 120 orders of magnitude larger than that observed. Steven Weinberg and a number of string theorists have invoked the 'weak anthropic principle': i.e. the reason that physicists observe a universe with such a small cosmological constant is that no physicists (or any life) could exist in a universe with a larger cosmological constant. Many cosmologists find this an unsatisfying explanation: perhaps because while the weak anthropic principle is self-evident (given that living observers exist, there must be at least one universe with a cosmological constant which allows for life to exist) it does not attempt to explain the context of that universe. For example, the weak anthropic principle alone does not distinguish between:

- Only one universe will ever exist and there is some underlying principle that constrains the CC to the value we observe.

- Only one universe will ever exist and although there is no underlying principle fixing the CC, we got lucky.

- Lots of universes exist (simultaneously or serially) with a range of CC values, and of course ours is one of the life-supporting ones.

Other possible explanations for dark energy include quintessence or a modification of gravity on the largest scales. The effect on cosmology of the dark energy that these models describe is given by the dark energy's equation of state, which varies depending upon the theory. The nature of dark energy is one of the most challenging problems in cosmology.

A better understanding of dark energy is likely to solve the problem of the ultimate fate of the universe. In the current cosmological epoch, the accelerated expansion due to dark energy is preventing structures larger than superclusters from forming. It is not known whether the acceleration will continue indefinitely, perhaps even increasing until a big rip, or whether it will eventually reverse.

Classical Electromagnetism

Classical electromagnetism or classical electrodynamics is a branch of theoretical physics that studies the interactions between electric charges and currents using an extension of the classical Newtonian model. The theory provides an excellent description of electromagnetic phenomena whenever the relevant length scales and field strengths are large enough that quantum mechanical effects are negligible. For small distances and low field strengths, such interactions are better described by quantum electrodynamics.

Fundamental physical aspects of classical electrodynamics are presented in many texts, such as those by Feynman, Leighton and Sands, Panofsky and Phillips, and Jackson.

History

The physical phenomena that electromagnetism describes have been studied as separate fields since antiquity. For example, there were many advances in the field of optics centuries before light was understood to be an electromagnetic wave. However, the theory of electromagnetism, as it is currently understood, grew out of Michael Faraday's experiments suggesting an electromagnetic field and James Clerk Maxwell's use of differential equations to describe it in his *A Treatise on Electricity and Magnetism* (1873). For a detailed historical account, consult Pauli, Whittaker, Pais, and Hunt.

Other Areas of Inquiry

Cosmologists also study:

- Whether primordial black holes were formed in our universe, and what happened to them.

- The GZK cutoff for high-energy cosmic rays, and whether it signals a failure of special relativity at high energies

- The equivalence principle, whether or not Einstein's general theory of relativity is the correct theory of gravitation, and if the fundamental laws of physics are the same everywhere in the universe.

- The increasing complexity of universal structures, an example being the progressively greater energy rate density.

Lorentz Force

The electromagnetic field exerts the following force (often called the Lorentz force) on charged particles:

$$\mathbf{F} = q\mathbf{E} + q\mathbf{v} \times \mathbf{B}$$

where all boldfaced quantities are vectors: F is the force that a particle with charge q experiences, E is the electric field at the location of the particle, v is the velocity of the particle, B is the magnetic field at the location of the particle.

The above equation illustrates that the Lorentz force is the sum of two vectors. One is the cross product of the velocity and magnetic field vectors. Based on the properties of the cross product, this produces a vector that is perpendicular to both the velocity and magnetic field vectors. The other vector is in the same direction as the electric field. The sum of these two vectors is the Lorentz force.

Therefore, in the absence of a magnetic field, the force is in the direction of the electric field, and the magnitude of the force is dependent on the value of the charge and the intensity of the electric field. In the absence of an electric field, the force is perpendicular to the velocity of the particle and the direction of the magnetic field. If both electric and magnetic fields are present, the Lorentz force is the sum of both of these vectors.

The Electric Field E

The electric field E is defined such that, on a stationary charge:

$$\mathbf{F} = q_0 \mathbf{E}$$

where q_0 is what is known as a test charge. The size of the charge doesn't really matter, as long as it is small enough not to influence the electric field by its mere presence. What is plain from this definition, though, is that the unit of E is N/C (newtons per coulomb). This unit is equal to V/m (volts per meter).

In electrostatics, where charges are not moving, around a distribution of point charges, the forces determined from Coulomb's law may be summed. The result after dividing by q_0 is:

$$\mathbf{E}(\mathbf{r}) = \frac{1}{4\pi\varepsilon_0} \sum_{i=1}^{n} \frac{q_i (\mathbf{r} - \mathbf{r}_i)}{|\mathbf{r} - \mathbf{r}_i|^3}$$

where n is the number of charges, q_i is the amount of charge associated with the ith charge, r_i is the position of the ith charge, r is the position where the electric field is being determined, and ε_0 is the electric constant.

If the field is instead produced by a continuous distribution of charge, the summation becomes an integral:

$$\mathbf{E}(\mathbf{r}) = \frac{1}{4\pi\varepsilon_0} \int \frac{\rho(\mathbf{r}')(\mathbf{r} - \mathbf{r}')}{|\mathbf{r} - \mathbf{r}'|^3} d^3\mathbf{r}'$$

where $\rho(\mathbf{r}')$ is the charge density and $\mathbf{r} - \mathbf{r}'$ is the vector that points from the volume element $d^3\mathbf{r}'$ to the point in space where E is being determined.

Both of the above equations are cumbersome, especially if one wants to determine E as a function of position. A scalar function called the electric potential can help. Electric potential, also called voltage (the units for which are the volt), is defined by the line integral

$$\varphi(\mathbf{r}) = -\int_C \mathbf{E} \cdot d\mathbf{l}$$

where $\varphi(r)$ is the electric potential, and C is the path over which the integral is being taken.

Unfortunately, this definition has a caveat. From Maxwell's equations, it is clear that $\nabla \times E$ is not always zero, and hence the scalar potential alone is insufficient to define the electric field exactly. As a result, one must add a correction factor, which is generally done by subtracting the time derivative of the A vector potential described below. Whenever the charges are quasistatic, however, this condition will be essentially met.

From the definition of charge, one can easily show that the electric potential of a point charge as a function of position is:

$$\varphi(\mathbf{r}) = \frac{1}{4\pi\varepsilon_0} \sum_{i=1}^{n} \frac{q_i}{|\mathbf{r} - \mathbf{r}_i|}$$

where q is the point charge's charge, r is the position at which the potential is being determined, and r_i is the position of each point charge. The potential for a continuous distribution of charge is:

$$\varphi(\mathbf{r}) = \frac{1}{4\pi\varepsilon_0} \sum_{i=1}^{n} \frac{q_i}{|\mathbf{r} - \mathbf{r}_i|}$$

where $\rho(\mathbf{r'})$ is the charge density, and $\mathbf{r} - \mathbf{r'}$ is the distance from the volume element $d^3\mathbf{r'}$ to point in space where φ is being determined.

The scalar φ will add to other potentials as a scalar. This makes it relatively easy to break complex problems down in to simple parts and add their potentials. Taking the definition of φ backwards, we see that the electric field is just the negative gradient (the del operator) of the potential. Or:

$$\mathbf{E}(\mathbf{r}) = -\nabla\varphi(\mathbf{r}).$$

From this formula it is clear that E can be expressed in V/m (volts per meter).

Electromagnetic Waves

A changing electromagnetic field propagates away from its origin in the form of a wave. These waves travel in vacuum at the speed of light and exist in a wide spectrum of wavelengths. Examples of the dynamic fields of electromagnetic radiation (in order of increasing frequency): radio waves, microwaves, light (infrared, visible light and ultraviolet), x-rays and gamma rays. In the field of particle physics this electromagnetic radiation is the manifestation of the electromagnetic interaction between charged particles.

General Field Equations

As simple and satisfying as Coulomb's equation may be, it is not entirely correct in the context of classical electromagnetism. Problems arise because changes in charge distributions require a non-zero amount of time to be "felt" elsewhere (required by special relativity).

For the fields of general charge distributions, the retarded potentials can be computed and differentiated accordingly to yield Jefimenko's Equations.

Retarded potentials can also be derived for point charges, and the equations are known as the Liénard–Wiechert potentials. The scalar potential is:

$$\varphi = \frac{1}{4\pi\varepsilon_0} \frac{q}{|\mathbf{r} - \mathbf{r}_q(t_{ret})| - \frac{\mathbf{v}_q(t_{ret})}{c} \cdot (\mathbf{r} - \mathbf{r}_q(t_{ret}))}$$

where q is the point charge's charge and r is the position. r_q and v_q are the position and velocity of the charge, respectively, as a function of retarded time. The vector potential is similar:

$$\mathbf{A} = \frac{\mu_0}{4\pi} \frac{q\mathbf{v}_q(t_{ret})}{\left|\mathbf{r} - \mathbf{r}_q(t_{ret})\right| - \dfrac{\mathbf{v}_q(t_{ret})}{c} \cdot (\mathbf{r} - \mathbf{r}_q(t_{ret}))}.$$

These can then be differentiated accordingly to obtain the complete field equations for a moving point particle.

Models

Branches of classical electromagnetism such as optics, electrical and electronic engineering consist of a collection of relevant mathematical models of different degrees of simplification and idealization to enhance the understanding of specific electrodynamics phenomena, cf. An electrodynamics phenomenon is determined by the particular fields, specific densities of electric charges and currents, and the particular transmission medium. Since there are infinitely many of them, in modeling there is a need for some typical, representative

(a) electrical charges and currents, e.g. moving pointlike charges and electric and magnetic dipoles, electric currents in a conductor etc.;

(b) electromagnetic fields, e.g. voltages, the Liénard–Wiechert potentials, the monochromatic plane waves, optical rays; radio waves, microwaves, infrared radiation, visible light, ultraviolet radiation, X-rays, gamma rays etc.;

(c) transmission media, e.g. electronic components, antennas, electromagnetic waveguides, flat mirrors, mirrors with curved surfaces convex lenses, concave lenses; resistors, inductors, capacitors, switches; wires, electric and optical cables, transmission lines, integrated circuits etc.;

all of which have only few variable characteristics.

References

- Biskamp, Dieter. Nonlinear Magnetohydrodynamics. Cambridge, England: Cambridge University Press, 1993. 378 p. ISBN 0-521-59918-0

- Davidson, Peter Alan (May 2001) An Introduction to Magnetohydrodynamics Cambridge University Press, Cambridge, England, ISBN 0-521-79487-0

- Lorrain, Paul ; Lorrain, François and Houle, Stéphane (2006) Magneto-fluid dynamics: fundamentals and case studies of natural phenomena Springer, New York, ISBN 0-387-33542-0

- Rosa, Richard J. (1987) Magnetohydrodynamic Energy Conversion (2nd edition) Hemisphere Publishing, Washington, D.C., ISBN 0-89116-690-4

- "Magnetohydrodynamics" In Zumerchik, John (editor) (2001) Macmillan Encyclopedia of Energy Macmillan Reference USA, New York, ISBN 0-02-865895-7

- Eckert, Michael (2006). The Dawn of Fluid Dynamics: A Discipline Between Science and Technology. Wiley. p. ix. ISBN 3-527-40513-5.

- Vilenkin, Alex (2007). Many worlds in one : the search for other universes. New York: Hill and Wang, A division of Farrar, Straus and Giroux. p. 19. ISBN 978-0-8090-6722-0.

- Jones, Mark; Lambourne, Robert (2004). An introduction to galaxies and cosmology. Milton Keynes Cambridge, UK New York: Open University Cambridge University Press. p. 232. ISBN 0-521-54623-0.

- e.g. Liddle, A. An Introduction to Modern Cosmology. Wiley. ISBN 0-470-84835-9. This argues cogently "Energy is always, always, always conserved."

- Ghose, Tia (26 February 2015). "Big Bang, Deflated? Universe May Have Had No Beginning". Live Science. Retrieved 28 February 2015.

- Ali, Ahmed Faraq (4 February 2015). "Cosmology from quantum potential". Physics Letters B. 741: 276–279. doi:10.1016/j.physletb.2014.12.057. Retrieved 28 February 2015.

- Overbye, Dennis (19 June 2014). "Astronomers Hedge on Big Bang Detection Claim". New York Times. Retrieved 20 June 2014.

- Amos, Jonathan (19 June 2014). "Cosmic inflation: Confidence lowered for Big Bang signal". BBC News. Retrieved 20 June 2014.

Permissions

Index

www.ingramcontent.com/pod-product-compliance
Lightning Source LLC
Chambersburg PA
CBHW061316190326
41458CB00011B/3819